Examining Optoelectronics in Machine Vision and Applications in Industry 4.0

Oleg Sergiyenko
Autonomous University of Baja California, Mexico

Julio C. Rodriguez-Quiñonez
Autonomous University of Baja California, Mexico

Wendy Flores-Fuentes
Autonomous University of Baja California, Mexico

A volume in the Advances in Computational
Intelligence and Robotics (ACIR) Book Series

Published in the United States of America by
        IGI Global
        Engineering Science Reference (an imprint of IGI Global)
        701 E. Chocolate Avenue
        Hershey PA, USA 17033
        Tel: 717-533-8845
        Fax: 717-533-8661
        E-mail: cust@igi-global.com
        Web site: http://www.igi-global.com

Library of Congress Cataloging-in-Publication Data

Names: Sergiyenko, Oleg, 1969- editor. | Rodríguez-Quiñonez, Julio C.,
    1985- editor. | Flores-Fuentes, Wendy, 1978- editor.
Title: Examining optoelectronics in machine vision and applications in
    industry 4.0 / Oleg Sergiyenko, Julio C. Rodriguez-Quiñonez, and Wendy
    Flores-Fuentes, editor.
Description: Hershey, PA : Engineering Science Reference, an imprint of IGI
    Global, [2021] | Includes bibliographical references and index. |
    Summary: "This book focuses on the examination of emerging technologies
    for the design, fabrication and implementation of optoelectronic
    sensors, devices and systems using a machine vision approach to support
    industrial, commercial and scientific applications"-- Provided by
    publisher.
Identifiers: LCCN 2020024121 (print) | LCCN 2020024122 (ebook) | ISBN
    9781799865223 (hardcover) | ISBN 9781799865230 (paperback) | ISBN
    9781799865247 (ebook)
Subjects: LCSH: Computer vision--Industrial applications. | Machine
    learning. | Optical pattern recognition--Equipment and supplies. |
    Optoelectronic devices.
Classification: LCC TA1634 .E977 2021 (print) | LCC TA1634 (ebook) | DDC
    006.3/7--dc23
LC record available at https://lccn.loc.gov/2020024121
LC ebook record available at https://lccn.loc.gov/2020024122

This book is published in the IGI Global book series Advances in Computational Intelligence and Robotics (ACIR) (ISSN: 2327-0411; eISSN: 2327-042X)

British Cataloguing in Publication Data
A Cataloguing in Publication record for this book is available from the British Library.

All work contributed to this book is new, previously-unpublished material. The views expressed in this book are those of the authors, but not necessarily of the publisher.

For electronic access to this publication, please contact: eresources@igi-global.com.

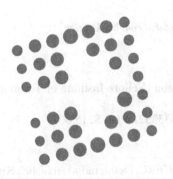

Advances in Computational Intelligence and Robotics (ACIR) Book Series

Ivan Giannoccaro
University of Salento, Italy

ISSN:2327-0411
EISSN:2327-042X

MISSION

While intelligence is traditionally a term applied to humans and human cognition, technology has progressed in such a way to allow for the development of intelligent systems able to simulate many human traits. With this new era of simulated and artificial intelligence, much research is needed in order to continue to advance the field and also to evaluate the ethical and societal concerns of the existence of artificial life and machine learning.

The **Advances in Computational Intelligence and Robotics (ACIR) Book Series** encourages scholarly discourse on all topics pertaining to evolutionary computing, artificial life, computational intelligence, machine learning, and robotics. ACIR presents the latest research being conducted on diverse topics in intelligence technologies with the goal of advancing knowledge and applications in this rapidly evolving field.

COVERAGE

- Cyborgs
- Computational Logic
- Algorithmic Learning
- Heuristics
- Neural Networks
- Intelligent control
- Artificial Life
- Cognitive Informatics
- Artificial Intelligence
- Synthetic Emotions

IGI Global is currently accepting manuscripts for publication within this series. To submit a proposal for a volume in this series, please contact our Acquisition Editors at Acquisitions@igi-global.com or visit: http://www.igi-global.com/publish/.

Titles in this Series

For a list of additional titles in this series, please visit: www.igi-global.com/book-series

Applications of Artificial Intelligence for Smart Technology
P. Swarnalatha (Vellore Institute of Technology, Vellore, India) and S. Prabu (Vellore Institute of Technology, Vellore, India)
Engineering Science Reference • © 2021 • 330pp • H/C (ISBN: 9781799833352) • US $215.00

Resource Optimization Using Swarm Intelligence and the IoT
Vicente García Díaz (University of Oviedo, Spain) Pramod Singh Rathore (ACERC, Delhi, India) Abhishek Kumar (ACERC, Delhi, India) and Rashmi Agrawal (Manav Rachna University, India)
Engineering Science Reference • © 2021 • 300pp • H/C (ISBN: 9781799850953) • US $225.00

Deep Learning Applications and Intelligent Decision Making in Engineering
Karthikrajan Senthilnathan (Revoltaxe India Pvt Ltd, Chennai, India) Balamurugan Shanmugam (Quants IS & CS, India) Dinesh Goyal (Poornima Institute of Engineering and Technology, India) Iyswarya Annapoorani (VIT University, India) and Ravi Samikannu (Botswana International University of Science and Technology, Botswana)
Engineering Science Reference • © 2021 • 332pp • H/C (ISBN: 9781799821083) • US $245.00

Handbook of Research on Natural Language Processing and Smart Service Systems
Rodolfo Abraham Pazos-Rangel (Tecnológico Nacional de México, Mexico & Instituto Tecnológico de Ciudad Madero, Mexico) Rogelio Florencia-Juarez (Universidad Autónoma de Ciudad Juárez, Mexico) Mario Andrés Paredes-Valverde (Tecnológico Nacional de México, Mexico & Instituto Tecnológico de Orizaba, Mexico) and Gilberto Rivera (Universidad Autónoma de Ciudad Juárez, Mexico)
Engineering Science Reference • © 2021 • 554pp • H/C (ISBN: 9781799847304) • US $295.00

Applications of Artificial Neural Networks for Nonlinear Data
Hiral Ashil Patel (Ganpat University, India) and A.V. Senthil Kumar (Hindusthan College of Arts and Science, India)
Engineering Science Reference • © 2021 • 315pp • H/C (ISBN: 9781799840428) • US $245.00

Analyzing Future Applications of AI, Sensors, and Robotics in Society
Thomas Heinrich Musiolik (Berlin University of the Arts, Germany) and Adrian David Cheok (iUniversity, Tokyo, Japan)
Engineering Science Reference • © 2021 • 335pp • H/C (ISBN: 9781799834991) • US $225.00

701 East Chocolate Avenue, Hershey, PA 17033, USA
Tel: 717-533-8845 x100 • Fax: 717-533-8661
E-Mail: cust@igi-global.com • www.igi-global.com

Editorial Advisory Board

List of Reviewers

Table of Contents

Wendy Flores-Fuentes, Autonomous University of Baja California, Mexico
Moises Rivas-Lopez, Autonomous University of Baja California, Mexico
Daniel Hernandez-Balbuena, Autonomous University of Baja California, Mexico
Oleg Sergiyenko, Autonomous University of Baja California, Mexico
Julio Cesar Rodriguez-Quiñonez, Autonomous University of Baja California, Mexico
Javier Rivera-Castillo, Autonomous University of Baja California, Mexico
Lars Lindner, Autonomous University of Baja California, Mexico
Luis C. Basaca-Preciado, Centro de Enseñanza Técnica y Superior (CETYS), Mexico
Fabian N. Murrieta-Rico, Autonomous University of Baja California, Mexico
Felix F. Gonzalez-Navarro, Autonomous University of Baja California, Mexico

Lars Lindner, Autonomous University of Baja California, Mexico
Oleg Sergiyenko, Autonomous University of Baja California, Mexico
Moisés Rivas-Lopez, Autonomous University of Baja California, Mexico
Wendy Flores-Fuentes, Autonomous University of Baja California, Mexico
Julio C. Rodríguez-Quiñonez, Autonomous University of Baja California, Mexico
Daniel Hernandez-Balbuena, Autonomous University of Baja California, Mexico
Fabian N. Murrieta-Rico, Autonomous University of Baja California, Mexico
Mykhailo Ivanov, Autonomous University of Baja California, Mexico

Tatyana Strelkova, Kharkiv National University of Radio Electronics, Ukraine
Alexander Strelkov, Kharkiv National University of Radio Electronics, Ukraine
Vladimir M. Kartashov, Kharkiv National University of Radio Electronics, Ukraine
Alexander P. Lytyuga, Kharkiv National University of Radio Electronics, Ukraine
Alexander S. Kalmykov, Kharkiv National University of Radio Electronics, Ukraine

Detailed Table of Contents

This chapter presents the application of optoelectronic devices fusion as the base for those systems with non-linear behavior supported by artificial intelligence techniques, which require the use of information from various sensors for pattern recognition to produce an enhanced output. It also included a deep survey to define the state of the art in industrial applications following this tendency to identify and recognize the most used optoelectronic sensors, interconnectivity, raw data collection, data processing and interpretation, data fusion, intelligent decision algorithms, software and hardware instrumentation and control. Finally, it exemplifies how these technologies implemented in the industry can also be useful for other kinds of sector applications.

One focus of present chapter is defined by the further development of the technical vision system (TVS). The TVS mainly contains two principal parts, the positioning laser (PL) and the scanning aperture (SA), which implement the optomechanical function of the dynamic triangulation. Previous versions of the TVS uses stepping motors to position the laser beam, which leads to a discrete field of view (FOV). Using stepping motors, inevitable dead zones arise, where 3D coordinates cannot be detected. One advance of this TVS is defined by the substitution of these discrete actuators by DC motors to eliminate dead zones and to perform a continuous laser scan in the TVS FOV. Previous versions of this TVS also uses a constant step response as closed-loop input. Thereby the chapter describes a new approach to position the TVS laser ray in the FOV, using a trapezoidal velocity profile as trajectory.

Chapter 3

Tatyana Strelkova, Kharkiv National University of Radio Electronics, Ukraine
Alexander Strelkov, Kharkiv National University of Radio Electronics, Ukraine
Vladimir M. Kartashov, Kharkiv National University of Radio Electronics, Ukraine
Alexander P. Lytyuga, Kharkiv National University of Radio Electronics, Ukraine
Alexander S. Kalmykov, Kharkiv National University of Radio Electronics, Ukraine

The chapter covers development of mathematical model of signals in optoelectronic systems. The mathematical model can be used as a base for detection algorithm development for optical signal from objects. Analytical expressions for mean values and signal and noise components dispersion are cited. These expressions can be used for estimating efficiency of the offered algorithm by the criterion of detection probabilistic characteristics and criterion of signal/noise relation value. The possibility of signal detection characteristics improvement with low signal-to-noise ratio is shown. The method is proposed for detection of moving objects and combines correlation and threshold methods, as well as optimization of the interframe processing of the sequence of analyzed frames. This method allows estimating the statistical characteristics of the signal and noise components and calculating the correlation integral when detecting moving low-contrast objects. The proposed algorithm for detecting moving objects in low illuminance conditions allows preventing the manifestation of the blur effect.

Chapter 4

Jesús Elias Miranda-Vega, Autonomous University of Baja California, Mexico
Javier Rivera-Castillo, Engineering Institute, Autonomous University of Baja California, Mexico
Moisés Rivas-López, Engineering Institute, Autonomous University of Baja California, Mexico
Wendy Flores-Fuentes, Faculty of Engineering, Autonomous University of Baja California, Mexico
Oleg Sergiyenko, Engineering Institute, Autonomous University of Baja California, Mexico
Julio C. Rodríguez-Quiñonez, Faculty of Engineering, Autonomous University of Baja California, Mexico
Daniel Hernández-Balbuena, Faculty of Engineering, Autonomous University of Baja California, Mexico

An application of landslide monitoring using optical scanner as vision system is presented. The method involves finding the position of non-coherent light sources located at strategic points susceptible to landslides. The position of the light source is monitored by measuring its coordinates using a scanner based on a 45° sloping surface cylindrical mirror. This chapter shows experiments of position light source monitoring in laboratory environment. This work also provides improvements for the optical scanner by using digital filter to smooth the opto-electronic signal captured from a real environment. The results of these experiments were satisfactory by implementing the moving average filter and median filter.

Chapter 5

Tatyana A. Strelkova, Kharkiv National University of Radio Electronics, Ukraine
Alexander P. Lytyuga, Kharkiv National University of Radio Electronics, Ukraine
Alexander S. Kalmykov, Kharkiv National University of Radio Electronics, Ukraine

The chapter is devoted to the creation of a comprehensive approach to the physical and mathematical description of signals in optoelectronics in machine vision, taking into account the phenomena of interaction of optical radiation with system elements. A new methodology for the study of the statistical properties of input and output signals in optoelectronic systems is proposed, taking into account the availability of grouped statistical properties that do not obey the Poisson statistics. The basis is the joint use of wave and corpuscular description of signals in systems, stochastic flow theories, and elements of statistical detection theory. Information and energetic technology have been developed that integrates the theoretical justification of signal description under various observation conditions and decision-making methods.

Chapter 6

Oleksandr Ryazantsev, Volodymyr Dahl East Ukrainian National University, Ukraine
Ganna Khoroshun, Volodymyr Dahl East Ukrainian National University, Ukraine
Andrii Riazantsev, Volodymyr Dahl East Ukrainian National University, Ukraine
Tatyana Strelkova, Kharkiv National University of Radio Electronics, Ukraine

The rapid development and use of optical systems in measurement, navigation, and space technology to obtain accurate and detailed information about an object of observation is accompanied by the problems of transmitting high quality information through the optical system and processing of the obtained data. Integration of artificial intelligence systems in industry requires the creation and improvement of objective assessment and self-assessment systems. This is especially designed for automated recognition and classification systems. The problem of the object movement registration also contains some peculiarities such as background and main signal separation, noise influence and main objects selecting. Information about data quality is a set of properties that reflects the degree of suitability of specific information. It contains the data about objects and their relationship to achieve the goals of user requirements.

Chapter 7

Fernando Merchan, Universidad Tecnologica de Panamá, Panama
Martin Poveda, Universidad Tecnologica de Panamá, Panama
Danilo E. Cáceres-Hernández, Universidad Tecnologica de Panamá, Panama
Javier E. Sanchez-Galan, Universidad Tecnologica de Panamá, Panama

This chapter focuses on the contributions made in the development of assistive technologies for the navigation of blind and visually impaired (BVI) individuals. A special interest is placed on vision-based systems that make use of image (RGB) and depth (D) information to assist their indoor navigation. Many commercial RGB-D cameras exist on the market, but for many years the Microsoft Kinect has been used as a tool for research in this field. Therefore, first-hand experience and advances on the use of Kinect for the development of an indoor navigation aid system for BVI individuals is presented. Limitations that can be encountered in building such a system are addressed at length. Finally, an overview of novel avenues of research in indoor navigation for BVI individuals such as integration of computer vision algorithms, deep learning for the classification of objects, and recent developments with stereo depth vision are discussed.

Oleksandr Poliarus, Kharkiv National Automobile and Highway University, Ukraine
Yevhen Poliakov, Kharkiv National Automobile and Highway University, Ukraine

Remote detection of landmarks for navigation of mobile autonomous robots in the absence of GPS is carried out by low-power radars, ultrasonic and laser rangefinders, night vision devices, and also by video cameras. The aim of the chapter is to develop the method for landmarks detection using the color parameters of images. For this purpose, the optimal system of stochastic differential equations was synthesized according to the criterion of the generalized variance minimum, which allows to estimate the color intensity (red, green, blue) using a priori information and current measurements. The analysis of classical and nonparametric methods of landmark detection, as well as the method of optimal estimation of color parameters jumps is carried out. It is shown that high efficiency of landmark detection is achieved by nonparametric estimating the first Hilbert-Huang modes of decomposition of the color parameters distribution.

Jesús Elias Miranda Vega, Polytechnic University of Baja California, Mexico
Anastacio González Chaidez, Polytechnic University of Baja California, Mexico
Cuauhtémoc Mariscal García, Faculty of Engineering, Autonomous University of Baja
California, Mexico
Moisés Rivas López, Polytechnic University of Baja California, Mexico
Wendy Flores Fuentes, Faculty of Engineering, Autonomous University of Baja California,
Mexico
Oleg Sergiyenko, Engineering Institute, Autonomous University of Baja California, Mexico

The systems based on image recognition play an important role in many cases where inspection methods are critical for industrial processes. Machine vision is required in industry for monitoring and detecting objects in real-time applications. Industrial robots are increasingly used in a variety of industries and applications such as manipulators and mobile robots. These devices are necessary for dangerous work conditions and tasks that humans cannot do; however, these industrial robots can also do some human activities such as transport material, select, and identify objects. In order to substitute human capabilities is no easy task. Nevertheless, the use of artificial intelligence has been replacing some human activities for increased productivity. By supporting machine learning technologies, this chapter presents a k nearest neighbors' algorithm for image classification of mobile robots to detect and recognize objects.

Chapter 10

The problems of highlighting the main informational aspects of images and creating their adequate models are discussed in the chapter. Vision systems can receive information about an object in different frequency ranges and in a form that is not accessible to the human visual system. Vision systems distort the information contained in the image. Therefore, to create effective image processing and transmission systems, it is necessary to formulate mathematical models of signals and interference. The chapter discusses the features of perception by the human visual system and the issues of harmonizing the technical characteristics of industrial systems for receiving and transmitting images. Methods and algorithms of pattern recognition are discussed. The problem of conjugation of the characteristics of the technical vision system with the consumer of information is considered.

Preface

Examining Optoelectronics in Machine Vision and Applications in Industry 4.0 is the 2nd edition of Optoelectronics and Machine Vision 1st edition. This new edition obey to the importance of examining the optoelectronics characteristics and principles to be applied in novel manufacturing of machine vision systems, which play one of the most important roles nowadays, as stated by Valerie C. Coffey "machine vision is the eyes of the industry 4.0" (Coffey, 2018).

The industry 4.0 involves the development of cyberphysical systems able to see and take decisions. Is a revolution to replace humans in an automated and interconnected process. Also called the Industrial Internet of Things (IIoT). A multidisciplinary integration of optoelectronics devices for sensors and cameras, artificial intelligence algorithms, embedded systems, robust control, robotics, interconnectivity, big data and cloud computing (Sergiyenko, et al., 2019).

For machine vision, the computing imaging is based on the optoelectronics, which is the study of technologies and physical phenomena's, which incur from the combination of optics and semiconductor electronics. Optoelectronics combine the benefits of electronic data processing with the advantages of fast, electromagnetic and electrostatic noiseless transmission of light. Thereby electrical signals are converted and processed into optical signals and vice versa, wherein the produced light can either propagate in free space or in solid light-transmissive mediums. Under this scheme examining optoelectronics is the core of this book, developed with the aim of their application on industry 4.0.

HISTORY

The *History of Optoelectronics* is closely connected to the historical development of light sensors and optical communication. Before the electrical measurement technology was used for light measurement, light was quantified by comparing visually the intensity of two light sources. Thereby the contrast sensitivity of the human eye was taken in advantage. This method had the great disadvantage that the results depended on the subjective impression of the measuring person. Especially when measuring light with different colors and at low intensities, the comparisons were very critical. In the mid-18th century the Ohm's Law and the temperature dependency of the electrical resistance of metals was discovered. Temperatures were measured using these resistor-type thermometers and by development of more sensitive measurement circuits, even smaller temperature differences could be determined. The first *Bolometer* was developed around 1880 by Samuel Pierpont Langley (1834-1906) and possessed thin blackened strip of platinum, through which small temperature differences could be measured, caused by optical radiation incidence. The bolometer was further developed in Germany at the "Physikalisch-Technische Reichsanstalt" among others by the physicist Otto Lummer (1860-1925) and used to examine the radiation of a black body for the first time. Even today, bolometers are used in particular for measuring infrared radiation. However, the resistance temperature dependence is not the only effect that can be used to measure light. In 1821 Thomas Johann Seebeck discovered (1770-1831) the so-called thermoelectricity, which explains the appearance of an electrical voltage, when two metals are brought together in two points using different temperatures. This effect can also be used for radiation measurements and based on *Thermocouples* the *Thermopile* was developed by Leopoldo Nobili (1784-1835) in 1830. Thermopiles are used today for heat flow or broadband radiation measurement. Another effect for measurement of light is given by the pyroelectric effect, which was 1824 defined by the Scot David Brewster (1781-1868) and named pyroelectricity. This effect describes the property of certain piezoelectric crystals to respond to a temperature change in time with a charge separation, which produces a measurable electrical voltage. Due to lack of technologies, that effect could take place for more intensive use by means of infrared sensors until the middle of the 20th century. There are other light sensors which base on the thermal effect of radiation. Marcel J. E. Golay (1902-1989) in 1947 developed the *Golay Cell*, in which a gas absorbs radiation, thus warms and expands itself. The expansion of gas is measured using optoelectronics and depends on the intensity of the incident light. Current detectors are mostly based on the *Photoelectric Effect*, which describes the interaction of photons with material. Thereby an electron is released from a bond (atomic bond, valence band or conduction bond of a solid) absorbing a photon, wherein the energy of the photon must be at least as large as the binding energy of the electron. It is usually distinguished between the inner and the outer photoelectric effect. The *Inner Photoelectric Effect* describes the conductance increase in semiconductors by formation of free charge carrier pairs in the form of electrons and holes using the energy of an absorbed photon. The electron is thereby increased from the valence band into the conduction band, which represents a higher energy level. Therefore, the energy of the single photon must be at least equivalent to the band gap of the irradiated semiconductor. Since the size of the bandgap depends on the material and the energy of a photon by the Planck constant is directly dependent on its frequency, there is a maximum light wavelength up to which the inner photoelectric effect occurs. The inner photoelectric effect is the physical principle of many modern light sensors, like photoresistors, photodiodes, phototransistors and CCD sensors. Based on this effect, Werner Siemens (1816-1892) built in 1875 the first photoresistor, with which he could detect light magnitudes quantitatively using a photometer. One year later in 1876 the English William Grylls Adams (1819-1892) found a related effect, which is the base of all modern solar cells. In 1930 Walter Schottky (1886-1976) was able to demonstrate at the German physicist conference the first selenium solar cell that could provide enough electrical energy to power a small electric motor. Beside selenium, the photovoltaic effect was also researched using germanium and silicon. The American engineer Russell Ohl (1898-1987) studied 1940 pure silicon and observed a photovoltaic effect, which was much stronger than in selenium. He discovered the p-n junction of semiconductors during resistance measurements of silicon wafers, whose resistance drastically altered when exposed to light. Thus, the technique of the photodiode was born and represent since the state of the art. The *Outer Photoelectric Effect* describes the process of an electron release from a semiconductor or metal surface using light irradiation. This effect was first discovered in 1839 by the Frenchman Edmond Becquerel (1820-1891) when working with galvanic cells, which provided an electrical voltage when irradiated with light. Heinrich Hertz (1857-1894) inspected in 1886 the stimulating effect of ultraviolet light on the formation of electrical sparks. The research of Hertz was continued by his student Wilhelm Hallwachs (1858-1922) in 1887. The Russian Alexander Grigorjewitsch Stoletow (1839-1896) explained the generation of an electric current under irradiation of UV light and use of the outer photoelectric effect. This electric current represents a measure of the intensity of the incident UV light. The first *Photocell* was developed in 1893 by Julius Elster (1854-1920) and Hans Friedrich Geitel (1855-1923). It consists of an evacuated glass envelope containing a photo-cathode and an anode. By incident light, electrons are released from the photocathode and discharged from the opposite anode, resulting in a measurable photocurrent. The photocell has been improved over time, at which the released electrons are accelerated by electric and magnetic fields and by the recovered energy so-called secondary electrons generated when hitting more electrons. Due to this amplifier effect the photomultiplier was developed in the 1930 years, with also single photons could be detected. These vacuum tubes have been long time the most reliable light sensors, but were later replaced by light sensors based on semiconductor technology and using the inner photoelectric effect. In the 1950 years the first machine vision applications concepts were developed. James J. Gibson (1904-1979) introduces the concept of optical flow, based on his research the mathematical models for optical signals were developed. In 1960 Larry Roberts wrote his PhD thesis where he discussed about the possibility of extract 3D geometrical information from 2D perspective (Actually this concept is the base of stereo vision). In 1970's MIT open a Machine vision course and researchers starts to work in "low-level" tasks as edge detection and segmentation. In 1980's the development of smart cameras or intelligent cameras generates a takeoff in machine vision. The OCR (optical character recognition) and face recognition popularize the practical applications of machine vision systems. In 1990's the applications of machine vision systems become common, and a massive growth of machine vision industries leads to new innovations. In this decade neural networks start to work in conjunction with machine vision systems. In 1995 the firs machine vision interface is standardized, the IIDC 1.04 (Fire Wire or IEEE1394) was created, IIDC was not created in the machine vision industry but it was adopted by the industry. In 2000's several machine vision interfaces born, Camera link 1.0 in 2000, GigE Vision 1.0 in 2006, Camera Link 1.2 in 2007, IIDC 1.32 in 2008 and GigE Vision 1.2 in 2009. Then, in 2011 GigE Vision 2.0 and USB3 Vision 1.0 in 2012.

Nowadays, this big and rush evolution, and the development of machine vision systems and their interconnections reinvented the manufacturing, it transformed in a smart manufacturing with the aim of automated and autonomous processes able to adapt to changes on the fly thanks to the capability of a fast track of end-to-end process. This in response that since the beginning of mass production, one of the human dreams have been that machines will automatically make everything after the humans have gone home, this gave origin to the industrial revolution, which brought as benefits the population and average length of life increase, as well a new lifestyle with everyday tasks easier and with an improved quality of life. The Industrial Revolution 1.0 happened from the 1760s to 1830s, where men's physical strength was replaced by machines powered by steam fed by coal. The production was mechanized and the technology began to emerge. During the period from 1840s till 1870s the Industrial Revolution 2.0, also known as the technological revolution was distinguished by the replacement of steam power by electrical energy and the use of iron and steel for the development of the heavy industry. Was in the 20th century that the Industrial Revolution 3.0 befall with the digital development, beginning with the construction of the Z1 computer designed by Konrad Zuse from 1936 to 1937, a binary electrically driven mechanical calculator with limited programmability, continuing with the development of supercomputers and the communication technologies. The industry 4.0 is already underway, it was in 2011 first mentioned by Bosch at the Hannover Trade Fair, sounding like science fiction that the machines began to manage themselves and the production process, so they no longer needed manpower. Replacing human vision with machine vision systems based on optoelectronics to monitor, measure, and analyze, the artificial intelligence and the interconnectivity and cloud computing to make decisions, and the virtual handling of robots to operate the industrial processes.

IMPORTANCE OF EXAMINING OPTOELECTRONICS IN MACHINE VISION AND APPLICATIONS IN INDUSTRY 4.0

The importance of *Optoelectronic Devices* cannot be neglected in today's life and business and the numerous applications of optoelectronics components can no longer be covered by a course, which is why these applications must be broken down into different areas of study. From the resulting different fields of applications for optoelectronic devices, the present book will inspect two special areas: The development of optoelectronic devices and the application of these optoelectronics devices in machine vision, especially for industry 4.0. The development of optoelectronics can be carried out at different levels, where the device level and the system level describe two most important one. The device level is represented by optoelectronics components, which can be classified by *transmitters* and *receptors*. Transmitters are optoelectronics actuators, which produce light from the electricity that is laser and light - emitting diodes. Thereby the emission spectrum of the transmitters can be located in both visible and invisible (UV or IR) spectral range. Receptors are optoelectronics detectors, which convert light into electrical signals, like photoresistors, photodiodes, phototransistors, photomultipliers, CCD sensors, CMOS sensors and more. These receptors can be further divided into light sensors and image sensors. An optoelectronic system, containing a transmitter and a receptor is called a photoelectric sensor, among which belong the optocouplers for example. Besides transmitters and receptors exist other components, which are needed for light-transmission, light-amplification or light–modulation. The transmission of optical signals can be carried out through free space or in connection with waveguides and optical circuits. Optical modulators are devices, which modules light with defined characteristic. This may be for example a time or spatial variation in amplitude or phase. Light receptors and optical modulators get described in detail in book chapter 4. On the device level, optoelectronics have made a major advance in the last two decades, especially in measurement precision and measurement time by use of new theoretical concepts. These advances allow a higher resolution of measured objects with shorter measurement time, which in turn facilitate the application of real-time methods in image processing. A mobile robot, for example, can thus detect its surroundings more accurate and correct the trajectory of his current path in real time (Sergiyenko, et al., 2017).

Besides pure data transmission, optoelectronics are also used by machines, to receive and process visual information's. *Machine Vision* is based on the abilities of the human visual system though, which allows machines to identify and classify clearly patterns or objects from their near surroundings. Especially machine vision systems are currently used in industrial processes, in automation technology and quality management. Further, fields of applications can be found for example in traffic technology and in security technology. Few current research projects are dealing with the understanding of the real meaning and content of an image. Instead, are developed algorithms for object recognition and description. Features of these objects are then measured, objects are classified and based on these results, decisions are taken or processes are controlled. Since image understanding is very much related with design and application of computational methods, it can be classified as a branch of computer science, which has strong relationships with photogrammetry, signal processing and artificial intelligence. The used methods of machine vision originate from mathematics, in particular geometry, linear and superior algebra, probability, statistics and functional analysis. Typical tasks of machine vision can be defined by object recognition, identification and measurement of object features and geometric structures and object movements tracking. Hereby are fall back on algorithms of image processing, for example segmentation and on algorithms of pattern recognition, for example for classifying objects, and for predictions. Machine vision is used mainly in *industrial* and *natural environments*. Industrial environments usually possess standardized and consistent conditions, such as camera position, lighting, speed of a conveyer, background color, or the location of the measured object. Thereby these techniques of machine vision nowadays can be successfully used in such environments. The situation is completely different in natural environments, where the conditions change continuously and may not even be predictable. Such environments provide far more difficult requirements on used machine vision algorithms and the conditions often cannot be manipulated. Therefore, the development of robust and error-free running programs in natural environments is far more difficult, than in industrial environments (Rivas-Lopez, et al., 2019).

Machine vision for the industry applications has been huge benefited by the developments of interconnected devices. These recent technologies and philosophies related to the automation, interconnectivity, artificial intelligence, cloud computing, collaborative smart robotics and autonomy of industrial processes are in the framework of the Industry Revolution 4.0. A futurist industry strategy that merges real and virtual worlds for a live production integrated by robotics and machine to machine collaboration where optoelectronics is the core of the machine vision, the eyes of the industrial processes.

IMPACT OF THE BOOK

The topic of this book *Examining Optoelectronics in Machine Vision and Applications in Industry 4.0* takes place in current research and application of devices that detect and measure their environment by optoelectronics sensors in machine vision systems that take the role of the eyes in industrial processes for monitoring, inspection, continuous improving the quality of products and processes, by a widespread interconnection of devices for robots collaboration in smart productions. Besides, optoelectronics sensors can be found in a wide range of modern life, be it motion sensors, solar cells, facial recognition security systems, light sensors in digital cameras, and in the latest smartphones, collaborative robots and automatic guided vehicles just to name a few. Also, there is a wide range of applications for light transmitters, such as laser light for measurement of objects, optical fiber for the transmission of light signals and in general lighting lamps with a special frequency band for the excitation of chemical processes. The impact of the book can be defined by endorsement in studies and development of complex machine vision systems using optoelectronics devices. Through methodical presentation and description of individual machine vision systems, the engineer, scientist or practitioner shall be supported with fundamental knowledge and applications examples, which differ in complexity.

By these reasons, nowadays machine vision can be found in most of manufactures industries. Machine vision is used by electronic industry to perform inspections of integrated circuits, inspections that can not be performed by the human eye without the aid of microscopy technology. Throughout the manufacturing industry, machine vision is used to provide information about automatic machine movements, bonding and wire bonding and visual inspections only to mention a few applications. On microelectronics industry is used to provide position feedback of the components and loose wires. The food industry find useful machine vision system for sorting, measure size, and analyze the content of packages. On the television industry machine vision is used in white balance applications, PCB tracking, etc. On the automobile industry machine vision systems are used to look the flushness of sheet metals, control the paint quality, visual feedback in arm robot tracking. Practically every nowadays industry has been adopted machine vision system in some way or another and their interconnectivity is taking place every day for a further industry 4.0 philosophy.

The book provides a basic introduction to physical principles of optoelectronics sensors and their electrical wiring, as well as examples of computer algorithms for object recognition and analysis in machine vision systems.

Thus exists a necessary connection between optoelectronics and machine vision, which represents the main constituent of the present book. The reader thereby is introduced to the development and implementation of different optoelectronics systems in various machine vision applications. The book shall serve as a reference work for students and researchers of applied physics and optics, as well as a reference book for engineers and technicians in today's industry of machine vision.

ORGANIZATION OF THE BOOK

The book contains ten chapters, which are briefly described in the following:

Chapter 1 presents the application of optoelectronics devices fusion as the base for those systems with non-linear behavior supported by artificial intelligence techniques, which require the use of information from different sensors for pattern recognition to produce an enhanced output. The chapter gives a wide introduction and presentation of optoelectronics sensor fusion in machine vision. It describes the theoretical approach to fuse the measurement data of photodiodes and CCD devices, to obtain a lower detection error in angle measurement, while deals with theoretically background of sensor and data fusion. It includes a deep survey to define the state-of-the-art in industrial applications following this tendency, to identify and recognize the most novelty used optoelectronics sensors, interconnectivity, raw data collection, data processing and interpretation, data fusion, intelligent decision algorithms, software & hardware instrumentation and control. Finally, it exemplifies how these technologies implemented in the industry can also be useful for other kinds of sector applications.

Chapter 2 gives an introduction and summary of actually laser scanners used as optical devices. A definition of laser scanner is given, some applications areas are described and laser scanners categorized by design principles, measurement values and measurement methods. The chapter describes in detail four commonly used methods for spatial coordinate's measurement, also provides a special example of a Technical Vision System prototype and specifies the mainly used mechanical principles for laser beam positioning, with the mechanical actuators for moving scanning mirrors. The TVS mainly contains two principal parts, the Positioning Laser (PL) and the Scanning Aperture (SA), which implement the opto-mechanical function of the Dynamic Triangulation. Previous versions of the TVS uses stepping motors to position the laser beam, which leads to a discrete field of view (FOV). Using stepping motors, inevitably leads to dead zones, where 3D coordinates cannot be detected. One recent advance of this TVS is defined by the substitution of these discrete actuators by DC Motors, to eliminate dead zones and to perform a continuous laser scan in the TVS FOV. Previous versions of this TVS also uses a constant step response as closed-loop input. Thereby present chapter describes a new approach to position the TVS laser ray in the FOV, using a trapezoidal velocity profile as trajectory. Furthermore, the chapter describes the class of light and image sensors used for laser beam measuring and it gives some technical implementations about how to use light sensors. The post-processing of the measured physical signals gets described in the last section of this chapter.

Chapter 3 introduces a wide theoretically frame of statistical signal properties in optoelectronic systems and discusses a derivation of the signaltonoise relation when measuring optical radiation. Analytical expressions for mean values and signal and noise components dispersion are cited. These expressions can be used for estimating efficiency of the offered algorithm by the criterion of detection probabilistic characteristics and criterion of signal/noise ratio value. The possibility of signal detection characteristics improvement with low signal-to-noise ratio is shown. The chapter develops a mathematical model of signals in output plane of optoelectronic systems and shows the possibility of signal detection characteristics improvement. The method is proposed for detection of moving objects and combines correlation and threshold methods, as well as optimization of the interframe processing of analyzed frames sequence. This method allows estimating the statistical characteristics of the signal and noise components; calculate the correlation integral when detecting moving low-contrast objects. Besides, the proposed algorithm for detecting moving objects in low illumination conditions allows preventing the manifestation of the blur effects.

Chapter 4 summarizes systems and techniques of machine vision optical scanners for landslide monitoring. Thereby different systems are reviewed, which use fixed triangulation sensors with CCD or PSD sensors, Polygonal Scanners, Pyramidal and Prismatic Facets, Holographic scanners, Galvanometer and Resonant scanners or scanners with off-axis parabolic mirrors. This chapter is focused on the use of active laser scanners systems for landslide monitoring and compares the use of incoherent and coherent light in scanners using triangulation. One of the proposed methods involves finding the position of non-coherent light sources located at strategic points susceptible to landslides. The position of the light source is monitored by measuring its coordinates using a scanner based on a 45° sloping surface cylindrical mirror. In this chapter are shown experiments of position light source monitoring in laboratory environment. This work also provides improvements for the optical scanner by using digital filter to smooth the optoelectronics signal captured from a real environment. The results of these experiments were satisfactory by implementing the moving average filter and median filter.

Chapter 5 describes that the monograph is devoted to the creation of a comprehensive approach to the physical and mathematical description of signals in Optoelectronics in Machine Vision, taking into account the phenomena of interaction of optical radiation with system elements. A new methodology for the study of the statistical properties of input and output signals in optoelectronic systems is proposed, taking into account the availability of grouped statistical properties that do not obey the Poisson statistics. The basis is the joint use of wave and corpuscular description of signals in systems, stochastic flow theories, and elements of statistical detection theory. Information and energetic technology have been developed that integrates the theoretical justification of signal description under various observation conditions and decision-making methods.

Chapter 6 discusses the rapid development and use of optical systems in measurement, navigation and space technology to obtain accurate and detailed information about an object of observation is accompanied by the problems of transmitting high quality information through the optical system and processing of the obtained data. Integration of artificial intelligence systems in industry requires the creation and improvement of objective assessment and self-assessment systems. This is specially designed for automated recognition and classification systems. The problem of the object movement registration also contains some peculiarities as background and main signal separation, noise influence and main objects selecting. Information about data quality is a set of properties that reflects the degree of suitability of specific information. It contains the data about objects and their relationship to achieve the goals of user requirements.

Chapter 7 focuses on the contributions made in the development of assistive technologies for the navigation of blind and visually impaired (BVI) individuals. A special interest is placed on vision based systems that make use of image (RGB) and depth (D) information to assist their indoor navigation. Many commercial RGB-D cameras exist on the market, but for many years the Microsoft Kinect has been used as a tool for research in this field. Therefore, first-hand experience and advances on the use of Kinect for the development of an indoor navigation aid system for BVI individuals, is presented. Limitations that can be encountered in building such a system are addressed at length. Finally, an overview of novel avenues of research in indoor navigation for BVI individuals such as: integration of computer vision algorithms, deep learning for the classification of objects, and recent developments with stereo depth vision are discussed.

Chapter 8 introduces remote detection of landmarks for navigation of mobile autonomous robots in the absence of GPS, this is carried out by low-power radars, ultrasonic and laser rangefinders, night vision devices and also by video cameras. The aim of the chapter is to develop the method for landmarks detection using the color parameters of images. For this purpose, the optimal system of stochastic differential equations was synthesized according to the criterion of the generalized variance minimum, which allows to estimate the color intensity (red, green, blue) using a priori information and current measurements. The analysis of classical and nonparametric methods of landmark detection, as well as the method of optimal estimation of color parameters jumps is carried out. It is shown that high efficiency of landmark detection is achieved by nonparametric estimating the first Hilbert-Huang modes of decomposition of the color parameters distribution.

Chapter 9 explains how nowadays the systems based on image recognition play an important role in many cases where inspection methods are critical for industrial processes, it describes that machine vision is required in industry for monitoring and detecting objects in real-time applications, that the industrial robots are increasingly used in a variety of industries and applications such as manipulators and mobile robots. It justifies why these devices are necessary for dangerous work conditions and tasks that humans cannot do; and however, these industrial robots can also do some human activities such as transport material, select and identify objects. It states that substitute human capabilities is not an easy task. Nevertheless, the use of artificial intelligence has been replacing some human activities for increased productivity. By supporting machine learning technologies in this chapter is presented a k-nearest neighbors' algorithm for image classification of mobile robots to detect and recognize objects.

Chapter 10 discusses the problems of highlighting the main informational aspects of images and creating their adequate models. It explains how vision systems can receive information about an object in different frequency ranges and in a form that is not accessible to the human visual system. Also, describes that vision systems distort the information contained in the image. Therefore, to create effective image processing and transmission systems, why it is necessary to formulate mathematical models of signals and interference. The chapter discusses the features of perception by the human visual system and the issues of harmonizing the technical characteristics of industrial systems for receiving and transmitting images. Methods and algorithms of pattern recognition are described. The problem of conjugation of the characteristics of the technical vision system with the consumer of information is considered.

REFERENCES

Coffey, V. C. (2018). Machine Vision: The Eyes of Industry 4.0. *Optics and Photonics News*, *29*(7), 42–49. doi:10.1364/OPN.29.7.000042

Rivas-Lopez, M., Sergiyenko, O., Flores-Fuentes, W., & Rodríguez-Quiñonez, J. C. (2019). *Optoelectronics in Machine Vision-Based Theories and Applications*. IGI Global.

Sergiyenko, O., Flores-Fuentes, W., & Mercorelli, P. (Eds.). (2019). *Machine Vision and Navigation*. Springer Nature.

Sergiyenko, O., Flores-Fuentes, W., & Tyrsa, V. (2017). *Methods to improve resolution of 3D Laser Scanning*. Lambert Academic Publisher.

Introduction

Optoelectronics is the combinational study of the physics of light and electronic devices that utilize light (Cardinale, 2003). The importance of optoelectronics devices cannot be understated; optoelectronics devices are merged in everyday lives and must be studied on deep. From the lecturer's perspective, the application of optoelectronics devices covers a wide area of opportunities that can be difficult to cover in only one course. As result, the area of optoelectronics devices must be studied by particular fields or applications. Due that particular need, this book is intended to study two particular fields: The development of optoelectronics devices, and the application of these optoelectronics devices in machine vision.

The development of optoelectronics can be understood in several ways, it can be analyzed at device level or at system level. At device level exists representative components as photodiodes, phototransistors, optoisolators, photoresistors, LED, CCD, etc., where in last decades the trend is picoseconds devices. Picosecond optoelectronics devices are electronic components that can record in real time the physical and electronic events that take place on picosecond time scales (Lee, 2012). Several picosecond sensors systems have been developed as: picosecond pulse generation, picosecond photoconductors even high power picosecond lasers, and all these advances are taken place in the current machine vision systems. On other hand, at system level, optoelectronics devices have been taken major breakthroughs in the last 20 years with the development of the further technologies and application of theoretical concepts developed since the early 70s that can now be applied due the current technology. The next section describes some of the most representative optoelectronics systems applied in machine vision, a review of machine vision systems is provided, trending topics in both areas are discussed and finally the scope of the book is presented.

3D OPTOELECTRONIC SCANNING TECHNOLOGIES

3D Optoelectronics Scanners plays an important role in machine vision applications; they are complete optoelectronics systems that analyze real world objects or scenes to collect digital information about the shape, and possibly appearance as color. Different technologies can be used to develop 3D Optoelectronics Scanners, each technology with its own advantages and limitations so, it is important to mention the most widely used systems.

Laser Triangulation

Laser triangulation scanner is a kind of active optoelectronic system. The principle of triangulation cares on the projection of a light pattern, i.e., a line or a dot is projected by the laser over an object and captured by the digital camera (or other type of photodetector). The distance from object to system can be calculated by trigonometry, as long as you know a priori distance between the scanning system and the camera/photodetector (Franĉa, et al., 2005), (Básaca, et al., 2010).

The laser triangulation concept can be further categorized into three categories: 1) static triangulations (Petrov, et al., 1998), (Marshall, 2004); 2) dynamic triangulations (Sergiyenko, et al., 2009), (Sergiyenko[a], et al., 2009) (Sergiyenko, 2010), (Sergiyenko, et al., 2011); 3) laser line projection systems with complex sensory part suggesting multiple polyhedrons analysis at the same time in parallel processing (Bin, 2016).

Laser static triangulation principle (Marshall, 2004) in general can be based on two schemes represented in Fig 1 (a and b). The first one uses a fixed angle of emission and variable distance; the second one, on the contrary, fixed triangulation base and variable scanning angle (Sergiyenko, et al., 2020).

Figure 1. Two principles of laser triangulation (a – "with fixed angle of emission"; b - "with fixed triangulation distance", consists of a laser dot or line generator and a 1D or 2D sensor located at a triangulation distance from the light source.

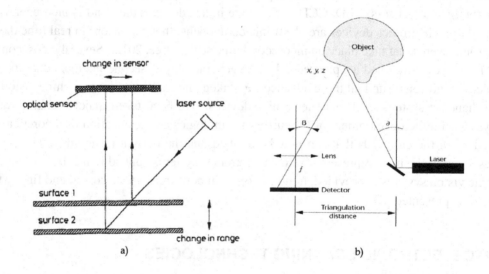

The laser line projection method in fact is based on the same principle, however to save operating time it uses the complex laser projecting head with several lasers, as the rule in pulsed mode, making n triangulations in parallel simultaneously. The general principle of such system (Bin, 2016) can be viewed on Fig.2.

Figure 2. Spatial angle measurement system.

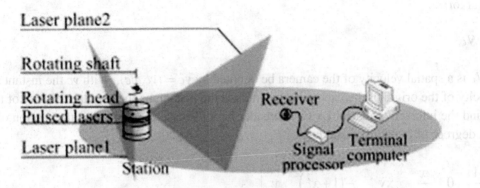

Stereo Vision

In Stereo Vision systems two or more cameras simultaneously capture the same scene. Every stereo vision system has their own function methodology, but in principle the stereo approach uses the following sequence of processing steps: Image acquisition, Camera Modeling, Feature Extractions, Correspondence Analysis and Triangulation. No further equipment and no special projections are required (Sansoni, Trebeschi, & Docchio, 2009). The advantages of the stereo approach are the simplicity and low cost; the major problem of stereo vision system is the correspondence problem. A variant of stereo vision system implies the use of multiple cameras or a single moving camera to obtain multiple images, in this case a constraint is that the scene doesn't contain moving parts (Sansoni, Trebeschi, & Docchio, 2009).

The general principle of such system (Chaumette, 2006) can be viewed on Fig.3. Any stereovision technical system is approach to a multicamera system (Rodriguez-Quiñonez, et al., 2017), (Ramírez-Hernández, L. R, et al., 2020). In the elementary case under consideration it is two cameras system. If a stereovision system is used, and a 3-D point is visible in both left and right images (see Figure 3), it is possible to use as visual features s (vector s contains the desired values of the scene/features):

$$\mathbf{s} = \mathbf{x}s = (\mathbf{x}l, \mathbf{x}r) = (xl, y\,l, xr, yr),$$

i.e., to represent the point by just stacking in **s** the x and y coordinates of the observed point in the left and right images.

For a 3-D point with coordinates $X = (X, Y, Z)$ in the camera frame, which projects in two images as a 2-D point with coordinates $x = (x, y)$, we have (Chaumette, 2006):

$$\begin{cases} x = X/Z = (u - c_u)/f\,\alpha \\ y = Y/Z = (v - c_v)/f, \end{cases} \tag{1}$$

where image measurements matrix $\mathbf{m} = (u, v)$ gives the coordinates of the image point expressed in pixel units, and $\mathbf{a} = (c\,u, c\,v, f, \alpha)$ is the set of camera intrinsic parameters: $c\,u$ and $c\,v$ are the coordinates of the principal point, f is the focal length, and α is the ratio of the pixel dimensions. In this case, we take $\mathbf{s} = \mathbf{x} = (x, y)$, the image plane coordinates of the point.

Taking the time derivative of the projection equations (1), we obtain the result which can be written in general form

$$\mathbf{X} = \mathbf{Lx} \, \mathbf{V}c, \tag{2}$$

where $\mathbf{V}c$ is a spatial velocity of the camera be denoted by $vc = (vc, \omega c)$, (with vc the instantaneous linear velocity of the origin of the camera frame and ωc the instantaneous angular velocity of the camera frame) and the interaction matrix $\mathbf{Lx}$ (we consider here the case of controlling the motion of a camera with six degrees of freedom) related to $\mathbf{x}$ is

$$L_x = \begin{bmatrix} \dfrac{-1}{Z} & 0 & \dfrac{x}{Z} & xy & -\left(1+x^2\right) & y \\[2mm] 0 & \dfrac{-1}{Z} & \dfrac{y}{Z} & 1+y^2 & -xy & -x \end{bmatrix} \tag{3}$$

Lx is an interaction matrix related to s, or feature Jacobian. In the matrix $\mathbf{Lx}$, the value Z is the depth of the point relative to the camera frame. Therefore, any control scheme that uses this form of the interaction matrix must estimate or approximate the value of Z.

Figure 3. A stereovision system

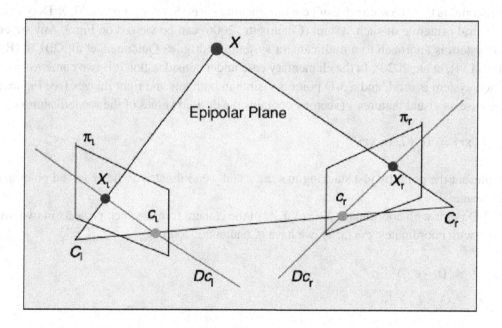

Structured Light

The structured light scanners project coded light patterns on the scene, such patterns can be: horizontal and vertical grids. However, more complex patterns as sine waves have been proposed (Scharstein & Szeliski, 2003). When locating such coded points in the image grabbed by a camera, the correspondence problem is solved with no requirement for geometrical constraint and the 3D structure of the scene is recovered by triangulation (Salvi, Pages, & Batlle 2004), (Gu, et al., 2013). Typical structured light 3D measurement system is made up of CCD cameras used for image capturing and a DLP projector used for digital fringe pattern image projection (Liu, et al., 2015).

Time of Flight

It is an active sensor that measures the travel time of infrared light, and consequently it does not interfere with the scene in the visual spectrum; its core components are a CMOS chip and an infrared light source (Cui, 2010). Optical range cameras use an emitted near-infrared light who is reflected back from the scene and capture by the camera. Depending on the camera model, the illumination can be modulated by sinusoidal or pseudonoise signal. The received signal is demodulated an the phase shift provides the time between emission and detection, indicating how far the light has traveled and thus allowing to indirectly measure the depth of the scene (Reynolds, et al., 2011).

CT Scan

Computed tomography uses a computer to combine many X-ray images taken from different angles, to produce a cross sectional tomographic. The CT scan refers to computer tomography and the most common used technology is X-ray, although, other types exists as positron emission tomography (PET) and single-photon emission computed tomography (SPECT). The uses of the mentioned technologies allow us to see inside of some objects without performing invasive procedures. The most common area of application is in medicine, in the detection of tumors, infarctions, hemorrhage, calcifications, bone trauma, etc..., (Hsieh, 2009).

MACHINE VISION

Machine vision is the technology and methods used to provide imaging-based automatic inspection and analysis for such applications as process control, and robot guidance in industry (Stanney & Hale, 2014), (Zou & Zhao, 2015). Machine vision is not a new topic, however, in the last 30 years the topic has moved on at accelerating rate, so, some topics that were considered tangential to machine vision fifteen or ten years ago, now need to be included on the literature (Rivas-Lopez, et al., 2018). An example of this topic is the role of artificial neural networks in machine vision. Artificial Neural Networks can be useful in circumstances where conventional methods do not provide desired outputs and the problem is strongly correlated to conditional probabilities; an example is the corner, edges/boundaries detection. To understand a complete machine vision system is important to know its main elements at hardware level (Image acquisition), and software level (Image processing) and the link between them (Miranda-Vega, et al, 2020), (Sanchez-Castro, et al., 2020).

Image Acquisition

The image acquisition process involves three steps; (1) energy reflected from the object of interest, (2) optical system which focuses the energy and (3) a sensor which measures the quantity of energy (Moeslund, 2012), (Rivas-Lopez, et al., 2013).

Energy

The source of the energy can be natural (as the sun light) or artificial. The purpose of lighting system is to provide some source of measurable energy. It is needed to consider a reliable source of light, as with the human eye, the vision system is affected by the quality and intensity of light (Zou & Zhao, 2015). Light sources may include incandescent, lasers, fluorescent, X-Ray, infrared and ultraviolet sources. Is important to bear in mind that the human eye is sensitive to the visible spectrum of light 400-700nm (nn = nanometer) and the visible electromagnetic spectrum of most of the cameras is different. The Lighting or Illumination is one of the most important parts of machine vision systems, as rule-of-thumb you can consider that illumination is 66% of the entire system design (Flores-Fuentes, et al., 2014), (Flores-Fuentes, et al., 2018).

Optical System and Image Sensor

The light reflected by the object of interest must be captured by a photodetector component; usually this component is a CCD or a camera; however in other cases the component can be a simple optical sensor. An optical system includes a diverse variety of components, a typical optical system in machine vision applications is composed of the sensor, lenses and filters; however, these components can vary according to the requirements. A first main component on an optical system is the Lens. A lens is an element which focuses the incoming light onto the sensor, a simple lens consists in a simple piece of material (typically glass) and compound lenses are made up of several simple lenses (Macnaughton, 2005), (Zou & Zhao, 2015). A second component to consider in the optical system is the Filter. Optical filters are devices most often used to selectively transmit certain wavelengths, (Hecht, 2015). There are different types of filters as Monochromatic filters who only allow a thin range of wavelengths, Dichroic filters allow to pass a selected wavelength and reflects the others, An absorptive filter absorbs some wavelengths and allow to pass others. Filters and Lenses work in conjunction to conditioning the image received by the photodetector (Miranda-Vega, et al., 2019).

The typical image sensor used in machine vision is the charge-coupled device CCD, however Active Pixel Sensors as complementary metal-oxide-semiconductor (CMOS) have acquired great popularity in the field of image sensors. CCD has higher light sensitivity, translates into better images in low-light conditions (Nilsson, 2008). CCD and CMOS sensors are often seen as competitors, however, each sensor has its own strengths and weaknesses. CCD sensors generally offer better light sensitivity than CMOS sensors; however CMOS sensors are less expensive than CCD. This can change in the following years since quality of modern CMOS sensors is rapidly increasing (Nilsson, 2008).

Image Processing

The image processing is the computer manipulation of image by mathematical operations or algorithms to enhance its quality or extract particular features. Typically image processing is implemented at software but with the development of current technology more algorithms of image processing are implemented at hardware. Image processing can be generally divided into three levels of processing: low level, intermediate level, and high-level. Low level processing performs image acquisition by transferring the electronic signal of the photodetector into a numeric form. Low level also realizes the pre-processing task, pre-processing of the image includes correction of distortion, noise removal, grey level correction and correction of blurring. Intermediate level processing performs image segmentation operations as: thresholding, edge-based segmentation and region-based segmentation. The intermediate level also includes the image representation, and description. High level processing performs recognition and interpretation, this can be done by classifiers or with multilayer neural networks (Brosnan, & Sun, 2004), (Zou & Zhao, 2015). The image processing application includes Classification, Feature extraction, Pattern recognition, Projection and Multi-Scale signal analysis (Mariscal-García, et al., 2020).

TRENDS OF OPTOELECTRONIC SYSTEMS AND MACHINE VISION

The trends in optoelectronics and machine vision can be divided into two areas, commercially and industrially. In both cases, as the computing capabilities increase, the new systems allow to perform more complex operations.

In the commercial field the use of smartphone allows the average person to acquire a whole machine vision system. The current advances in visual odometry in conjunction with the inertial measurement unit (IMU) of a smartphones enable to create novel systems (Castro-Toscano, et al., 2018), (Castro-Toscano, et al., 2017). Both visual odometry and the inertial navigation system (obtained from the IMU) can determine the movements and positions of a robot or a pedestrian according to the world coordinate system (Tomažič & Škrjanc, 2015). With camera-enabled mobile phones, the industry of application developers can create a new range of imaging applications, applications as automatic face detection for cataloging photos, user recognition for social media, room measurement for remodeling, camera based gesture control and augmented reality navigation are only a few trending applications for smartphones that support its technology in machine vision (Durini, 2014). It is noteworthy that vision systems that are based on statistical tools for classification as neural networks are already using these technologies in their algorithms.

Otherwise, in the industrial area, machine vision systems have acquired great acceptability in inspection and test process. The software trend in this area is to incorporate learning based methods as neural networks, Neuro-fuzzy systems and support vectors machine to machine vision systems to increase their classification capabilities (e. g. the classification of conforming products, and automatic optical inspection) (Bosque & Echanobe, 2014), (Shanmugamani, Sadique, & Ramamoorthy, 2015), (Zhang, et al., 2014). In the research area, the scientists are working with Biologically-Inspired new visual classifications (Tang & Qiao, 2014), (Kerr, et al., 2015), and new approaches as dynamic triangulation and fusion of different scanning technologies have been implemented (Rodríguez, 2011), (Rodríguez-Quiñonez, 2014), (Budzan & Kasprzyk, 2016), (Ivanov, et al., 2020), (Trujillo-Hernández, et al., 2019).

SCOPE OF THE BOOK

The book is aimed to acquaint the reader with technical issues and commonly used in the area of machine vision; with chapters focused on visual inspection and recognition, such as "Recognition system by using machine vision tools and machine learning techniques for mobile robots", where is introduced a supporting machine learning technologies, a k nearest neighbors' algorithm for image classification of mobile robots to detect and recognize objects. Like the chapter "Technical vision model of the visual systems for industry application" where the highlighting of the main informational aspects of images and their adequate models are described for discussing the features of perception by the human visual system and the issues of harmonizing the technical characteristics of industrial systems for receiving and transmitting images. And the chapter "Detection of landmarks by mobile autonomous robots based on estimating the color parameters of the surrounding area", and the chapter "Indoor navigation aid systems for the blind and visually impaired based on depth sensors" that focuses on the contributions made in the development of assistive technologies for the navigation of blind and visually impaired (BVI) individuals, with a special interest placed on vision based systems that make use of image (RGB) and depth (D) information to assist their indoor navigation. This book has wide covered areas of machine vision, considering not only passive systems but also active systems as Laser scanners, providing a review of different scanning technologies and realizes a comparison between them. Different coordinate measuring methods as Time-of-flight, Phasing and triangulation are compared; these methods are described in terms of content and mathematically. The chapter "Advances in laser scanners" provides an analysis of dynamic triangulation, and numerate several components of a full developed technical vision system. Additionally, the chapter includes a post-processing of measured values section, where the authors explain different approaches to improve the obtained 3D measurements. The chapter "Reducing the optical noise of machine vision optical scanners for landslide monitoring" provides a summary of the systems and techniques for this common application of machine vision. The reviewed systems covers scanners with fixed triangulation sensors using CCD or PSD, Polygonal Scanners, Pyramidal and Prismatic Facets, Holographic scanners, Galvanometer and Resonant scanners, Scanners with off-axis parabolic mirrors. This chapter is focused on the use of active laser scanners systems for Landslide monitoring, and compares the use of incoherent vs coherent light in scanners for scanners that use the triangulation method with landslide monitoring applications. Also, here are reviewed common CMOS Image sensors used in optoelectronics systems, this book explains some of the most important parameters such as noise, dynamic range, sensitivity and resolution in relation to the analog-mixed- signal circuit design, technology option/type of CMOS process, some of design methodologies used in the designing the CMOS Image sensor and update some of well-known circuit technique/architecture, such as in chapters "Statistical characteristics of optical signals and images in machine vision a systems", and "Informational model of optical signals and images in machine vision systems", that are devoted to the creation of a comprehensive approach to the physical and mathematical description of signals in Optoelectronics in Machine Vision, taking into account the phenomena of interaction of optical radiation with system elements.

Finally, the book pays some attention to very important aspect of machine vision and optical scanning: the methods applicable for rectification of optical signals with the aim to improve the quality of surrounding perception. For example, the chapters "Methods of Reception and Signal Processing in Machine Vision Systems", and "Optoelectronic devices fusion in machine vision applications" introduce us into the problem of such techniques applications, on the example of optical ray axis search, both hardware and applying mathematical formalism.

This is a complete book about Examining Optoelectronics in Machine Vision and Applications in Industry 4.0; it provides an accessible, well-organized overview of machine vision applications and properties that emphasizes basic principles. Coverage combines an optional review from key concepts, up to the progress gradually through more advanced topics. This book includes the recent developments on the field, emphasizes fundamental concepts and analytical techniques, rather than a comprehensive coverage of different devices, so readers can apply them to all current, and even future, devices. This book discusses important properties for different types of applications, such as analog or digital links, the theoretical methods of image processing in optoelectronic systems; optoelectronic devices integration for spatial coordinate measurement; machine vision applications on science, industry, agriculture and analysis of optoelectronic devices.

This book generally combines the optical, electrical and software behavior of a whole machine vision system, including interwoven properties as interconnections to external components and timing considerations. It provides the key concepts and analytical techniques that readers can apply to current and future devices. It also emphasizes the importance of time-dependent interactions between electrical and optical signals.

This is an ideal reference for graduate students and researchers in electrical engineering and applied physics departments, as well as practitioners in the machine vision industry.

REFERENCES

Básaca, L. C., Rodríguez, J., Sergiyenko, O. Y., Tyrsa, V. V., Hernández, W., Hipólito, J. I. N., & Starostenko, O. (2010, July). 3D laser scanning vision system for autonomous robot navigation. In *Industrial Electronics (ISIE), 2010 IEEE International Symposium on* (pp. 1773-1778). IEEE.

Bin, X., Xiaoxia, Y., & Jigui, Z. (2016). Architectural stability analysis of the rotary-laser scanning technique. *Optics and Lasers in Engineering, 78*, 26–34. doi:10.1016/j.optlaseng.2015.09.005

Bosque, G., del Campo, I., & Echanobe, J. (2014). Fuzzy systems, neural networks and neuro-fuzzy systems: A vision on their hardware implementation and platforms over two decades. *Engineering Applications of Artificial Intelligence, 32*, 283–331. doi:10.1016/j.engappai.2014.02.008

Brosnan, T., & Sun, D. W. (2004). Improving quality inspection of food products by computer vision—a review. *Journal of Food Engineering, 61*(1), 3–16. doi:10.1016/S0260-8774(03)00183-3

Budzan, S., & Kasprzyk, J. (2016). Fusion of 3D laser scanner and depth images for obstacle recognition in mobile applications. *Optics and Lasers in Engineering, 77*, 230–240. doi:10.1016/j.optlaseng.2015.09.003

Cardinale, G. (2003). *Optoelectronics: introductory theory and experiments*. Cengage Learning.

Castro-Toscano, M. J., Rodríguez-Quiñonez, J. C., Hernández-Balbuena, D., Lindner, L., Sergiyenko, O., Rivas-Lopez, M., & Flores-Fuentes, W. (2017, June). A methodological use of inertial navigation systems for strapdown navigation task. In *2017 IEEE 26th International Symposium on Industrial Electronics (ISIE)* (pp. 1589-1595). IEEE. 10.1109/ISIE.2017.8001484

Castro-Toscano, M. J., Rodríguez-Quiñonez, J. C., Hernández-Balbuena, D., Rivas-Lopez, M., Sergiyenko, O., & Flores-Fuentes, W. (2018). Obtención de Trayectorias Empleando el Marco Strapdown INS/KF: Propuesta Metodológica. *Revista Iberoamericana de Automática e Informática Industrial, 15*(4), 391–403. doi:10.4995/riai.2018.8660

Chaumette, F., & Hutchinson, S. (2006). Visual servo control PART I: Basic approaches. *IEEE Robotics & Automation Magazine, 13*(4), 82–90. doi:10.1109/MRA.2006.250573

Cui, Y., Schuon, S., Chan, D., Thrun, S., & Theobalt, C. (2010, June). 3D shape scanning with a time-of-flight camera. In *Computer Vision and Pattern Recognition (CVPR), 2010 IEEE Conference on* (pp. 1173-1180). IEEE.

Durini, D. (Ed.). (2014). *High performance silicon imaging: fundamentals and applications of cmos and ccd sensors*. Elsevier.

Flores-Fuentes, W., Miranda-Vega, J. E., Rivas-López, M., Sergiyenko, O., Rodríguez-Quiñonez, J. C., & Lindner, L. (2018). Comparison between different types of sensors used in the real operational environment based on optical scanning system. *Sensors (Basel), 18*(6), 1684. doi:10.339018061684 PMID:29882912

Flores-Fuentes, W., Rivas-Lopez, M., Sergiyenko, O., Gonzalez-Navarro, F. F., Rivera-Castillo, J., Hernandez-Balbuena, D., & Rodríguez-Quiñonez, J. C. (2014). Combined application of power spectrum centroid and support vector machines for measurement improvement in optical scanning systems. *Signal Processing, 98*, 37–51. doi:10.1016/j.sigpro.2013.11.008

França, J. G. D., Gazziro, M. A., Ide, A. N., & Saito, J. H. (2005, September). A 3D scanning system based on laser triangulation and variable field of view. In *Image Processing, 2005. ICIP 2005. IEEE International Conference on* (Vol. 1, pp. I-425). IEEE.

Gu, J., Nayar, S. K., Grinspun, E., Belhumeur, P. N., & Ramamoorthi, R. (2013). Compressive structured light for recovering inhomogeneous participating media. Pattern Analysis and Machine Intelligence. *IEEE Transactions on, 35*(3), 1–1.

Hecht, J. (2015). *Understanding fiber optics*. Jeff Hecht. doi:10.1117/3.1445658

Hsieh, J. (2009, November). *Computed tomography: principles, design, artifacts, and recent advances*. Bellingham, WA: SPIE.

Ivanov, M., Sergiyenko, O., Tyrsa, V., Lindner, L., Flores-Fuentes, W., Rodríguez-Quiñonez, J. C., ... Mercorelli, P. (2020). Influence of data clouds fusion from 3D real-time vision system on robotic group dead reckoning in unknown terrain. *IEEE/CAA Journal of Automatica Sinica, 7*(2), 368-385.

Kerr, D., McGinnity, T. M., Coleman, S., & Clogenson, M. (2015). A biologically inspired spiking model of visual processing for image feature detection. *Neurocomputing, 158*, 268–280. doi:10.1016/j.neucom.2015.01.011

Lee, C. H. (Ed.). (2012). *Picosecond optoelectronic devices*. Elsevier.

Liu, X., Sheng, H., Zhang, Y., & Xiong, Z. (2015, May). A Structured Light 3D Measurement System Based on Heterogeneous Parallel Computation Model. In *Cluster, Cloud and Grid Computing (CCGrid), 2015 15th IEEE/ACM International Symposium on* (pp. 1027-1036). IEEE.

Macnaughton, J. (2005). *Low vision assessment*. Elsevier Health Sciences.

Mariscal-García, C., Flores-Fuentes, W., Hernández-Balbuena, D., Rodríguez-Quiñonez, J. C., Sergiyenko, O., González-Navarro, F. F., & Miranda-Vega, J. E. (2020, June). Classification of Vehicle Images through Deep Neural Networks for Camera View Position Selection. In *2020 IEEE 29th International Symposium on Industrial Electronics (ISIE)* (pp. 1376-1380). IEEE.

Miranda-Vega, J. E., Rivas-López, M., & Flores-Fuentes, W. (2020). k-Nearest Neighbor Classification for Pattern Recognition of a Reference Source Light for Machine Vision System. *IEEE Sensors Journal*, 1. doi:10.1109/JSEN.2020.3024094

Miranda-Vega, J. E., Rivas-Lopez, M., Flores-Fuentes, W., Sergiyenko, O., Rodríguez-Quiñonez, J. C., & Lindner, L. (2019). Methods to reduce the optical noise in a real-world environment of an optical scanning system for structural health monitoring. In *Optoelectronics in machine vision-based theories and applications* (pp. 301–336). IGI Global. doi:10.4018/978-1-5225-5751-7.ch011

Moeslund, T. B. (2012). *Introduction to video and image processing: Building real systems and applications*. Springer Science & Business Media. doi:10.1007/978-1-4471-2503-7

Nilsson, F. (2008). *Intelligent network video: Understanding modern video surveillance systems*. CRC Press. doi:10.1201/9781420061574

Petrov, M., Talapov, A., Robertson, T., & Lebedev, A. Zhilyaev, a., and Polonskiy, L. (1998). Optical 3D Digitizers: Bringing Life to the Virtual World. *IEEE Computer Graphics and Applications, 18*(3), 28-37.

Ramírez-Hernández, L. R., Rodríguez-Quiñonez, J. C., Castro-Toscano, M. J., Hernández-Balbuena, D., Flores-Fuentes, W., Rascón-Carmona, R., Lindner, L., & Sergiyenko, O. (2020). Improve three-dimensional point localization accuracy in stereo vision systems using a novel camera calibration method. *International Journal of Advanced Robotic Systems, 17*(1), 1729881419896717. doi:10.1177/1729881419896717

Reynolds, M., Doboš, J., Peel, L., Weyrich, T., & Brostow, G. J. (2011, June). Capturing time-of-flight data with confidence. In *Computer Vision and Pattern Recognition (CVPR), 2011 IEEE Conference on* (pp. 945-952). IEEE. 10.1109/CVPR.2011.5995550

Rivas-Lopez, M., Flores, W., Rivera, J., Sergiyenko, O., Hernández-Balbuena, D., & Sánchez-Bueno, A. (2013). *A method and electronic device to detect the optoelectronic scanning signal energy centre*. Optoelectronics-Advanced Materials and Devices.

Rivas-Lopez, M., Sergiyenko, O., Flores-Fuentes, W., & Rodríguez-Quiñonez, J. C. (Eds.). (2018). *Optoelectronics in Machine Vision-Based Theories and Applications*. IGI Global.

Rodríguez, J. C., Sergiyenko, O. Y., Tyrsa, V. V., Basaca, L. C., & Hipólito, J. I. N. (2011, March). Continuous monitoring of rehabilitation in patients with scoliosis using automatic laser scanning. In Health Care Exchanges (PAHCE), 2011 Pan American (pp. 410-414). IEEE. doi:10.1109/PAHCE.2011.5871941

Rodríguez-Quiñonez, J., Sergiyenko, O., Hernandez-Balbuena, D., Rivas-Lopez, M., Flores-Fuentes, W., & Basaca-Preciado, L. (2014). Improve 3D laser scanner measurements accuracy using a FFBP neural network with Widrow-Hoff weight/bias learning function. *Opto-Electronics Review*, 22(4), 224–235. doi:10.247811772-014-0203-1

Rodríguez-Quiñonez, J. C., Sergiyenko, O., Flores-Fuentes, W., Rivas-Lopez, M., Hernandez-Balbuena, D., Rascón, R., & Mercorelli, P. (2017). Improve a 3D distance measurement accuracy in stereo vision systems using optimization methods' approach. *Opto-Electronics Review*, 25(1), 24–32. doi:10.1016/j.opelre.2017.03.001

Salvi, J., Pages, J., & Batlle, J. (2004). Pattern codification strategies in structured light systems. *Pattern Recognition*, 37(4), 827–849. doi:10.1016/j.patcog.2003.10.002

Sanchez-Castro, J. J., Rodríguez-Quiñonez, J. C., Ramírez-Hernández, L. R., Galaviz, G., Hernández-Balbuena, D., Trujillo-Hernández, G., . . . González-Navarro, F. F. (2020, June). A Lean Convolutional Neural Network for Vehicle Classification. In *2020 IEEE 29th International Symposium on Industrial Electronics (ISIE)* (pp. 1365-1369). IEEE. 10.1109/ISIE45063.2020.9152274

Sansoni, G., Trebeschi, M., & Docchio, F. (2009). State-of-the-art and applications of 3D imaging sensors in industry, cultural heritage, medicine, and criminal investigation. *Sensors (Basel)*, 9(1), 568–601. doi:10.339090100568 PMID:22389618

Scharstein, D., & Szeliski, R. (2003, June). High-accuracy stereo depth maps using structured light. In *Computer Vision and Pattern Recognition, 2003. Proceedings. 2003 IEEE Computer Society Conference on* (Vol. 1, pp. I-195). IEEE. 10.1109/CVPR.2003.1211354

Sergiyenko, O., Flores-Fuentes, W., & Mercorelli, P. (Eds.). (2020). *Machine Vision and Navigation.* Springer Nature. doi:10.1007/978-3-030-22587-2

Sergiyenko, O., Hernandez, W., Tyrsa, V., Devia Cruz, L., Starostenko, O., & Pena-Cabrera, M. (2009). Remote Sensor for Spatial Measurements by Using Optical Scanning. MDPI. *Sensors*, 9(7), 5477-5492.

Sergiyenko, O., Tyrsa, V., Basaca Preciado, L., Rodriguez-Quinonez, J., Hernandez, W., Nieto-Hipolito, J., & Starostenko, O. (2011). Electromechanical 3D optoelectronic scanners: resolution constraints and possible ways of its improvement. In Optoelectronic Devices and Properties. In-Tech.

Sergiyenko, O. Yu. (2010). Optoelectronic System for Mobile Robot Navigation. *Optoelectronics, Instrumentation and Data Processing*, 46(5), 414-428.

Sergiyenko, O. Y., Tyrsa, V. V., Devia, L. F., Hernandez, W., Starostenko, O., & Rivas-Lopez, M. (2009). Dynamic Laser Scanning method for Mobile Robot Navigation. *Proceedings of ICCAS-SICE, ICROS-SICE International Joint Conference*, 4884-4889.

Shanmugamani, R., Sadique, M., & Ramamoorthy, B. (2015). Detection and classification of surface defects of gun barrels using computer vision and machine learning. *Measurement*, 60, 222–230. doi:10.1016/j.measurement.2014.10.009

K. Stanney, & K. S. Hale (Eds.). (2014, July). Advances in Cognitive Engineering and Neuroergonomics. *AHFE Conference.*

Tang, T., & Qiao, H. (2014). Exploring biologically inspired shallow model for visual classification. *Signal Processing*, *105*, 1–11. doi:10.1016/j.sigpro.2014.04.014

Tomažič, S., & Škrjanc, I. (2015). Fusion of visual odometry and inertial navigation system on a smartphone. *Computers in Industry*, *74*, 119–134. doi:10.1016/j.compind.2015.05.003

Trujillo-Hernández, G., Rodríguez-Quiñonez, J. C., Ramírez-Hernández, L. R., Castro-Toscano, M. J., Hernández-Balbuena, D., Flores-Fuentes, W., . . . Mercorelli, P. (2019, October). Accuracy Improvement by Artificial Neural Networks in Technical Vision System. In *IECON 2019-45th Annual Conference of the IEEE Industrial Electronics Society* (Vol. 1, pp. 5572-5577). IEEE. 10.1109/IECON.2019.8927596

Zhang, B., Huang, W., Li, J., Zhao, C., Fan, S., Wu, J., & Liu, C. (2014). Principles, developments and applications of computer vision for external quality inspection of fruits and vegetables: A review. *Food Research International*, *62*, 326–343. doi:10.1016/j.foodres.2014.03.012

Zou, X., & Zhao, J. (2015). *Machine Vision Online Measurements. In Nondestructive Measurement in Food and Agro-products*. Springer Netherlands.

Chapter 1
Optoelectronic Devices Fusion in Machine Vision Applications

Wendy Flores-Fuentes
https://orcid.org/0000-0002-1477-7449
Autonomous University of Baja California, Mexico

Javier Rivera-Castillo
Autonomous University of Baja California, Mexico

Moises Rivas-Lopez
Autonomous University of Baja California, Mexico

Lars Lindner
https://orcid.org/0000-0002-0623-6976
Autonomous University of Baja California, Mexico

Daniel Hernandez-Balbuena
Autonomous University of Baja California, Mexico

Luis C. Basaca-Preciado
Centro de Enseñanza Técnica y Superior (CETYS), Mexico

Oleg Sergiyenko
https://orcid.org/0000-0003-4270-6872
Autonomous University of Baja California, Mexico

Fabian N. Murrieta-Rico
https://orcid.org/0000-0001-9829-3013
Autonomous University of Baja California, Mexico

Julio Cesar Rodriguez-Quiñonez
Autonomous University of Baja California, Mexico

Felix F. Gonzalez-Navarro
Autonomous University of Baja California, Mexico

ABSTRACT

This chapter presents the application of optoelectronic devices fusion as the base for those systems with non-linear behavior supported by artificial intelligence techniques, which require the use of information from various sensors for pattern recognition to produce an enhanced output. It also included a deep survey to define the state of the art in industrial applications following this tendency to identify and recognize the most used optoelectronic sensors, interconnectivity, raw data collection, data processing and interpretation, data fusion, intelligent decision algorithms, software and hardware instrumentation and control. Finally, it exemplifies how these technologies implemented in the industry can also be useful for other kinds of sector applications.

DOI: 10.4018/978-1-7998-6522-3.ch001

INTRODUCTION

With the continual technology grow due to innovations in sensors, devices, and systems, especially those based in a machine vision approach and also enhanced by the internet of things, and in consequence considered smart sensors, devices or systems, the industrial sector is in a revolution. Also, quotidian and modern human life habits have drastically changed with the incorporation of those devices and interconnectivity. Machine vision is supported and enhanced by optoelectronic devices, traditional signal, and image processing methods, as well as by novelty artificial intelligence algorithms, and the fusion of all these. The output from a machine vision system is information about the content of the optoelectronic signal, it is the process whereby a machine, usually a digital computer and/or electronic hardware automatically processes an optoelectronic signal and reports what it means. Machine vision methods to provide spatial coordinates measurement has developed in a wide range of technologies for multiples fields of applications such as robot navigation, medical scanning, structural health monitoring and industrial process. Each technology has specified properties that could be categorized as an advantage and disadvantage according to its utility to the application purpose. With these technologies implementations, most of the industrial processes can be automated and optimized, resulting in recognition, popularity and high revenue for the industrial trademark, and in a continuous lifestyle evolution.

The present chapter surged in the research continuity of a 3D Vision System for a mobile robot navigation application (Basaca-Preciado, 2014), a 3D medical laser scanner (Rodriguez-Quiñonez J. C., 2014), a structural and environmental health monitoring system (Rivera-Castillo, 2017), a measurement of vegetation vitality system (Lindner L. S.-L.-B.-F.-Q.-P., 2017), and an industrial measurement system (Lindner, 2016). With the objective of increasing the accuracy of the systems, digital and analog processing signals methodologies have been developed in order to find the energetic center of the optoelectronic signal handled by these systems. Into the task of systems overall robustness, its measurement data has been submitted to statistical analysis, finding a non-linear behavior of the systems, leading to the need of artificial intelligence application such as neuronal network (NN) (Rodriguez-Quiñonez J. a.-B.-L.-F.-P., 2014), k-Nearest Neighbor (k-NN) (Real-Moreno, 2018) and support vector machine regression (SVMR) (Flores-Fuentes W. a.-L.-N.-C.-B.-Q., 2014), in a modern approach to the prediction of the non-linear measurement error of the systems to compensate it. In the process of obtaining enough information from a measurement system to extract from it a model to predict its measurement error. It has been done a search of attributes to build the training dataset and test dataset. It has been found that pattern recognition can be enhanced by the sensor fusion and redundancy theory. This theory refers to the synergistic use of information from various sensors to achieve the task required by the system. Input data (attributes) are combined, fused and grouped for proper quality and integrity of the decisions to be taken by the artificial intelligent algorithm. Besides, the benefits can be extracted from the redundant data, the reduction of uncertainty and the increasing of precision reliability. For these reasons, the photodiodes and charge-coupled devices (CCD) are fusion in the task of robust systems building for machine vision by Spatial Coordinates Measurement (Weckenmann, 2009), (Elfes, 1992), (ZHANG, 2008), (Shih, 2015). The specific properties of both, their advantages and limitations have been considered, since, the photodiode is the sensor who gives place to the laser-scanning and the CCD is the sensor which gives place to the close-range photogrammetry. The energetic center of the laser optoelectronic signals from the photodiode and the energetic center of the image signal from the CCD sensor are detected to combine these sensors outputs, and to exploit their natural synergy, experimental results are presented to demonstrate the increase of systems accuracy.

BACKGROUND

Optoelectronics is the study of any devices that produce an electrically-induced optical output or an optically-induced electrical output and the techniques for controlling such devices (Marston, 1999), it includes generation, transmission, routing, and detection of optoelectronic signals in a widespread of applications (Dagenais, 1995). Wherever light is used to transmit information, tiny semiconductor devices are needed to transfer electrical current into optical signals and vice versa. Examples include light-emitting diodes, photodetectors and laser diodes (Piprek, 2003).

Most optoelectronics devices applications have focused on single sensors and relatively simple processes to extract specific information from the sensor. However, the use of multiple sensors by an optoelectronic device fusion technology delivers more advanced information and enable to develop intelligent and sophisticated optoelectronic systems, in special for machine vision applications (Yallup, 2014). More than one optoelectronic sensor may be needed to fully monitor the observation space at all times. Methods of combining multiple sensor data are in developing due to the availability and computational power of communications devices that support algorithms needed to reduce the raw sensor data from multiple sensors to convert it to the information needed by the system user (Klein, 2003).

Optoelectronic systems with only one sensor are not recommended for those optoelectronics signals with non-linear behavior, the observations made by one sensor could be uncertain and occasionally incorrect, a single sensor could not detect all the critical characteristics of the optoelectronic signal as could be done with multiple sensors, which could provide different information each one, in special if they are from different technologies, they can provide different information regarding the target under observation. Besides, if a unique sensor is used, and it fails, the optoelectronic system is shut down, while this sensor is not working. Several sensors deliver redundant information to increase reliability in case of errors or failure of a sensor. The redundant information can reduce uncertainty and increases the accuracy with which the characteristics are perceived by the optoelectronic system. Further information from various sensors is obtained to characterize the environment, perceived in a way that would be impossible using only the information for each sensor separately (Castanedo, 2013).

Machine vision is supported and enhanced by optoelectronic devices, the output from a machine vision system is information about the content of the optoelectronic signal, it is the process whereby a machine, usually a digital computer and/or electronic hardware automatically processes an optoelectronic signal and reports what it means. That is, it recognizes the content, it not only locate the content but inspects it as well. Machine vision includes two components, measurement of features and pattern classification based on those features. Defining the pattern classification as the process of making a decision about measurement, with a set of measurements knowledge about the possible classes could be obtained, leading to making a decision (Snyder, 2010).

The machine vision technology has a widespread of applications, as is the automatic target recognition, 3D vision, vehicle vision, and the industrial inspection, that make use of it for replace or complement manual inspections and measurements, to automate the production, in order to increase production speed and yield, as well as to improve product quality.

In special, machine vision methods to provide spatial coordinates measurement has been developed in a wide range of technologies for multiples fields of applications such as robot navigation, medical scanning, structural and environmental health monitoring, measurement of vegetation vitality, and industrial measurement. Each technology has specific properties that could be categorized as advantages and disadvantages according to its utility to the application purpose.

The computer vision has guide the machine vision to the tendency of duplicate and/or enhance the abilities of human vision by electronically perceiving and understanding an image with a high a dimensional data, increasing the data storage and the time processing requirements, due to the complexity of algorithms to extract important patterns and trends to understand what data says (learning data). For some applications it is not necessary to acquire the whole vision of a scene as makes the human vision to extract the features of interest, it is possible to only extract the significant characteristics from the scene by the use of a positioning laser (PL) (Rodriguez-Quiñonez J. a.-B.-L.-F.-P., 2014), a rotatory mirror scanner and dynamic triangulation and/or by the extraction of important features from images captured through the use of camera with CCD (Charge Coupled Device) (Hou, 2011).

Researches previously published has found that the 3D modeling of objects and complex scenes constitutes a field of multi-disciplinary challenges and difficulties, in the going-over accuracy, reliability, quality, portability, low cost and automatization of the whole procedure. For these reasons, although there is a wide variety of sensors, optical scanning and digital cameras play the main role in this application, because even though these two types of sensors can work in a separate fashion, it is when they are fusion together when the best results are attained. The sensor fusion, in particular concerning these both technologies appears as a promising possibility to improve the data acquisition and the geometric and radiometric processing of these data (Gonzalez-Aguilera, 2010).

Optoelectronic Devices Fusion in Machine Vision Applications is expected to play an increasingly important role in the near future, by its desired characteristics, such as: optimization, deduction, robustness, flexibility, inference, accuracy, reliability, speed and cost-effectiveness.

By this motivation it is presented a review of modern optoelectronic devices, including light-emitting diodes, edge-emitting lasers, vertical-cavity lasers, electro-absorption modulators, a novel combination of amplifier and photodetector (Piprek, 2003), and CCD, to know the diverse optoelectronic devices, and which are the most suitable according to the needs in numerous machine vision applications.

The available sensor fusion techniques are classified into three nonexclusive categories: a) data association, b) state estimation, and c) decision fusion. Also, architectures and algorithms are introduced.

Machine vision operation is described, an introduction to the methods for imaging, processing, analysis, communication and actions are developed.

The current optoelectronic scanning technology and image processing for 3D reconstruction techniques are fusion, and a brief introduction to the optimal hardware architecture is reviewed.

OPTOELECTRONIC SENSORS FUSION IN MACHINE VISION

Spatial Coordinate Measurement

The spatial coordinate measurements have been developed through a lot of technologies during history. Between the artifacts, devices or systems we can find the theodolite, used for measuring horizontal and vertical angles; the inertial navigation system (INS), based on measurements obtained from the inertial measurement unit (IMU); the global positioning system (GPS), based on the calculation of time consumed in the communication of a receptor with at least four satellites; the radar, based on the measurement of electromagnetic wave traveling distances versus time; the stereoscopic vision systems (SVS), based on the identification of similarities between digital images acquired by cameras; and the optical scanning systems (OSS), based on the sensing of optical signals, such as the laser imaging detection and ranging (LiDAR), the laser doppler velocimetry (LVD) and the optoelectronic system proposed in this chapter.

Machine vision methods to provide spatial coordinates measurement has developed in a wide range of technologies, and their outputs data fusion. The computer vision has guided the machine vision to the tendency of duplicate the abilities of human vision by electronically perceiving and understanding an image with a high a dimensional data, increasing the data storage requirement, and the time processing due to the complexity of algorithms to extract important patterns and trends and understand what data says (learning data). For some applications it is not necessary to acquire the whole vision of a scene as makes the human vision to extract the features of interest (Sonka, 2014), it is possible to only extract the significant characteristics from the scene by the use of an OSS based on a positioning laser (PL), a rotary mirror scanner and dynamic triangulation, for detailed system overall see (Basaca-Preciado, 2014), (Rodriguez-Quiñonez J. C., 2014) (Flores-Fuentes W. a.-L.-N.-C.-B.-Q., 2014), (Flores-Fuentes W. a.-L.-Q.-B.-C., 2014).

1. Optoelectronic Devices

Optoelectronic devices take advantage of sophisticated interactions between electrons and light, it combines electronics and photonics. Photons are the smallest energy packets of light waves and their interaction with electrons, the energy exchange between them is the key physical mechanism in optoelectronic devices. Optoelectronics brings together optics and electronics within a single device, a single material, semiconductors designed to allow for the transformation of light into current and vice versa.

a. Photodetector Devices

The purpose of any photodetector is to convert light (photons) into electric current (electrons) when electromagnetic waves inside over photodetector material surface. The principle of operation of these devices is based on the transition of an electron from the valence band to the conduction band by the absorption of a photon. The condition in (1) should be accomplished to produce the photo-current process, where h is the Planck constant, v is the photon frequency and E_g is the prohibited band (band-gap) of the material to go from the valence band to the conduction band. Substituting (2) in (1) is obtained (3), where c is the light speed and λ is the light wavelength.

$$hv \geq E_g. \tag{1}$$

$$v = \frac{c}{\lambda}. \tag{2}$$

$$\lambda \leq \frac{hc}{E_g}. \tag{3}$$

Photodetectors are among the most common optoelectronic devices; they are used to automatic open doors, detect the signal from infrared remote controls, and record pictures in modern cameras. Photodetectors are fabricated from various semiconductor materials since the bandgap needs to be smaller than the energy of the photons detected. Photon absorption generates electron–hole pairs which are subsequently separated by the applied electrical field. Depending on the desired use, photodetectors are designed in many different ways and can be classified in photodiodes, phototransistors, infrared-detectors and charge-coupled devices (CCD), as shown in Figure 1. Modern photodetectors include an amplifier; the amplification photodetector simultaneously amplifies and absorbs the incoming light.

Figure 1. Photodetectors classification

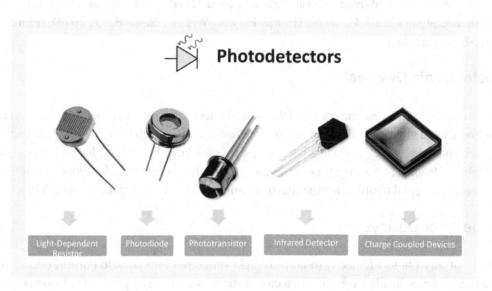

i. Light-Dependent Resistor

The Light-Dependent Resistor (LDR) is a photo-conductive cell with a spectral response similar to the human eye spectral response, although exists photoresistors with a wide spectral range that includes infrared and ultraviolet. The resistance of this cell increases and decreases according to the intensity of the light, that inside on it; when the light insides the LDR the resistance of this decreases, and the opposite happens when the light does not inside or decrease, the resistance of this device increases so it is also known as a photoresistor. They are commonly constructed with cadmium sulfide (CdS) and cadmium selenide (CdSe). The response time of this element ranges between 1 and 10 tenths of a second, this characteristic can be used to filter these small variations of light for other photodetectors that are more sensitives to small changes of light, as in the case of photodiodes and phototransistors.

ii. Photodiode

Photodiodes are light detecting devices fabricated in semiconductor materials when a photon of sufficient energy strikes a diode, it creates an electron-hole pair. This mechanism is also known as the inner photoelectric effect. If the absorption occurs in the junction's depletion region, or one diffusion length away from it, these carriers are swept from the junction by the built-in electric field of the depletion region. Thus holes move toward the anode, and electrons toward the cathode and a photocurrent is produced. The total current through the photodiode is the sum of the dark current (current that is generated in the absence of light) and the photocurrent, so the dark current must be minimized to maximize the sensitivity of the device. Photodiodes are similar to regular semiconductor diodes except that they may be either exposed or packaged with a window or optical fiber connection to allow light to reach the sensitive part of the device (Tavernier, 2011). The most common photodiodes are the P-N photodiode, PIN photodiode, avalanche photodiode, waveguide photodiode and amplification photodiode (Syms, 1992), (Piprek, 2003).

iii. Phototransistor

The phototransistor is an extension of a photodiode. Due to the photodiode current is very small, for some applications it requires to be amplified by a transistor. The phototransistor can be defined as a transistor with three external connections, collector (C), base (B) and emitter (E), two junctions (CB & BE), with a window that allows the light to inside over its sensitive material (CB junction), generating carriers in to activate it. It combines the properties of photodiodes and transistors, it is the light detection and the gain. The collector current can be regulated through the base current, but it is also affected by the light inside in its junctions. The incident photons generate electron-hole pairs near the large CB junction. The reverse polarization voltages of the CB junction, bring the holes to the surface of the base and electrons to the collector. The directly polarized BE junction causes the holes to circulate from base to emitter while electrons flow from the emitter to the base. The main function of the phototransistor is carried out with the electrons injected from the emitter crossing the small region of the base and reaching the collector that is more positive. This flow of electrons constitutes a collector current induced by light. The photo-induced electron-hollow pairs contribute to the base current and if the phototransistor is connected in a common emitter configuration, the light-induced base current appears as a collector current multiplied by β .or hfe. One of the main advantages of phototransistors over photodiodes is that they are more sensitive given the gain effect of transistors, whose sensitivity is around 100 to 1000 times higher. These phototransistors are also made of semiconductors and can usually be fabricated in bipolar NPN and PNP junction, although there are also phototransistors based on field-effect transistors (FET) whose advantage is that the offset voltage is less than the bipolar junction transistor (BJT). Another of the main advantage of phototransistors is ist responds speed.

iv. Infrared-Detectors

This detector is sensible to infrared radiation (IR). Thermal and photonic photodetectors are the main types of infrared-detectors. The effects of the incident IR radiation can be followed through many temperature-dependent phenomena; a thermal detector type has no wavelength dependence. Major characteristics indicating infrared detector performance are photosensitivity, the photodetector is wavelength dependent. In comparison with visible and ultraviolet rays, infrared radiation has small energy. To increase infrared detection efficiency, the detector should be cooled. The signal output from a detector is generally quite small and needs to be amplified. When designing a preamplifier, it is considered the impedance match between the detector and the amplifier, if the incident light is modulated by a chopper, the use of a tuned amplifier is necessary to reduce noise. If the detector is cooled, it is also practical to cool the amplifier. IR photodetectors are classified in intrinsic type and extrinsic type (Ge:Au, Ge:Hg, Ge:Cu, Ge:Zn, Si:Ga, Si:As), the intrinsic type is classified on Photoconductive type (PbS, PbSe, InSb, HgCdTe) and Photovoltaic type (Ge, InGaAs, Ex. InGaAs, InAs, InSb, HgCdTe) (Rogalski, 2002). The most common infrared detectors are InGaAs PIN photodiode and linear image sensor, PbS and PbSe photoconductive detector, InAs and InSb photovoltaic detector,MCT (HgCdTe) and InSb photoconductive detector, and two-color detector (McClure, 2003) (Tribolet, 2008).

v. Light-Emitting Diode (LED) as Photodetector

Light-emitting diodes (LED´s) are devices designed to emit radiant energy in a very narrow wavelength spectral region, however, these devices also can be used as photosensors, as described at (Brooks, 2001), and (MIMS, 2000). The photo-current generated by the light incidence into the LED is very small, so it has to be amplified by a high impedance circuit. According to the LED principle of operation when it is polarized or subjected to an electric field between its terminals, the electrons acquire enough energy and move from the valence band to the conduction band, however, these electrons can only be maintained for a very short time in a state of greater energy and when they return to the lower energy valence band they emit radiant energy (light) at a certain wavelength that depends on the LED manufacturing design. To use the LED as a photodetector the process is inverted when the LED is exposed to a radiant energy of the same or greater wavelength, electron-hollow pairs or charge carriers are generated resulting in a decrease in the resistance of the semiconductor material, allowing the fluid of an electrical current called a photocurrent, due to the energy of the photons are ceded to the electrons.

vi. Charge-Coupled Devices (CCD)

CCD is the abbreviation for charge-coupled device, also known as CCD image sensor. CCD sensors are typically built with silicon technology. Silicon-based integrated circuits (ICs), consisting of a dense matrix of photodiodes or photogates that operate by converting light energy, in the form of photons, into electronic charges. When a photon strikes a silicon atom in or near a CCD photosite, the photon usually produces a free electron and a hole via the photoelectric effect.

The primary function of the CCD is to collect the photogenerated electrons in its potential wells (or pixels) during the CCD's exposure to radiation, and the hole is then forced away from the potential well and is eventually displaced into the Si substrate. While more light that is incident on a particular pixel, the higher the number of electrons that accumulate on that pixel. The output signal is then transferred to the computer for image regeneration or image processing (Li F. a., 2006).

The pixels in a CCD sensor can be arranged in various configurations, the linear and area array are the classical architectures. A linear or line scan CCD sensor consists of a single line of pixels, adjacent to a CCD shift register that is required for the read-out of the charge packets. The isolation between the pixels and the CCD register is achieved by a transfer gate. Typically, the pixels of a linear CCD are formed by photodiodes. Two-dimensional imagining is possible with area array CCD sensors; the entire image is captured with one exposure, eliminating the need for any movement by the sensor or the scene. As charge-coupled device (CCD) technology matures it dominates digital imaging, for commercial and industrial manufacturing applications (Janesick, 2001).

Conventional CCDs are designed for the front-illuminated mode of operation. Front illuminated CCDs are quite economical to manufacture by using standard wafer fabrication procedures and are popular in consumer imagining applications, as well as industrial-grade applications. However, front-illuminated CCDs are inefficient at short wavelengths (e.g., blue and UV) due to the absorption of photons by the polysilicon layers if photogates are used as the pixel element. Typically, conventional front-illuminated CCDs are adequate for low-end applications and for consumer electronics. But, for large, professional observatories and high-end industrial inspection systems that demand that require extremely sensitive detectors, conventional thick frontside-illuminated chips are rarely used. Thinned back-illuminated CCDs offer a more compatible solution for these applications. Backside-thinned back-illuminated chips are rarely used. Thinned back-illuminated CCDs offer a more compatible solution for these applications. Backside-thinned back-illuminated CCDs exhibit a superior responsively, remarkably in the shorter wavelength region (Li F. M., 2003).

b. Conditioners for Photodetectors

For conditioning of the opto-electronic signal output from photodetectors, active devices such as Junction Field-Effect Transistor (JFET) and operational amplifiers (OPAM´s) can be used. The JFET is an element whose main function is to control the current flowing through a circuit. The magnitude that controls these elements is the application of a voltage at the gate terminal. One of the main characteristics of JFET is that it has a high input impedance, unlike BJT that has a low input impedance. Typically the terminals are called source (S), drain (D) and gate (G). The OPAM is a differential amplifier that has a high gain with high input impedance and low output impedance. This electronic element is a very versatile device since they have different uses such as an amplifier, peak detector, voltage comparator, adder, subtractor. An application that stands out such as its use for the design of analog active filters. Since OPAM´s are integrated circuits, there are different models that use BJT such as UA741 and others that are designed based on JFET like TL081.

While for conditioning the optical signal from photodetectors (without converting the optical signal into an electrical signal), the semiconductor optical amplifiers (SOA) and the fiber optic amplifiers are the ones that perform this task. The principle of operation of the SOA is based on the operation of the semiconductor laser, with the exception of the SOA has no feedback. When light travels through the active region of SOA, it causes electrons to lose energy in the form of photons and return to the elemental state. These stimulated photons have the same wavelength as the optical input signal, which causes this signal to be amplified at the output terminal. One of the most modern optical amplifiers used in optical communications systems is the erbium-doped fiber amplifier, between its advantages, are that they have cylindrical geometry which is compatible with fiber transmission, so the coupling is facilitated. Another advantage is that this device is insensitive to the polarization of the input and normal variations in the real environment. The third advantage is that they provide high gain with low noise

c. Light Emitting Devices (LED´s)

The LED´s are based on the electroluminescence optical-electrical phenomenon, electrical generation of light through spontaneous emission. It is when a material emits light in response to the passage of an electric current, radiative recombination of electrons and holes are done in the material, usually a semiconductor. In semiconductor electrons and holes are separated by doping the material to form a p-n junction, then recombination of electrons and holes are done, and it is when the excited electrons release their energy as photons (light). As an isolated lump of material will not emit significant quantities of light, in thermal equilibrium (room temperature) the number of downward electron transitions is extremely small. To improve the optical output, the material should be moved far from equilibrium, so that the rate of spontaneous emission is considerably increased. This might be done by taking (for example) a p-type material, which already contains a large holes density, and pouring electrons into it by a forward-biased p-n junction. Spontaneous emission generates photons that travel in random directions, so the emission is isotropic, the emission is also un-polarized, and the output is incoherent, with a spectrum consisting of a broad range of wavelengths, consequently, the LED cannot be used as a source for high-speed, long-distance optical communications; the dispersion caused by such an extended-spectrum would be far too large. LED transmitters are therefore restricted to short-haul applications.

A single photon is able to generate an identical second photon by stimulating the recombination of an electron-hole pair. This photon multiplication is the key physical mechanism of lasing. The second photon exhibits the same wavelength and the same phase as the first photon, doubling the amplitude of their monochromatic wave. Subsequent repetition of this process leads to strong light amplification. However, the competing process is the absorption of photons by the generation of new electron-hole pairs. Stimulated emission prevails when more electrons are present at the conduction band level than at the valence band level. This carrier inversion is the first requirement of lasing and it is achieved at pn-junctions by providing conduction band electrons from the n-doped side and valence band holes from the p-doped side. The photon energy is given by the band gap, which depends on the semiconductor material. Continuous current injection into the device leads to continuous stimulated emission of photons, but only if enough photons are constantly present in the device to trigger this process. Therefore, optical feedback and the confinement of photons in an optical resonator is the second basic requirement of lasing.

The absorption and stimulated emission process produced in a LED is the principle of light amplification by stimulated emission of radiation (LASER), the exponentially growing wave produces an amplification ought to offer a method of light generation. To achieve this, the electron density is raised far above the equilibrium level by the injection of electrons across a forward-biased diode. The laser is a device based on the stimulated emission of electromagnetic radiation. It emits light through a process of optical amplification; it emits spatially coherently light as a tight spot, achieving a very high irradiance, or they can have very low divergence in order to concentrate their power at a great distance. The laser enables applications such as cutting, welding and lithography for optical disk drives, laser printers, barcode scanners, fiber-optic, optical communication, surgery, skin treatments, device enforcement for marking targets and measuring range and speed, and lighting display.

The first semiconductor lasers consisted of heavily doped p-n junctions with cleaved facets. This construction is inefficient, as there is no defined region in which recombination can take place. Carriers can be lost to diffusion before recombination occurs. As a result, these early devices required a lot of current to reach the threshold. Threshold currents were reduced with the developments of the double-heterostructure laser, which has a thin region of semiconductor with a smaller energy gap sandwiched between two oppositely doped semiconductors with wide bandgap energy. When forward biased, carriers flow into the active region and recombine more efficiently because of the potential barriers of the heterostructure confine the carriers to the active region. There is an added advantage of guided the laser light because the refractive index of the cladding layers is less than the active region. See Figure 2.

Figure 2. Light-emitting devices classification

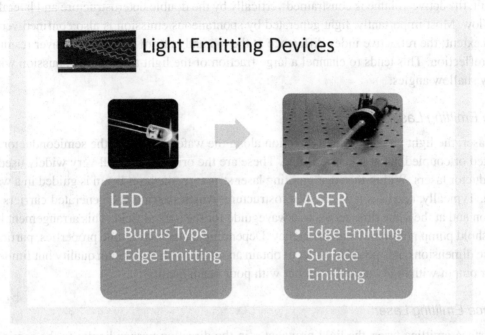

i. Burrus-Type LED

In place of the simple p-n junction, a more complicated structure is used for the Burrus-type LED. Typically, this might be a P-n-N double heterostructure, containing three layers: a p-type layer with a wide energy gap, a narrow-gap n-type layer, and a wide-gap n-type layer. A single heterojunction provides potential barriers of different heights for electrons and holes; the double heterostructure provides a high barrier at different positions for electrons and holes. In this way, the recombination region may be limited to a defined region of space, resulting in an increase in the radiative recombination efficiency. This is a high-efficiency device, suitable for use with multimode fiber. To fabricate the device, a double heterostructure is first grown on a substrate, and a SiO2 isolation layer is then deposited on the lower layer of the heterojunction. This layer is etched, to expose the heterojunctionn over a small window. Metallization layers are then added to the upper and lower surfaces; clearly, the lower of these make contact with the heterojunction only in the region of the window. A deep well, aligned with this window, is then etched right through the substrate to the upper layer of the heterojunction, and multimode optical fiber is epoxied into the well in contact with the LED surface. When a forward current is passed through the LED, it flows mainly through the region immediately below the fiber. The recombination region (or active region) is therefore confined vertically by the double heterojunction and laterally by the distribution of current flow.

ii. Edge Emitting LED

This Edge Emmiting LED also called the superluminescent LED, presents an alternative geometry to the Burrus-type LED (Fukuda, 1999). This is also based on a double heterostructure, but now the emission is taken from the edge of the junction rather than its surface. Generally, the current is forced to flow through a narrow strip down the device center line by the introduction of a current-blocking silica layer. As a result, the active volume is constrained vertically by the double heterostructure and laterally by the current flow. Most importantly, light generated by spontaneous emission is also confined vertically to a certain extent; the refractive index difference at the top and bottom of the active layer results in total internal reflection. This tends to channel a large fraction of the light towards the emission windows at relatively shallow angles.

iii. Edge Emitting Laser

In this Laser, the light propagates in a direction along the wafer surface of the semiconductor chip and is reflected or coupled out at a cleaved edge. These are the original and still very widely used form of semiconductor lasers. Within the edge-emitting laser structure, the laser beam is guided in a waveguide structure. Typically, one uses a double heterostructure, which restricts the generated carriers to a narrow region and at the same time serves as a waveguide for the optical field. This arrangement leads to a low threshold pump power and high efficiency. Depending on the waveguide properties, particularly its transverse dimensions, it is possible either to obtain an output with high beam quality but limited output power, or output with high output power but with poor beam quality.

iv. Surface Emitting Laser

In the surface-emitting laser, the light propagates in the direction perpendicular to the semiconductor wafer surface. These are further subdivided into monolithic and external-cavity devices: Monolithic means that the laser resonator is realized in the form of two semiconductor Bragg mirrors with the quantum well section in between. Such a device is called VCSEL (vertical-cavity surface-emitting laser) and is electrically pumped in most cases (Lorenser, 2003).

2. Sensors and Data Fusion

Data fusion refers to the synergistic use of information from different sensors for proper quality and integrity of the decision making to accomplish a task required by a system. All tasks that demand any type of parameter estimation from multiple sources can benefit from the use of data/information fusion methods. The terms information fusion and data fusion are typically employed as synonyms; but in some scenarios, the term data fusion is used for raw data (obtained directly from the sensors) and the term information fusion is employed to define already processed data. In this sense, the term information fusion implies a higher semantic level than data fusion. Other terms associated with sensor fusion that typically appear in the literature include decision fusion, data combination, data aggregation, multi-sensor data fusion, and data fusion (Castanedo, 2013).

Systems with only one sensor are not recommended for those with non-linear behavior, the observations made by one sensor could be uncertain and occasionally incorrect, a single sensor could not detect all the critical characteristics of the system as could be done with multiple sensors, which could provide different information each one, in special if they are from different technologies and can provide different information regarding the system. Besides if a unique sensor is used, and it fails the system is shutdown while this sensor is not working. The redundant information can reduce uncertainty and increases the accuracy with which the characteristics are perceived by the system. Several sensors delivering redundant information increase reliability in case of errors or failure of a sensor. Further information from various sensors to characterize the environment perceived in a way that would be impossible using only receive the information for each sensor separately.

The goal of using sensor fusion in multi-sensor environments is to obtain a lower detection error probability and higher reliability by using data and information from multiple distributed sources, as shown in Figure 3.

Figure 3. Sensors and data fusion applied to systems with non-linear behavior

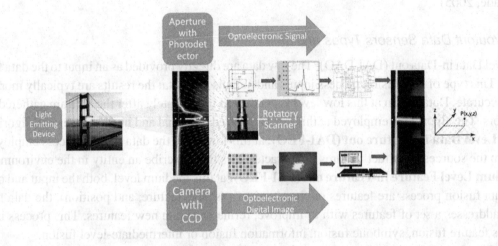

a. Sensors and Data Fusion Classification

Fusion techniques and methods are difficult to classify due to they involve a big variety of application fields. General classification is based on data criteria, as described in Figure 4.

Figure 4. Sensors and data fusion classification

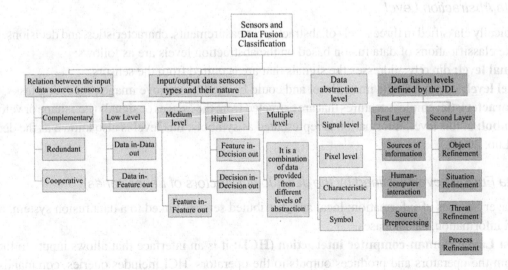

i. Relation Between the Input Data Sources (sensors)

Complementary: Each sensor provides complementary information that can be combined together, the information provided by the sensors represents different parts of the scene and could thus be used to obtain more complete global information (Vauhkonen, 2014).

Redundant: Various sensors providing information about the same target or scene (Abidi, 1992).

Cooperative: Two or more sensors are independently used to perform detection, providing information that is combined into new information that is typically more complex than the original information (Labayrade, 2005).

ii. Input/output Data Sensors Types and their Nature

Low Level Data in-Data out (DAI-DAO): The raw data are directly provided as an input to the data fusion process. This type of data fusion process inputs and outputs raw data; the results are typically more reliable or accurate. Data fusion at this low level is conducted immediately after the data are gathered from the sensors. The algorithms employed at this level are based on signal and image processing algorithms.

Low Level Data in-Feature out (DAI-FEO): at this low level, the data fusion process employs raw data from the sources to extract features or characteristics that describe an entity in the environment;

Medium Level Feature in-Feature out (FEI-FEO): at this medium level, both the input and output of the data fusion process are features or characteristics (shape, texture, and position), the data fusion process addresses a set of features with to improve, refine or obtain new features. This process is also known as feature fusion, symbolic fusion, information fusion or intermediate-level fusion.

High Level Feature in-Decision out (FEI-DEO): takes symbolic representations as sources and combines them to obtain a more accurate decision, this high level obtains a set of features as input and provides a set of decisions as output. Most of the classification systems that perform a decision based on a sensor's inputs fall into this category of classification.

High Level Decision in-Decision out (DEI-DEO): This type of classification is also known as decision fusion. It fuses input decisions to obtain better or new decisions (Dasarathy, 1997).

Multiple Level: It is a combination of data provided from different levels of abstraction.

Iii. Data Abstraction Level

It is typically classified in three levels of abstraction: measurements, characteristics, and decisions. Other possible classifications of data fusion based on the abstraction levels are as follows:

Signal level: directly addresses the signals that are acquired from the sensors.

Pixel level: operates at the image level and could be used to improve image processing tasks.

Characteristic: employs features that are extracted from the images or signals (i.e., shape or velocity),

Symbol: at this level, information is represented as symbols; this level is also known as the decision level (Luo, 2002).

iv. Data Fusion Levels Defined by the JDL (Joint Directors of Laboratories).

First Layer Sources of information: local and distributed sensors linked to a data fusion system, and or a priori information, as databases.

First Layer Human-computer interaction (HCI): it is an interface that allows inputs to the system from the operators and produces outputs to the operators. HCI includes queries, commands, and information on the obtained results and alarms. It incorporates not only multimedia methods for human interaction (graphics, sound, tactile interface, etc.), but also methods to assist humans in direction of attention, and overcoming human cognitive limitations.

First Layer Source Preprocessing: an initial process allocates data to appropriate processes and performs data pre-screening. Source preprocessing reduces the data fusion system load by allocating data to appropriate processes. The database management system stores the provided information and the fused results. This system is a critical component because of the large amount of highly diverse information that is stored.

Second Layer Level 1 Processing (Object Refinement): transforms sensor data into a consistent set of units and coordinates, refines and extends in time estimates of an object's position, kinematics, or attributes, assigns data to objects to allow the application of statistical estimation techniques, and refines the estimation of an object's identity or classification.

Second Layer Level 2 Processing (Situation Refinement): formal and heuristic techniques are used to examine, in a conditional sense, the meaning of Level 1 processing results.

Second Layer Level 3 Processing (Threat Refinement): it is an impact assessment, at this level the impact of techniques used in level 2 is evaluated to obtain a proper perspective. And a future projection is performed to identify possible risks, vulnerabilities, and operational opportunities. This level includes an evaluation of the risk or threat and a prediction of the logical outcome;

Second Layer Level 4 Processing (Process Refinement): it is a process concerned about other processes, monitors the data fusion process performance to provide information about real-time control and long-term performance, identifies what information is needed to improve the multilevel fusion product (inferences, positions, identities, etc.), determines the source-specific requirements to collect relevant information (i.e., which sensor type, which specific sensor, which database), and allocates and directs the sources to achieve mission goals (Luo, 2002).

b. Sensors and Data Fusion Methods

Most current sensors data fusion methods algorithms can be broadly classified as illustrating at Figure 5.

Figure 5. Sensors and data fusion methods

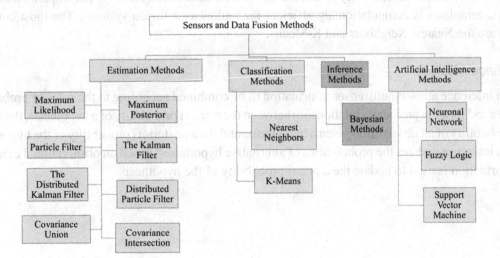

i. Estimation Methods

These methods take the weighted average of redundant information coming from a group of sensors and use it as the value of the fusion sensors. These methods determine the state of the target under movement (typically the position) given the observation or measurements. State estimation techniques are also known as tracking techniques with a final goal to obtain a global target state from the observations.

The estimation problem involves finding the values of the vector state (e.g., position, velocity, and size) that fits as much as possible with the observed data. From a mathematical perspective, we have a set of redundant observations, and the goal is to find the set of parameters that provides the best fit to the observed data. In general, these observations are corrupted by errors and the propagation of noise in the measurement process. State estimation methods fall under level 1 of the JDL classification and could be divided into two broader groups:

Linear dynamics and measurements: here, the estimation problem has a standard solution. Specifically, when the equations of the object state and the measurements are linear, the noise follows the Gaussian distribution, do not refer to it as a cluttered environment; in this case, the optimal theoretical solution is based on the Kalman filter;

Nonlinear dynamics: the state estimation problem becomes difficult, and there is not an analytical solution to solve the problem in a general manner. In principle, there are no practical algorithms available to solve this problem satisfactorily.

This method allows real-time processing of data. Some of them are: Maximum Likelihood and Maximum Posterior, Particle Filter, The Kalman Filter, The Distributed Kalman Filter, Distributed Particle Filter, and Covariance Consistency Methods: Covariance Intersection/Union. Being The Kalman filter the most popular.

ii. Classification Methods

A data association is performed to determine the set of measurements that correspond to each target. The multidimensional feature space can be partitioned into distinct regions, each representing an identification or identity class. The location of a feature vector is compared to pre-specified locations in feature space. A similarity measure must be computed, and each observation is compared to a priori class. A feature space may be partitioned by geometrical or statistical boundaries. Therefore, the templating approach may declare a unique identity or an identity with associated uncertainty. The implementation of parametric templates is computationally efficient for multisensor fusion systems. The most common methods are the Nearest Neighbors and K-Means.

iii. Inference Methods

Bayesian inference allows multisensor information to be combined according to the rules of probability theory. Bayes' formula provides a relationship between the a priori probability of a hypothesis, the conditional probability of an observation given a hypothesis, and the a posteriori probability of the hypothesis. Bayesian inference updates the probabilities of alternative hypotheses, based on observational evidence. New information is used to update the a priori probability of the hypothesis.

iv. Artificial Intelligence Methods

In these methods decisions are typically taken based on the knowledge of the perceived situation, high-level inference is required, as the human reasoning, for pattern recognition, planning, induction, deduction, and learning.

The artificial intelligence methods are based on an initial data set and the rule-based knowledge base comprising rules, frames and scripts. The inference process uses the a priori data set, and searches the complete set of rules to identify the applicable rule. The rule selection strategies from among multiple applicable rules include refraction, actuation, rule ordering, specificity, and random choice. Examples of these methods are Neuronal Network, Fuzzy Logic, Support Vector Machines, etc.

Artificial intelligence methods can be seen as the advanced versions of the estimation, the classification, and the inference methods. Artificial intelligence methods can be model-free, rather than model-specific, and have a sufficient degree of freedom to fit complex nonlinear relationships, with the necessary precautions to properly generalize. As a result, artificial intelligence methods can effectively conduct sensor fusion at different levels (Hall, 1997).

c. Sensors and Data Fusion Architectures

This describes where the sensors and data fusion process will be performed, as shown in Figure 6.

Figure 6. Sensors and data fusion architectures

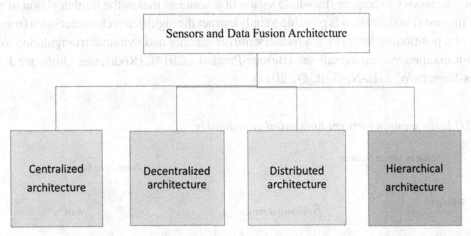

i. Centralized Architecture

All of the fusion processes are executed in a central processor that uses the provided raw measurements from the sources.

ii. Decentralized Architecture

It is composed of a network of nodes in which each node has its own processing capabilities and there is no single point of data fusion. Therefore, each node fusion its local information with the information that is received from its peers. Data fusion is performed autonomously, with each node accounting for its local information and the information received from its peers.

iii. Distributed Architecture

Measurements from each source node are processed independently before the information is sent to the fusion node; the fusion node accounts for the information that is received from the other nodes. Therefore, each node provides an estimation of the object state based on only their local views, and this information is the input to the fusion process, which provides a fused global view.

iv. Hierarchical Architecture

Other architectures comprise a combination of decentralized and distributed nodes, generating hierarchical schemes in which the data fusion process is performed at different levels in the hierarchy.

3. Machine Vision for Spatial Coordinate Measurement

Machine vision methods to provide spatial coordinates measurement has developed in a wide range of technologies, the computer vision has guided the machine vision to the tendency of duplicate the abilities of human vision by electronically perceiving and understanding an image with a high a dimensional data, increasing the data storage requirement, and the time processing due to the complexity of algorithms to extract important patterns and trends and understand what data says (learning data). For some applications it is not necessary to acquire the whole vision of a scene as makes the human vision to extract the features of interest (Luo, 2002), it is possible to only extract the significant characteristics from the scene by the use of a positioning laser (PL), a rotatory mirror scanner and dynamic triangulation as shown at Figure 7, for detailed system overall see (Basaca-Preciado, 2014), (Rodriguez-Quiñonez J. C., 2014) and (Flores-Fuentes W. a.-L.-N.-C.-B.-Q., 2014).

Figure 7. 3D laser scanning system for spatial coordinates

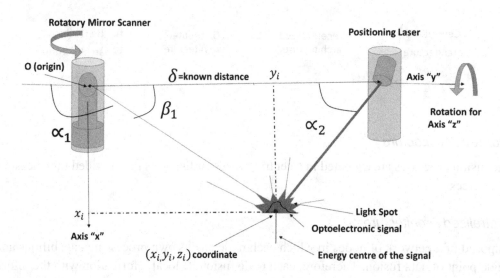

a. Machine Vision System Overall

The main sensor is a light scanning system, it can be a coherent light emitter source such as a laser or an incoherent light source such as a bulb like the ones used in motor vehicles. We assume that for any light emitter source there is only one energy centre that represents its point position.

A rotating optical aperture, designed as a 45° slanted mirror surface attached to a cylindrical rod that deflects the light beam into a double convex lens provided with an interference filter and a photodiode. The cylindrical rod is mounted onto the shaft of a dc electric motor, and as it rotates, an electronic signal is generated. Figure 8 shows the main elements of the optical scanning aperture. When the mirror starts spinning, the sensor "s" is synchronized when the origin generates a pulse indicating the 0° position and the starting of a cycle of 360° that finishes immediately before the "s" sensor generates the next beginning pulse. These pulses are used to calculate the scanning frequency and the zero reference, which is used to measure the angle where the light emitter source is positioned.

Figure 8. Optical scanning aperture

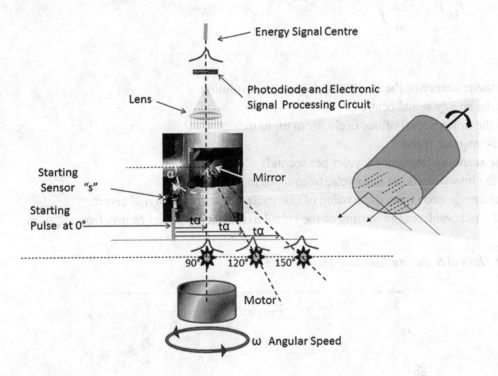

The signal timing diagram in Figure 9 shows the starting pulse and the optoelectronic signal relation used to calculate the light emitter energy signal centre position as described in the equations below. The interval $T_{2\pi}$ is the time for one motor revolution, defined as the time between m_1 and m_1 as in (4), which are expressed by the code $I_{2\pi}$ as defined in (5). As for m_1, it is the reference to the starting of one motor revolution (0°).

$$T_{2\pi} = \frac{2\pi}{\omega} = \frac{2\pi}{2\pi f} = \frac{1}{f}.$$ (4)

$$I_{2\pi} = T_{2\pi} f.$$ (5)

The time $t\alpha$ is the time between the reference of the starting of one motor revolution and the energy signal centre, that is, the interval between m_1 and m_2. This can be expressed by the code I_α is defined in (6). As for m_2, it is the energy signal centre. The angle under measurement is then calculated by (7).

$$I_\alpha = t_\alpha f.$$ (6)

$$\alpha = \frac{2\pi I_\alpha}{I_{2\pi}} = \frac{2\pi t_\alpha f}{T_{2\pi} f} = \frac{2\pi t_\alpha f}{\frac{2\pi}{\omega} f} = t_\alpha \omega = t_\alpha 2\pi f.$$ (7)

Where:

   m_1 is the reference of the starting of one motor revolution
   m_2 is the energy signal centre
   $T_{2\pi}$ is the time interval of one cycle, from m_1 to m_1
   ω is the angular speed
   f is the scanning frequency (cycles per second).
   $I_{2\pi}$ is the interval code of one cycle, from m_1 to m_1
   I_α is the range code from the starting of one cycle to the energy signal centre
   t_α is the interval from the starting of the signal to the energy signal centre, from m_1 to m_2

Figure 9. Optical Scanning Aperture Timing

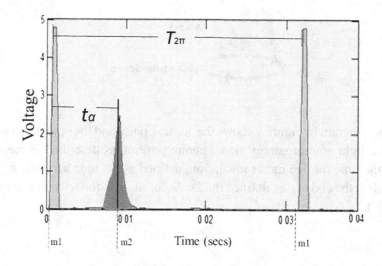

b. Applications of Machine Vision for Spatial Coordinate Measurement.

The present chapter surged in the research continuity of a 3D Vision System for mobile robot navigation application, a 3D medical laser scanner, a structural and environmental health monitoring system, a measurement of vegetation vitality system, and an industrial measurement system shown in Figure 10.

Figure 10. Applications of machine vision for spatial coordinate measurement

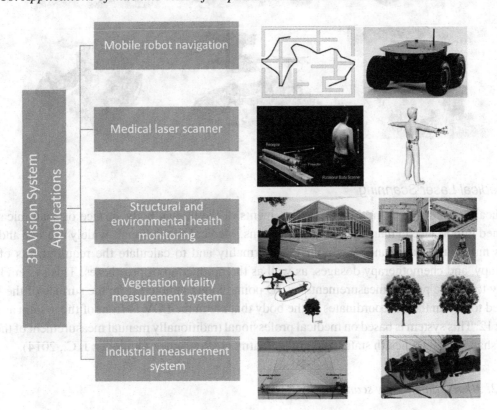

i. Mobile Robot Navigation

Mobile Robot Navigation, also known as Navigation System (NS) utilizes Technical Vision System (TVS) to obtain the characteristics of obstacles appearing in the field of view (FOV) of a mobile robot; with that information, the NS calculates a new trajectory to reach its programmed goal. The mobile robot needs to have a goal programmed to begin operation, then an initial trajectory is calculated by calculating the shortest path between the current position and goal. The mobile robot will follow the initial trajectory until an obstacle is detected by an array of infrared (IR) sensors. The IR sensors signal will trigger the TVS to perform a scan and obtain the characteristics (3D coordinates) of the detected obstacles. If no obstacles are detected, the mobile robot will reach its goal and stop advancing.

This NS is based in the Pioneer 3-AT (Figure 11) which is a small, four-wheel, four-motor skid-steer intelligent robotics platform, with all the characteristics to be a very suitable research development platform to implement the navigation and 3D vision system for autonomous mobile robots (Basaca-Preciado, 2014).

Figure11. Pioneer 3-AT mobile robotic platform

ii. 3D Medical Laser Scanning

For Medical Scanning System, the 3D measurements of the human body surface or anatomical areas have gained importance in many medical applications. Medical professionals widely use size and shape to assess nutritional status and developmental normality and to calculate the requirements of drug, radiotherapy, and chemotherapy dosages, as well as the production of prostheses. This system has the capability to realize precise measurements by 3D point clouds sampled from the surface of the human body, used to obtain the 3D coordinates of the body that are in the FOV in front of the system as shown in Figure 12. This system is based on medical professional traditionally manual measurement of the body size and shape to assess health status and guide treatment (Rodriguez-Quiñonez J. C., 2014).

Figure 12. 3D rotational body scanner

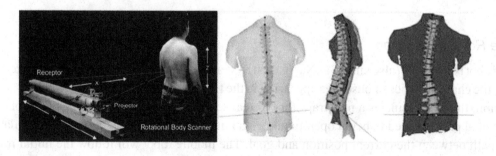

iii. Structural and Environmental Health Monitoring System

Important engineering constructions as shown in Figure 13, require geometrical monitoring to predict their structural health during their lifetime. Structure experiences deterioration and damage due to environmental conditions and excessive load conditions. Optical Scanning Systems (OSS) provides position measurements for structural health monitoring (SHM) task byspatial coordinates measurements. This system is based on an initial prototype consisting of an incoherent light source emitter (non-rotating) mounted on the structure under monitoring, a passive rotating optical aperture sensor designed with a 45°-sloping mirror, and embedded into a cylindrical micro rod. A photodiode to capture the beam of light while the cylindrical micro rod mounted on a dc electrical motor shaft is rotating, and a signal

Figure 13. Structures (historical center, skyscraper, suspension bridge, dam water retention, antenna, and industrial warehouses).

energetic centre detector by digital and analog processing (Flores-Fuentes W. a.-L.-N.-C.-B.-Q., 2014), (Flores-Fuentes W. a.-L.-Q.-B.-C., 2014).

iv. Measurement of Vegetation Vitality System

Machine vision systems can be used to inspect the vegetation vitality, such as a human can do it. Of course, with a limited resolution, optimized to evaluate specific parameters with the lower resource inversion possible. The principle of functioning of 3D laser scanning systems for spatial coordinates makes possible the measurement of vegetation vitality. By the scanning of vegetation with a laser ray, and an optical aperture to sense the degree of light reflected from vegetation, the normalized differenced vegetation index (NDVI) can be defined. Since the NDVI is related to the vegetation chlorophyll, the vegetation vitality of an area can be defined by this system. In (Lindner L. S.-L.-B.-F.-Q.-P., 2017) is presented a TVS mounted on an unmanned aerial vehicle (drone) to support the measurement of the vegetation NDVI. Line-by-line scanning has been performed to measure the reflected laser energy from the vegetation as shown in Figure 14, where five signals with different energy levels, corresponding to five different color saturation found in the vegetation area scanned are illustrated.

Figure 14. Vegetation vitality signals (laser intensity reflected). a) Level a, b) Level b, c) Level c, d) Level d, e) Level e.

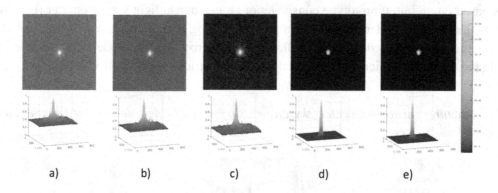

a)　　　b)　　　c)　　　d)　　　e)

v. Industrial Measurement System

Robot vision systems are fundamental for industrial tasks automation, they are required for robot navigation, mechanism positioning as well as for the measurement of materials and products characteristics. TVS based on laser and dynamic triangulation make possible to measure surface topography by physically scanning the laser light spot across a surface. The 3D coordinates of the scanned surface are provided by the TVS to realize its shape and dimensions. For this specific application, the proposed measurement methodology can be defined as a non-contact surface measurement technique by surface measurement system (SMS). Non-contact SMS´s are very important on those cases where the object of interest cannot have direct contact with the measurement system to obtain the surface information by different factors: its size, the shape, the difficulty to reach it or the nature of the material, due to it could be damage or degraded. Some industrial process where the non-contact surface measurements techniques applies are corrosion inspection, soldering inspection, material warpage inspection, products laser engraver (marked), etc.

Figure 15. Industrial surface measurement applications. a) Corrosion, b) Soldering, c) Warpage, d) Laser engraver.

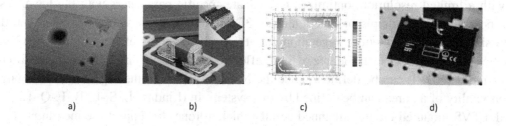

a)　　　b)　　　c)　　　d)

In process machine vision monitoring is in the state of art at the industry, it is possible thanks to Cyber-Physical Systems (CPSs), due to they are composed of networks of software and hardware components controlling physical parametres and processes, making decentralized decisions and triggering actions. CPSs are able to communicate, cooperate each other and also interact with humans in real-time. These are heterogeneous systems which contain different kinds of networks, interconnected sensors, actuators, and controllers. CPSs applied to manufacturing systems improve agility, responsiveness and guarantes the quality of the products. Therefore, CPSs have been introduced into manufacturing systems, which are called Cyber-Physical Production Systems (CPPSs) (Lins, 2020). Some examples of industrial applications following this tendency are the automated fabric inspection (Vargas, 2020), mineral sorting process (Gülcan, 2020), gear and gearboxes quality inspection (Moru, 2020), printed circuit board inspections (Dolezel, 2020), drill pipe emission inspection (Zhang C. &., 2019), automatic nondestructive spinach freshness detection (Huang, 2019), defect inspection for codes on complex backgrounds of plastics containers (Liang, 2019), automated visual inspection of glass bottle bottom (Zhou, 2019).

b. Optoelectronic Scanning and Digital Image Processing through the Energetic Center of Laser Spot

Optoelectronic scanning systems generate different shapes of signals, depending on the light source and the sensor of the scanner. A typical position measuring process includes an emitter source of light, like a laser diode or an incoherent light lamp and the position-sensitive detector like photodiode and CCD as a receiving device, which collects a portion of the back-reflected light from the target.

When a photodiode is used as a sensor on a scanner with rotating mirror it can also originate a similar Gaussian-like shape with some noise and deformation, the spatial distribution function of light has an Airy-function-like shape, by the other side the CCD produces output currents related to the "centre of mass" of light incident on the surface of the device, CCD use the light quantity distribution of the entire beam spot entering the light-receiving element to determine the beam spot centre or centroid and identifies this as the target position.

Concluding that, the photodiode signal originates a similar function to a CCD, consequently, it is possible to enhance the accuracy measurements in optoelectronic scanners with a rotating mirror, using a method for improving centroid accuracy by taking measurement in the energy centre of the signal (Sonka, 2014), (Li X. a., 2014).

i. Energetic Center of Laser Optoelectronic Signal Detected by Photodiode

The photodiode is the sensor, which is used for laser-scanning (Beiser, 1995). As SHM is an upcoming tendency of determining the integrity of structures and development of strategies to prevent undesirable damage, the OSS was designed, it has the necessity off detect a light emitter mounted on the structure under monitoring and calculate its energy centre localization. The OSS generates the targeted signals to be analyzed by the proposed methods previously assessed (Rivas-Lopez, 2014).

These six Energy Centre Localization Methods (ECLM), (Geometric Centroid, Power Spectrum Centroid, Analog Processing by Electronic Circuit, Saturation and Integration, Rising Edge, and Peak Detection), Figure 16 shows MATLAB code for Power Spectrum Centroid y graphical results of six ECLM. ECLM were assessed based on the assumption that the OSS signal from light emitter scanning is a Gaussian-like shape signal, the light emitter is an incoherent light, considered a punctual light source and due to the fact that with distance the light source expands its radius a cone-like or an even more complex shape is formed depending on the properties of the medium through which the light is traveling. And to reduce errors in position measurements, the best solution is taking the measurement in the energy centre of the signal generated by the OSS.

Figure 16. ECLM graphic representation

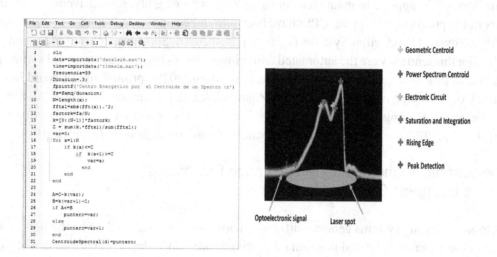

According to the results from a previous assessment made with an incoherent light source the most accurately ECLM are the Geometric Centroid and the Power spectrum centroid, by this reason the Geometric Centroid ECLM was selected to perform the energetic centre localization of a laser spot to measure laser beam deviation.

ii. Energetic Center of Laser Image Signal Detected by CCD

The CCD is the sensor who gives place to close-range photogrammetry (Franca). Photogrammetry gives information about the geometry of an object or surface through the use of photographs, provides less dense surface information but with high quality, especially along object space discontinuities, edges and borders, it can be defined as the automation of information extraction from digital images, based on image analysis methods.

A camera is located parallel to the OSS photodiode to obtain a second source of information regarding angle $\beta1$ by using a CCD sensor to acquire an image of the FOV at the same time as the OSS photodiode. The use of two methods has been proposed, one focused on the determination of edges and the shape centroid and other focused on the density of points measured on the surface of laser reflection. Both of them calculate the energetic center (image matrix centroid) applying morphological operations. Mathematical morphology is a tool for extracting image components that are useful in representation and description of region shape, it works directly on spatial domain, is a non-linear approach for detection of edges, by image transformations based on simple expanding and shrinking operations. Edge detection identifies and locates discontinuities in an image. An edge may be regarded as the boundary between two dissimilar regions in an image, classical methods of edge detection involve the convolution of an image with an operator (like Canny and Prewitt), which is developed to be sensitive to gradients in the image and returning values of zero in uniform areas.

In the method focused in the determination of edge and the shape centroid, a laser light beam is projected in a desired coordinate to create an edge, then the laser light scattered reflection is used to generate a minimum illumination difference (MID), to isolate only the laser light reflection from the image and to detect the edges between the two images (with/without laser light beam), a continuity the edge location is performed by an operator like Canny or Prewitt. Second, the laser light reflection energetic center is localized by the application of several morphological operations as dilation, erosion, boundaries location, fill image regions and remove open area (Li L. a., 2013). MATLAB code and results in Figure 17.

Figure 17. Graphic image processing stages by method focused on the determination of edge and the shape centroid.

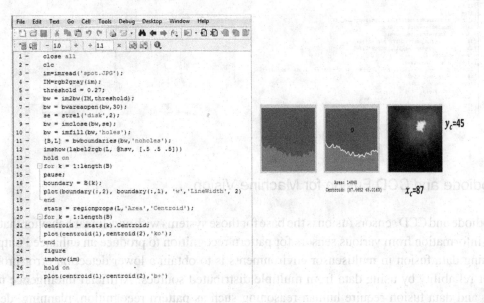

In the method focused in the density of points of laser reflection, the first step the image is captured, and the red color spectrum is obtained, a Gaussian blur filter (5x5 averaging matrix) is applied to smooth the captured image. By calculating the convolution of the original image and a 5x5 averaging matrix, is obtained the desired smoother image which allows finding the center of the laser spot more precisely. Straightforward formal implementation of the two-dimensional convolution equation in spatial form is used, calculated for finite intervals, defined by image frame limits.

After the image has been blurred the maximum values of red color are calculated for each row (x) and column (y) of the 2D matrix representing the image; the brighter the red color in the image, the higher the red value.

The final results are the x and y coordinates of the center of the laser spot inside the captured image (i.e. these coordinates instantly represented in pixels and further, converted to X and Y) (Damelin, 2012), (Zhang, 2012). MATLAB code and graphical results are shown in Figure 18.

These two Energy Centre Localization Methods (ECLM) for CCD Images have not been formerly evaluated to compare them, due to the measurement quality depends on surface reflection properties and lighting conditions, the surface reflection properties are dictated by a number of factors, as a) angle of the laser ray hitting, b) surface material, and c) roughness. Although for this fusion sensor purpose, the method focused on the density of points of laser reflection has been selected to be used during the experimentation.

Figure 18. Graphic Image Processing Stages by method focused on the density of points of laser reflection.

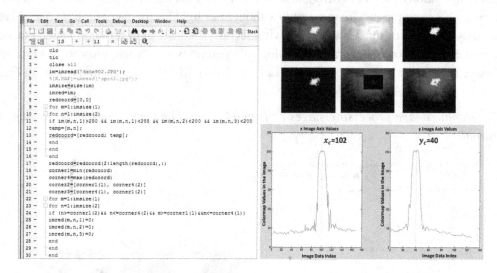

c) Photodiode and CCD Fusion for Machine Vision

The photodiode and CCD sensors fusion is the base for those systems with non-linear behavior that require the use of information from various sensors for pattern recognition to produce an enhanced output. The goal of using data fusion in multisensor environments is to obtain a lower detection error probability and higher reliability by using data from multiple distributed sources. Artificial intelligence methods for sensor and data fusion require human reasoning such as pattern recognition, planning, deduction and learning. The processes of inference used by expert systems begin with a group of initial data (data prior) and some basic rules. Machine Learning algorithms are investigated to fuse the sensors data, such as Support Vector Machine (SVM). The best model generated by each algorithm is called estimator. It is shown that the employment of estimators based on artificial intelligence can improve significantly the performance achieved by each sensor alone (Faceli, 2004), (Banerjee, 2012), (Waske, 2007).

In previous OSS experimentation angle $\beta 1$ measurements with photodiode have been done through an optical table, to observe the measurement error through the FOV, a total of 6020 measurements built the dataset that was used to train and to test an SVM algorithm in the error prediction, to measurement corrections of angle $\beta 1$ obtaining a satisfactory improvement, increasing from 53.39% to 97.60% the percentage of measurements with an error lower than 2.9°. Corrected angles $\beta 1$ were used to calculate the spatial coordinates by dynamic triangulation. At the same time, Navigation and Medical Scanning Systems experimentations were applying the Feed Forward backpropagation neural networks (NN) Levenberg−Marquardt algorithm with Widrow-Hoff Height/bias learning function to predict the error of spatial coordinates previously calculated with measurement of angle $\beta 1$. N_N provided an acceptable approximation of the measurement error up to 99.97%, however, the validation of the method shows only a 96% of performance.

For sensors and data fusion experimentation, the designed process explained in Figure 19 was followed. A total of 80 measurements built the new experimentation database of 80 instances (n rows) and 6 variables (p columns), the database values corresponding for column p=1 are the real objective angle under measurement (β1R, for p=2 are the real objective distance under measurement, for p=3 are the system scanning motor frequency, for p=4 are the angle calculation from photodiode output ($\beta_{1 photodiode}$, for p=5 are the angle calculation from CCD output (β1CCD, and for p=6 are the error measurement (E) calculated by (5).

$$E = \beta_{1R} - \beta_{1M} .$$ (8)

Where: E is the measurement error, representing how far the measurement from the real value is. β1R is the angle real value. β1M is the angle measured by the system (estimated fusion of photodiode and CCD by weighted average).

Figure 19. Sensors and Data Fusion Experimentation Flowchart

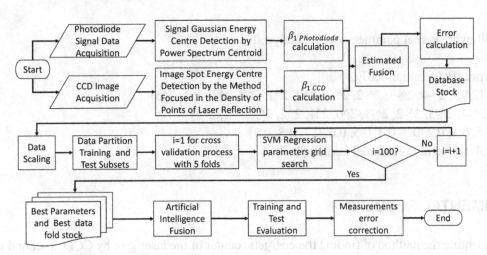

A scaling step on each column took place (from -1 to 1). The scaled database was partitioned in two sets, 80% training dataset and 20% test dataset, obtaining a training dataset of 64 rows with 6 columns and a test dataset of 16 rows with 6 columns. The 5 first columns are indicated as attributes and the last column as a target.

In order to fusion data by artificial intelligence, a properly validate the SVM regression performance was executed, the well-known k-fold cross-validation method was executed, repeating 100 times a 5-fold and in each run performing a grid search parameters, as illustrated at Figure 20.

Figure 20. Flowchart critical steps illustration

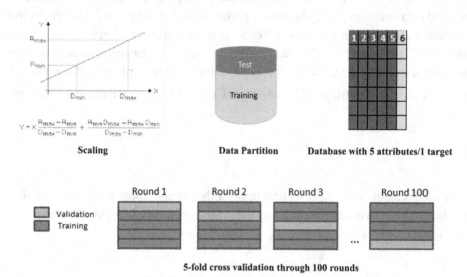

The following kernel parameters were analyzed:

Kernel type: [Polynomial, RBF],
Gamma: [2^-5, 2^-4, 2^-3, 2^-2, 2^-1, 2^0, 2^1, 2^2, 2^3, 2^4, 2^5]
C: [2^-5, 2^-4, 2^-3, 2^-2, 2^-1, 2^0, 2^1, 2^2, 2^3, 2^4, 2^5]
NU: [0.015, 0.1, 0.2, 0.3, 0.4, 0.5, 0.6, 0.7, 0.8, 0.9]
Polynomial degree in Polynomial kernel: [1, 2, 3]

EXPERIMENTAL RESULTS

By implementing the method of finding the energetic center of the laser spot by CCD, a second value of the $\beta1$ angle was obtained, the average of that value and the previously corrected angle by artificial intelligence algorithm generated a new value for the $\beta1$ angle, being the first fusion sensor experimentation.

Since the first experimentations for $\beta1$ angle error correction included the known values of frequency (scanner rotation), angle and distance of the scanning aperture (SA) and the angle of incidence $\beta1$ of the laser light on the phototransistor as input data (attributes) for the intelligence algorithm, this new experimentation, utilizes the intelligence algorithm SVMR as a fusion sensor technique, to include as attributes for the training process the $\beta1$ angle measured for both sensors, photodiode and CCD in order to obtain the best sensors characteristics fusion. Showing results illustrated on Table 1.

Table 1. Photodiode and CCD fusion sensor

Target	Photodiode Result	CCD Result	SVM Prediction	Fusion Result Error
90°	89.4°	91.1°	90.35°	0.35°
95°	98.1°	99.9	95.60°	0.60°
100°	99.4°	101.0°	100.08°	0.08°
105°	103.2°	104.0°	104.60°	0.40°
110°	108.0°	108.0°	109.80°	0.20°

The SVMR algorithm has demonstrated been an efficient intelligent tool for prediction, the change of attributes for the target prediction to include the CCD sensor, has improved the $\beta 1$ a${}_n$gle measurement accuracy (with an error lower than 1°), for the spatial coordinate dynamic triangulation calculation.

CONCLUSION

The goal of using sensor fusion in Machine Vision for Spatial Coordinates Measurement is to obtain a lower measurement error probability and higher reliability by using data from multiple distributed sources (signal energetic center of sensors, previously corrected by artificial intelligence). This method implements the use of a second value of a measured variable to be used as an attribute in SVMR prediction, which directly affects the overall system accuracy and consequently the monitoring, navigation and scanning systems performance.

Due to the measurement quality depends on surface reflection properties and lighting conditions, the surface reflection properties are dictated by angle of the laser ray hitting, and surface roughness material, further experimentation will be realized scanning over different surface materials like wood, skin, paper, paperboard, egg carton, etc, to build a database to train the system to be able to operate in almost any environment, installed in an optimal hardware architecture in order to further increase the system accuracy and expand the optical scanning applications.

The author's project consists of three main research stages (Laboratory, Outdoor, and Field). The design of experiment and measurements in this chapter describes the results of the Laboratory research stage, and takes them as background for Outdoor research stage, where the fusion of 2D/3D Image and Range Scanner Datasets will be done with data from real environment measurements, while the field research stage summarize future challenges for accomplishing the project scope of SHM for heritage buildings in Autonomous University of Baja California at Mexicali city. In the Laboratory Research Stage, we started with the build of the one OSS for 3D measurements, the use of one OSS provides of polar coordinates in a plane, the use of two OSS separated at a known distances, provide of a spatial coordinate in a plane by dynamic triangulation calculations, while the use of three OSS provides the full scope of 3D spatial coordinates measurement. In this stage, one OSS was built and characterized at laboratory environment. During tests has been observed that it presents a non-linear behavior throughout all the scanning scene, and in order to obtain an accuracy increase the requirement of computational intelligence techniques support surged, for pattern recognition to produce an enhanced output in a modern approach to the prediction of the non-linear measurement error of the OSS to compensate it. The outcome of our project has been a prototypal system for supporting an expert user in conducting structural analyses on building by extracting spatial coordinate information from light emitters mounted over a structure as indicator, in order to evaluate any structural displacement. Users may thus monitor the state of previous interventions, reason about the stability and integrity of the structure, identify potential risks, and plan reinforcement activities accordingly. Besides the specific scenario considered here, such a system may find application in several areas, as for instance in a museum scenario, bridge, damp, industrial structures or any other historical building monitoring projects. Concluding that this chapter outlines the proposed strategy by Artificial Intelligent Algorithms, and the capability to understand images and range scanner data through their fusion.

Future challenges for accomplishing the project scope of SHM for buildings are:

- Incorporate a third aperture to the OSS, the use of three apertures in the OSS provides the full scope of 3D spatial coordinates measurement.
- Perform real time measurements by the integration of the signal processing stages (acquisition, energy signal centre calculations, error compensations, fusion and triangulation), (Rivera-Castillo, 2017).
- Integrate the OSS signal processing by hardware, such as Field Programmed Gate Array (FPGA) to satisfy the monitoring system requirements to continue their tasks with safety and high accuracy on temporary occlusion conditions (Miranda-Vega, 2018).
- Include adaptive elements to the signal processing task in order to enhance the system to be capable of performance in an unknown environment.
- Implement a communication system for measurements record by wireless (Flores-Fuentes W. R.-L.-B.-Q.-P.-C., 2016).
- Evaluate the requirement of time processing reduction.
- Reduce the energy consumption of the OSS.
- Perform measurement analysis to identify structures natural patterns from damages.

REFERENCES

Abidi, M. A. (1992). *Data fusion in robotics and machine intelligence*. Academic Press Professional, Inc.

Banerjee, T. P., & Das, S. (2012). Multi-sensor data fusion using support vector machine for motor fault detection. *Information Sciences*, *217*, 96–107. doi:10.1016/j.ins.2012.06.016

Basaca-Preciado, L. C.-Q.-L.-B., Sergiyenko, O. Y., Rodríguez-Quinonez, J. C., García, X., Tyrsa, V. V., Rivas-Lopez, M., Hernandez-Balbuena, D., Mercorelli, P., Podrygalo, M., Gurko, A., Tabakova, I., & Starostenko, O. (2014). Optical 3D laser measurement system for navigation of autonomous mobile robot. *Optics and Lasers in Engineering*, *54*, 159–169. doi:10.1016/j.optlaseng.2013.08.005

Beiser, L. (1995). Fundamental architecture of optical scanning systems. *Applied Optics*, *31*(34), 7307–7317. doi:10.1364/AO.34.007307 PMID:21060601

Brooks, D. R., & Mims, F. M. III. (2001). Development of an inexpensive handheld LED-based Sun photometer for the GLOBE program. *Journal of Geophysical Research, D, Atmospheres, 106*(D5), 4733–4740. doi:10.1029/2000JD900545

Castanedo, F. (2013). A review of data fusion techniques. *TheScientificWorldJournal, 2013*, 1–19. doi:10.1155/2013/704504 PMID:24288502

Dagenais, M. a. (1995). *Integrated optoelectronics*. Academic Press.

Damelin, S. B. (2012). *The mathematics of signal processing*. Cambridge University Press.

Dasarathy, B. V. (1997). Sensor fusion potential exploitation-innovative architectures and illustrative applications. *Proceedings of the IEEE, 1*(85), 24–38. doi:10.1109/5.554206

Dolezel, P. &. (2020). Neural Network for Smart Adjustment of Industrial Camera-Study of Deployed Application. In *International Conference on Advanced Engineering Theory and Applications* (pp. 101-113). Springer. 10.1007/978-3-030-14907-9_11

Elfes, A. (1992). Multi-source spatial data fusion using Bayesian reasoning. *Data fusion in robotics and machine intelligence*, 137-163.

Faceli, K., de Carvalho, A. C. P. L. F., & Rezende, S. O. (2004). Combining intelligent techniques for sensor fusion. *Applied Intelligence, 3*(20), 199–213. doi:10.1023/B:APIN.0000021413.05467.20

Flores-Fuentes, W. L.-N.-C.-B.-Q., Rivas-Lopez, M., Sergiyenko, O., Gonzalez-Navarro, F. F., Rivera-Castillo, J., Hernandez-Balbuena, D., & Rodríguez-Quiñonez, J. C. (2014). Combined application of power spectrum centroid and support vector machines for measurement improvement in optical scanning systems. *Signal Processing, 98*, 37–51. doi:10.1016/j.sigpro.2013.11.008

Flores-Fuentes, W. L.-Q.-B.-C., Rivas-Lopez, M., Sergiyenko, O., Rodriguez-Quinonez, J. C., Hernandez-Balbuena, D., & Rivera-Castillo, J. (2014). Energy Center Detection in Light Scanning Sensors for Structural Health Monitoring Accuracy Enhancement. *Sensors Journal, IEEE, 7*(14), 2355–2361. doi:10.1109/JSEN.2014.2310224

Flores-Fuentes, W. R.-L.-B.-Q.-P.-C. (2016). Online shm optical scanning data exchange. In *2016 IEEE 25th International Symposium on Industrial Electronics (ISIE)* (pp. 940-945). IEEE.

Franca, J. G. (2005). A 3D scanning system based on laser triangulation and variable field of view. *Image Processing, 2005. ICIP 2005. IEEE International Conference on*. IEEE.

Fukuda, M. (1999). *Optical semiconductor devices*. John Wiley and Sons.

Gonzalez-Aguilera, D. L.-G. (2010). *Camera and laser robust integration in engineering and architecture applications*. INTECH Open Access Publisher. doi:10.5772/9959

Gülcan, E. (2020). A novel approach for sensor based sorting performance determination. *Minerals Engineering, 146*, 106130. doi:10.1016/j.mineng.2019.106130

Hall, D. L., & Llinas, J. (1997). An introduction to multisensor data fusion. *Proceedings of the IEEE, 1*(85), 6–23. doi:10.1109/5.554205

Hou, B. (2011). Charge-coupled devices combined with centroid algorithm for laser beam deviation measurements compared to a position-sensitive device. *Optical Engineering (Redondo Beach, Calif.)*, *50*(3), 033603–033603. doi:10.1117/1.3554379

Huang, X. Y., Yu, S., Xu, H., Aheto, J. H., Bonah, E., Ma, M., Wu, M., & Zhang, X. (2019). Rapid and nondestructive detection of freshness quality of postharvest spinaches based on machine vision and electronic nose. *Journal of Food Safety*, *39*(6), e12708. doi:10.1111/jfs.12708

Janesick, J. R. (2001). *Scientific charge-coupled devices.* SPIE Press.

Klein, L. A. (2003). Sensor and data fusion: a tool for information assessment and decision making. Bellingham, WA: Spie Press.

Labayrade, R., Royere, C., Gruyer, D., & Aubert, D. (2005). Cooperative fusion for multi-obstacles detection with use of stereovision and laser scanner. *Autonomous Robots*, *2*(19), 117–140. doi:10.100710514-005-0611-7

Li, F. a. (2006). *CCD image sensors in deep-ultraviolet: degradation behavior and damage mechanisms.* Springer Science and Business Media.

Li, F. M. (2003). Degradation behavior and damage mechanisms of CCD image sensor with deep-UV laser radiation. *Electron Devices. IEEE Transactions on*, *12*(51), 2229–2236.

Li, L., Ma, G., & Du, X. (2013). Edge detection in potential-field data by enhanced mathematical morphology filter. *Pure and Applied Geophysics*, *4*(179), 645–653. doi:10.100700024-012-0545-x

Li, X., Zhao, H., Liu, Y., Jiang, H., & Bian, Y. (2014). Laser scanning based three dimensional measurement of vegetation canopy structure. *Optics and Lasers in Engineering*, *54*, 152–158. doi:10.1016/j.optlaseng.2013.08.010

Liang, Q. Z., Zhu, W., Sun, W., Yu, Z., Wang, Y., & Zhang, D. (2019). In-line inspection solution for codes on complex backgrounds for the plastic container industry. *Measurement*, *148*, 106965. doi:10.1016/j.measurement.2019.106965

Lindner, L. S.-L.-B.-F.-Q.-P. (2017). Exact laser beam positioning for measurement of vegetation vitality. *Industrial Robot. International Journal (Toronto, Ont.)*.

Lindner, L. S.-Q.-L.-B.-F. (2016). Mobile robot vision system using continuous laser scanning for industrial application. *Industrial Robot. International Journal (Toronto, Ont.)*.

Lins, R. G., de Araujo, P. R. M., & Corazzim, M. (2020). In-process machine vision monitoring of tool wear for Cyber-Physical Production Systems. *Robotics and Computer-integrated Manufacturing*, *61*, 101859. doi:10.1016/j.rcim.2019.101859

Lorenser, D. a. (2003). Towards wafer-scale integration of high repetition rate passively mode-locked surface-emitting semiconductor lasers. *Applied Physics. B, Lasers and Optics*, *8*(79), 927–932.

Luo, R. C.-C., & Chih-Chen Yih. (2002). Multisensor fusion and integration: Approaches, applications, and future research directions. *Sensors Journal, IEEE*, *2*(2), 107–119. doi:10.1109/JSEN.2002.1000251

Marston, R. M. (1999). *Optoelectronics circuits manual.* Butterworth-Heinemann.

McClure, W. F. (2003). 204 years of near infrared technology: 1800-2003. *Journal of Near Infrared Spectroscopy*, *6*(11), 487–518. doi:10.1255/jnirs.399

Mims, F. M. (2000). Solar radiometer with light-emitting diodes as spectrally-selective detectors. *Applied Optics*, *39*(34), 6517–6518. doi:10.1364/AO.39.006517

Miranda-Vega, J. E.-F.-L.-Q.-B., Flores-Fuentes, W., Sergiyenko, O., Rivas-López, M., Lindner, L., Rodríguez-Quiñonez, J. C., & Hernández-Balbuena, D. (2018). Optical cyber-physical system embedded on an FPGA for 3D measurement in structural health monitoring tasks. *Microprocessors and Microsystems*, *56*, 121–133. doi:10.1016/j.micpro.2017.11.005

Moru, D. K., & Borro, D. (2020). A machine vision algorithm for quality control inspection of gears. *International Journal of Advanced Manufacturing Technology*, *106*(1-2), 105–123. doi:10.100700170-019-04426-2

Piprek, J. (2003). *Semiconductor optoelectronic devices: introduction to physics and simulation*. Academic Press.

Real-Moreno, O. C.-T.-O.-B.-F.-L. (2018). Implementing k-nearest neighbor algorithm on scanning aperture for accuracy improvement. *IECON 2018-44th Annual Conference of the IEEE Industrial Electronics Society*, 3182-3186.

Rivas-Lopez, M. C.-F.-Q.-B.-B. (2014). Scanning for light detection and Energy Centre Localization Methods assesment in vision systems for SHM. In *2014 IEEE 23rd International Symposium on* (pp. 1955-1960). Industrial Electronics (ISIE).

Rivera-Castillo, J. F.-F.-L.-N.-Q.-P., Flores-Fuentes, W., Rivas-López, M., Sergiyenko, O., Gonzalez-Navarro, F. F., Rodríguez-Quiñonez, J. C., Hernández-Balbuena, D., Lindner, L., & Básaca-Preciado, L. C. (2017). Experimental image and range scanner datasets fusion in shm for displacement detection. *Structural Control and Health Monitoring*, *24*(10), e1967. doi:10.1002tc.1967

Rodriguez-Quiñonez, J. B.-L.-F.-P., Sergiyenko, O., Hernandez-Balbuena, D., Rivas-Lopez, M., Flores-Fuentes, W., & Basaca-Preciado, L. (2014). Improve 3D laser scanner measurements accuracy using a FFBP neural network with Widrow-Hoff weight/bias learning function. *Opto-Electronics Review*, *4*(22), 224–235. doi:10.247811772-014-0203-1

Rodriguez-Quiñonez, J. C., Sergiyenko, O. Y., Preciado, L. C. B., Tyrsa, V. V., Gurko, A. G., Podrygalo, M. A., Lopez, M. R., & Balbuena, D. H. (2014). Optical monitoring of scoliosis by 3D medical laser scanner. *Optics and Lasers in Engineering*, *54*, 175–186. doi:10.1016/j.optlaseng.2013.07.026

Rodríguez-Quiñonez, J. C.-B.-L.-F.-P. (2016). Improve 3D laser scanner measurements accuracy using a FFBP neural network with Widrow-Hoff weight/bias learning function. *Opto-Electronics Review*, *22*(4), 224–235.

Rogalski, A. (2002). Infrared detectors: An overview. *Infrared Physics & Technology*, *3*(43), 187–210. doi:10.1016/S1350-4495(02)00140-8

Shih, H.-C. P., Hwang, C., Barriot, J.-P., Mouyen, M., Corréia, P., Lequeux, D., & Sichoix, L. (2015). High-resolution gravity and geoid models in Tahiti obtained from new airborne and land gravity observations: Data fusion by spectral combination. *Earth, Planets, and Space, 1*(67), 1–16. doi:10.118640623-015-0297-9

Snyder, W. E. (2010). *Machine vision*. Cambridge University Press. doi:10.1017/CBO9781139168229

Sonka, M. a. (2014). *Image processing, analysis, and machine vision*. Cengage Learning.

Syms, R. R. (1992). *Optical guided waves and devices*. McGraw-Hill.

Tavernier, F. a. (2011). *High-speed optical receivers with integrated photodiode in nanoscale CMOS*. Springer Science and Business Media. doi:10.1007/978-1-4419-9925-2

Tribolet, P. a. (2008). Advanced HgCdTe technologies and dual-band developments. In *SPIE Defense and Security Symposium*. International Society for Optics and Photonics. 10.1117/12.779902

Vargas, S. S., Stivanello, M. E., Roloff, M. L., Stiegelmaier, É., & Stemmer, M. R. (2020). Development of an Online Automated Fabric Inspection System. *Journal of Control. Automation and Electrical Systems, 31*(1), 73–83. doi:10.100740313-019-00514-6

Vauhkonen, J. a. (2014). Tree species recognition based on airborne laser scanning and complementary data sources. In Forestry applications of airborne laser scanning (pp. 135-156). Springer. doi:10.1007/978-94-017-8663-8_7

Waske, B. a. (2007). Fusion of support vector machines for classification of multisensor data. *Geoscience and Remote Sensing. IEEE Transactions on, 122*(45), 3858–3866.

Weckenmann, A. D.-R. (2009). Multisensor data fusion in dimensional metrology. *CIRP Annals-Manufacturing Technology, 2*(58), 701–721.

Yallup, K. a. (2014). *Technologies for smart sensors and sensor fusion*. CRC Press.

Zhang, F.-m. a.-h.-h. (2008). Multiple sensor fusion in large scale measurement. *Optics and Precision Engineering*, (7), 18.

Zhang, C. (2019). Automatic drill pipe emission control system based on machine vision. *Journal of Petroleum Exploration and Production Technology, 9*(4), 2737–2745.

Zhang, Q. A. (2012). 3-D shape measurement based on complementary Gray-code light. *Optics and Lasers in Engineering, 4*(50), 574–579.

Zhou, X. W. (2019). Automated visual inspection of glass bottle bottom with saliency detection and template matching. *IEEE Transactions on Instrumentation and Measurement, 68*(11), 4253–4267.

Chapter 2
Advances in Laser Scanners

Lars Lindner
 https://orcid.org/0000-0002-0623-6976
Autonomous University of Baja California,
Mexico

Oleg Sergiyenko
 https://orcid.org/0000-0003-4270-6872
Autonomous University of Baja California,
Mexico

Moisés Rivas-Lopez
Autonomous University of Baja California,
Mexico

Wendy Flores-Fuentes
 https://orcid.org/0000-0002-1477-7449
Autonomous University of Baja California,
Mexico

Julio C. Rodríguez-Quiñonez
Autonomous University of Baja California,
Mexico

Daniel Hernandez-Balbuena
Autonomous University of Baja California,
Mexico

Fabian N. Murrieta-Rico
 https://orcid.org/0000-0001-9829-3013
Autonomous University of Baja California,
Mexico

Mykhailo Ivanov
Autonomous University of Baja California,
Mexico

ABSTRACT

One focus of present chapter is defined by the further development of the technical vision system (TVS). The TVS mainly contains two principal parts, the positioning laser (PL) and the scanning aperture (SA), which implement the optomechanical function of the dynamic triangulation. Previous versions of the TVS uses stepping motors to position the laser beam, which leads to a discrete field of view (FOV). Using stepping motors, inevitable dead zones arise, where 3D coordinates cannot be detected. One advance of this TVS is defined by the substitution of these discrete actuators by DC motors to eliminate dead zones and to perform a continuous laser scan in the TVS FOV. Previous versions of this TVS also uses a constant step response as closed-loop input. Thereby the chapter describes a new approach to position the TVS laser ray in the FOV, using a trapezoidal velocity profile as trajectory.

DOI: 10.4018/978-1-7998-6522-3.ch002

INTRODUCTION

Laser scanners are optical devices, which use lasers to obtain certain information about surface topography, superficial coordinates or other characteristics, by physical sensing a light spot displacement across a surface. In opposition to stylus instruments, laser scanners' measuring is contactless and thereby has higher scanning speeds.

Applications for contactless measurement of 3D coordinates use mostly optical signals with CCD cameras or laser signals in Laser Scanning Systems (Toth & Zivcak, 2014). Cameras have the advantage that they resemble the way of human vision (Sergiyenko O., Optoelectronic System for Mobile Robot Navigation, Optoelectronics, 2010), which makes it easy to implement algorithms for different scenario detection in an unknown environment. Also, the scanning results do not depend on the examined object surface properties, when using cameras. On the other side, cameras are not preferable for single coordinate measurements (e.g. distance), due to their large amount of data generation. Another disadvantage is their dependency from the condition and existence of visible light and from atmospheric effects. Laser scanning systems however are suited for accurate coordinate measurements, which they can perform from objects in long distances and independent of ambient light. They also have the advantage of a fast measuring speed and a simple optical arrangement among low cost (Zhongdong, Peng, Xiaohui, & Changku, 2014). On the other hand, it must be noted, that for laser scanning systems the measurement readings depend on the scanning surface and that post-processing is required, due to large and high resolution 3D data sets.

One application, where measurement of 3D coordinates is absolutely needed, can be found in the movement control of Autonomous and Mobile Robots (AMR). The environment of a robot is typically measured with CCD cameras and / or laser scanning systems. In (Ohnishi & Imiya, 2013) for example, a robot is navigated using a „visual potential", which is computed using a sequence-capturing of various images by a camera mounted on the robot. Paper (Correal, Pajares, & Ruz, 2014) uses an automatic expert system for 3D terrain reconstruction, which captures his environment with two cameras in a stereoscopic way, similar to the human binocular vision. Laser scanning systems, as remote sensing technology, instead are known as Light Detection and Ranging (Lidar) systems, which are widely used in many areas, as well as in mobile robot navigation. Paper (Kumar, McElhinney, Lewis, & McCarthy, 2013) for example uses an algorithm and terrestrial mobile Lidar data, to compute the left and right road edge of a route corridor. In (Hiremath, van der Heijden, van Evert, Stein, & ter Braak, 2014), a mobile robot is equipped with a Lidar-system, which, using the time-of-flight principle, navigates in a cornfield.

However, other sensors and methods are also used to navigate mobile robots. Paper (Benet, Blanes, Simo, & Perez, 2002) for example uses infrared (IR) and ultrasonic sensors (US) for map building and object location of a mobile robot prototype. One ultrasonic rotary sensor is installed on the top and a ring of 16 infrared sensors are distributed in eight pairs around the perimeter of the robot. These IR sensors are based on the direct measurement of the IR light magnitude that is back-scattered from a surface placed in front of the sensor. The typical response time of these IR sensors for a distance measurement is about $2ms$. Distance measurement with this sensor can be realized from a few centimeters to $1m$. which represents one limitation of this approach. The range for coordinate measurements by triangulation can be far over $1m$. Paper (Volos, Kyprianidis, & Stouboulos, 2013) even experiments with a chaotic controlled mobile robot, which only uses an ultrasonic distance sensor for short-range measurement to avoid obstacle collision. The experimental results show applicability of chaotic systems to real autonomous mobile robots.

An optical 3D laser scanning system for navigation of autonomous mobile robots, called Technical Vision System (TVS), was developed at the Laboratory of Optoelectronics and Automated Measurement of the Autonomous University of Baja California (Sergiyenko O., Optoelectronic System for Mobile Robot Navigation, Optoelectronics, 2010). This TVS consists mainly of a laser scanning system, which uses the *Dynamic Triangulation* Method, to obtain 3D coordinates of any object under observation. This developed autonomous robot navigation system has his main task in prevention of obstacle collisions in an unknown environment. The TVS is examined with more detail in the section Technical Vision System. Laser scanners can be categorized by *Design Principles* and *Measurement Methods*, which are described briefly in the following two sections.

Design Principles

Design principles describe the characteristics by which a laser scanner is composed and define its construction. Three different main design principles can be determined for laser scanners:

- Hand-held laser scanners
- Mobile laser scanners
- Stationary laser scanners

Hand-held laser scanners are small and compact devices, which increasingly find applications, where flexibility and mobility play a central role for coordinate determination. In opposite to stationary laser scanners, they have a shorter scanning range. Hand-held laser scanners are often used as barcode readers or hand-held coordinate measurement device. Figure 1 shows an image of a barcode reader brand Motorola® and Figure 2 an image of a hand-held scanner brand Faro®.

Figure 1. Motorola LI4278 Barcode Reader [1]

Figure 2. Faro Hand-held Scanner Freestyle [2]

Mobile laser scanners are related to hand-held laser scanners, which are mainly used for autonomous mobile robot applications. They serve for navigation and orientation of the autonomous robot in a typically unknown environment and are part of the robot total system. Mobile laser scanners can also be installed on vehicles and even designed as portable measuring systems for man. Figure 3 shows a mobile laser scanner brand Riegl®.

Figure 3. Riegl VMX 450 mobile laser scanner [3]

In *stationary laser scanners* (also called terrestrial laser scanner) the surface geometry of objects is recorded digitally. Thereby a set of discrete sample points is produced, referred to as a point cloud. This point cloud is then converted into user data by post-processing using mathematical methods. The stationary laser scanners, such as cameras, are operated on tripods and are also found permanently installed on buildings or in institutions, for control and supervision purposes. Figure 4 shows a stationary laser scanner brand Faro®.

Figure 4. Faro Focus 3D stationary laser scanner [4]

Measuring Methods

Laser scanners measure 3D spatial coordinates from objects or 2D coordinates from surfaces, by determining the following measurement values, which corresponds to a specific coordinate measurement method. Table 1 gives a brief summary over the mostly used measurement methods, which are compared in following Table 2.

Table 1. Coordinate measurement methods

Measurement value	Measurement method
Flight time of the laser beam	Time-of-flight (TOF)
Phase difference between the transmitted and received laser beam	Phasing
Incident angle of the reflected laser beam	Static and Dynamic Triangulation
Intensity of the reflected laser beam	Imaging

Thereafter, these methods are described in terms of content and mathematically. In addition, current research references are given, which are using the respective methods.

Table 2. Advantages / Disadvantages of coordinate measurement methods

Method	Advantages	Disadvantages	Range
Time-of-flight	Short reaction time. No optical aperture angle.	Expensive for high resolution.	1m – some kilometers
Phasing	Less technical effort for high resolution.	Less measuring distance. Ambiguous for distance more than half wavelength.	Depends on the frequency
Triangulation	Cost-efficient and robust.	Depends on the surface.	1μm – 100m

Time-of-Flight

By the time-of-flight (TOF) method, the laser scanner measures the absolute flight time Δt of an emitted and received laser pulse. Thereby, the laser pulse must arrive and be reflected perpendicular on the measured surface, so the transmitter and receiver can be combined in the same measuring unit. Figure 5 shows the principle of TOF with the emitted (red) and the reflected laser pulse (blue).

Figure 5. Time-of-flight principle

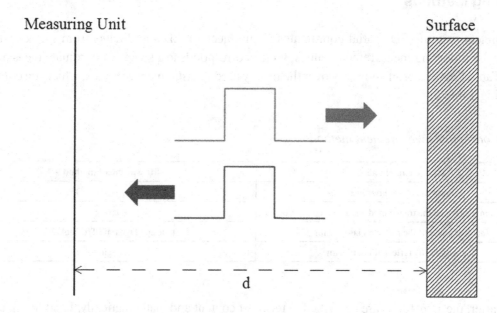

With the known laser pulse speed c the distance d between the measuring unit and the surface can be determined using:

$$d = c \cdot \frac{\Delta t}{2}. \tag{1}$$

The laser pulse speed c depends on the refractive index n of the medium and Δt denotes the measured time. Since the laser beam propagates with approximate speed of light, short reaction times must be measured. This results either in expensive sensors for time measurement or in a low resolution of the measured time.

Phasing

For the phasing measuring method, a harmonic laser signal is sent and overlays with its reflected signal. The distance between the measuring unit and the measured surface is determined using the phase shift between the emitted (red) and the reflected laser pulse train (blue), see Figure 6.

Figure 6. Reflection of a pulse train

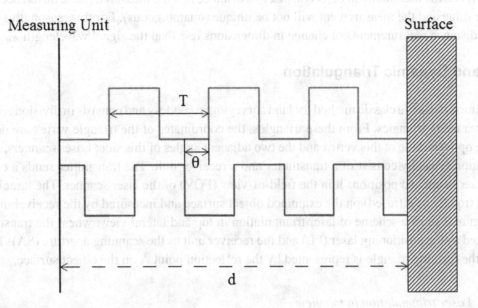

With the period of the laser signal T and the measured phase delay time Δt the phase difference angle is expressed by:

$$\theta = \frac{\Delta t}{T} 2\pi \,.$$

(2)

The wavelength of a harmonic signal λ is defined by its propagation speed c and oscillation frequency f.

$$\lambda = \frac{c}{f} = c \cdot T \,.$$

(3)

The double distance $2d$ is calculated by the sum of n wavelengths and the phase difference length, the wavelength λ is substituted using (3):

$$2d = n\lambda + \frac{\Delta t}{T}\lambda = ncT + \Delta tc \,.$$

(4)

Substitution of Δt from (2), the final expression for the distance d is received:

$$d = \frac{cT}{4\pi}\left[\theta + 2\pi n\right].$$

(5)

With accurate measurement of the phase angle θ the distance between the laser scanner and the examined object is determined. The most disadvantage of this method is the detection of the n signal periods. When the measurement object displaces a distance of λ the measured phase difference θ remains the same value and the measurement will not be unique (unambiguous). For this reason, this method is used for distance measurements of change in dimensions less than the signal wavelength λ.

Static and Dynamic Triangulation

Triangulation defines a classic method for land surveying in geodesy and consists in division and measuring an area using triangles. From these triangles, the coordinates of the triangle vertex are determined using the opposite side of this vertex and the two adjacent angles of this side. Laser scanners, which use triangulation typically consist of a transmitter and a receiver unit. The transmitter sends a constant or pulsed laser beam and positions it in the field-of-view (FOV) of the laser scanner. The laser beam gets reflected (totally or diffused) on the examined object surface and measured by the receiver unit. Figure 7 and Figure 8 show a scheme of laser triangulation in top and lateral view, where the transmitter unit is indicated as the positioning laser (PL) and the receiver unit as the scanning aperture (SA). The vertex point of the forming triangle is represented by the reflection point A *on* the object surface.

Figure 7. Laser triangulation in top view

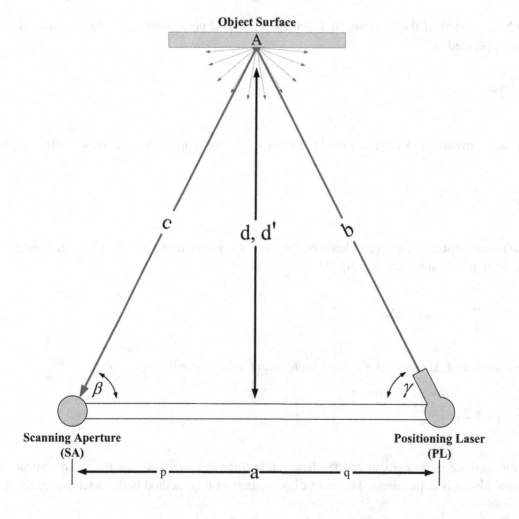

The use of the triangulation method in laser scanner systems can be of *static* or *dynamic* nature (Sergiyenko, et al., 2011). Static nature is referred to an operation mode without moving the positioning laser or the scanning aperture, which results in a statically behavior of the triangulation. The principle of dynamic triangulation varies the emitter and receiver angle, to eliminate this statically behavior and to increase the FOV of the laser scanner.

Figure 8. Laser triangulation in lateral view

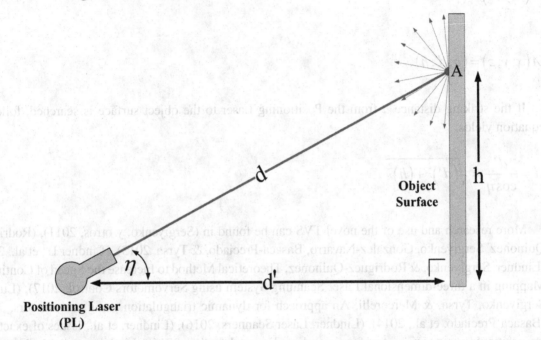

Trigonometric functions are used for calculation of 3D spatial coordinates using triangulation measuring methods. To measure distances using triangulation, consider the triangle in top (Figure 7) and lateral view (Figure 8). Here *a, b,* and *c* represent the sides of the triangle, γ is the angle of the PL, β is the angle of the SA and d' represents the projection of the distance d onto the top plane. The distance a between PL and SA is determined constructively and the angles β, γ, and η are measured using different technical measurement methods (Lindner, Sergiyenko, & Rodriguez-Quinonez, Theoretical Method to Increase the Speed of Continuous Mapping in a three-dimensional Laser Scanning System using Servomotors Control, 2017). By using the law of sines, the 3D coordinates of the reflection point $A(x;y;z)$ in a Cartesian coordinate system can be determined (Sergiyenko, et al., 2011):

$$d' = a \cdot \frac{\sin \beta \cdot \sin \gamma}{\sin(\beta + \gamma)}. \tag{6}$$

$$p = \frac{d'}{\tan \beta} = a \cdot \frac{\cos \beta \cdot \sin \gamma}{\sin(\beta + \gamma)}. \tag{7}$$

$$h = d' \cdot \tan\eta = a \cdot \frac{\sin\beta \cdot \sin\gamma}{\sin(\beta+\gamma)} \cdot \tan\eta . \tag{8}$$

Depending, where the origin of the Cartesian coordinate system is defined, either p or $q=a-p$ is taken as the x coordinate for the reflection point A.

$$A(x;y;z) = (p;d';h) . \tag{9}$$

$$A(x;y;z) = (q;d';h) . \tag{10}$$

If the striking distance d from the Positioning Laser to the object surface is searched, following equation yields:

$$d = \frac{d'}{\cos\eta} = \sqrt{(d')^2 + (h)^2} . \tag{11}$$

More research and use of the novel TVS can be found in (Sergiyenko, y otros, 2011), (Rodriguez-Quinonez, Sergiyenko, Gonzalez-Navarro, Basaca-Preciado, & Tyrsa, 2013), (Lindner L., et al., 2015), (Lindner, Sergiyenko, & Rodriguez-Quinonez, Theoretical Method to Increase the Speed of Continuous Mapping in a three-dimensional Laser Scanning System using Servomotors Control, 2017), (Lindner, Sergiyenko, Tyrsa, & Mercorelli, An approach for dynamic triangulation using servomotors, 2014), (Basaca-Preciado, et al., 2014), (Lindner, Laser Scanners, 2016), (Lindner, et al., Issues of exact laser ray positioning using DC motors for vision-based target detection, 2016), (Lindner, et al., Machine vision system errors for unmanned aerial vehicle navigation, 2017), (Lindner, et al., Machine vision system for UAV navigation, 2016), (Lindner L., et al., Mobile robot vision system using continuous laser scanning for industrial application, 2016) and (Lindner, et al., UAV remote laser scanner improvement by continuous scanning using DC motors, 2016).

Imaging

A laser scanner, which detects the intensity of the reflected laser beam, is called a laser imaging scanner. This scanner measures the surface reflectance by processing the returned signal energy from the laser beam. With this measuring method, the true reflectance can be determined, as these are not distorted by shadows or darkening effects (Hug & Wehr, 1997). The result is similar to a grey-scale photo image of the surface. (Hinks, Carr, Gharibi, & Laefer, 2015). Figure 9 shows an example of a grayscale panoramic image captured by a laser scanner.

One measuring device which uses this method is the Scanning Laser Altitude and Reflectance Sensor (ScaLARS) developed by the Institute of Navigation of the University of Stuttgart in Germany in the mid-1990s. The ScaLARS uses the phase measuring technique with a signal frequency of 1 MHz and 10 MHz, to obtain spatial data of a geographical area and to find the slant range from an airborne platform to the ground (Shan & Toth, 2008). Thereby this laser scanner belongs to the type of terrestrial laser scanners (TLS), which are used to survey and map a specific geographic area from an aircraft (Fang, Huang, Zhang, & Li, 2015). Another application of the imaging laser scanning can be found in (Garcia-Talegon, et al., 2015). Here, the intensity values of the reflected laser beam are used to assess pathologies of a complex building facade.

Figure 9. Grayscale panoramic image [5]

TECHNICAL VISION SYSTEM (TVS)

A laser scanning system must fulfill two main tasks: positioning a laser beam in the FOV and measuring the reflected laser beam on the scanned surface using optical sensors. The TVS predecessor prototype No.2 was developed at the Laboratory of Optoelectronics and Automated Measurement of University UABC and is shown in Figure 10:

Figure 10. Technical Vision System (TVS) No.2 (Rodriguez-Quinonez, Sergiyenko, Gonzalez-Navarro, Basaca-Preciado, & Tyrsa, 2013)

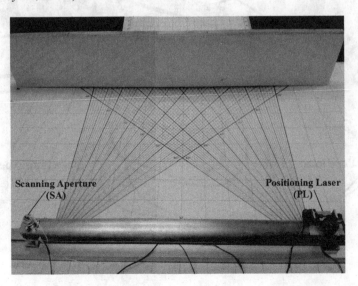

The TVS consists mainly of a laser scanning system, which uses dynamic triangulation, to obtain 3D coordinates of any object under observation. The TVS is composed of two principal parts: the Positioning Laser (PL) and the Scanning Aperture (SA). Using the measured angles and the formulas (6) – (10), the prototype determines three-dimensional coordinates in space of examined objects in front of the system. Although the TVS has replaced the disadvantages of camera light sensors by use of a single-sensor detector, it still operates using stepping motors to positioning the emitted laser. Stepping motors have a discontinuous behavior, due to their stepwise rotation of the motor shaft. Modern stepping motors in certain operating modes have step sizes less than one degree, but will always have unreachable angles in their field-of-motion (FOM). By using stepping motors in the previously developed TVS, the FOV is discretized and its resolution depends on the distance of the examined scanning object. Targets, which have dimensions less than the FOV resolution in a certain distance, cannot be scanned or captured. These targets are located within the dead zones of the TVS and thereby not detectable by the same.

As novel approach of the successor prototype No.3, these stepping motors were substituted by Direct Current Motors (DC Motors), to eliminate the disadvantage of a discrete FOV and replace it with a continuous behavior of scanning element. These DC motors will be controlled in closed- and open-loop configuration. By replacing the measurement principle of a discrete FOV with a continuous one, the discretization of the FOV for 3D coordinate measurement is no longer a physical limitation. Thereby applications become possible, which were previously limited by the fixed step size or by the distance to the examined object. The developed prototype TVS No.3 is shown in Figure 11.

Figure 11. Technical Vision System (TVS) No.3 (Lindner, Sergiyenko, Tyrsa, & Mercorelli, An approach for dynamic triangulation using servomotors, 2014)

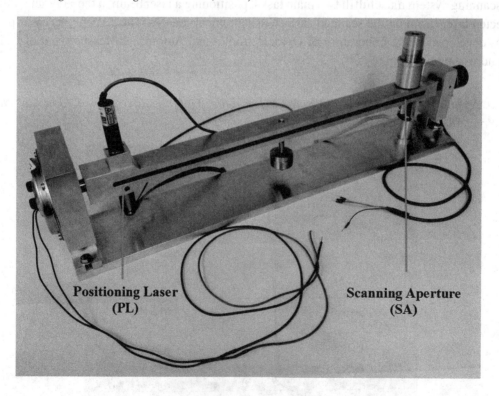

The new TVS prototype No.3 contains the new PL and SA, which comprises the following advantages compared with the TVS prototype No.2:

- Using DC motors instead of stepping motors, to substitute the discrete FOV by a continuous one and thereby elimination of dead zones in the TVS FOV.
- Using DC motors directly without gear, to increase speed and accuracy of the 3D coordinate measurement.
- Minimization of the PL relative positioning error by simplification of mechanical design and optimized control.
- By lowering the TVS mechanical degree of complexity, possible causes of systematic errors are eliminated.
- Opening the possibility for implementation of new scanning algorithms, e.g. variable step scanning method.

The designs and main characteristics of the PL and SA of both prototypes are explained in detail below. Extensive preliminary research about the TVS can be found in (Lindner, et al., Issues of exact laser ray positioning using DC motors for vision-based target detection, 2016), (Lindner, et al., Machine vision system for UAV navigation, 2016), (Lindner L., et al., Mobile robot vision system using continuous laser scanning for industrial application, 2016), (Lindner, et al., UAV remote laser scanner improvement by continuous scanning using DC motors, 2016), (Flores-Fuentes, et al., 2016), (Flores-Fuentes, et al., 2016), (Flores-Fuentes, et al., 2015), (Murrieta Rico, et al., 2016), (Murrieta-Rico, et al., 2015), (Murrieta-Rico, et al., 2016) and (Murrieta-Rico, et al., 2016). Another application of the TVS was explored in the medical field, where it can be used to monitor scoliosis (Rodriguez-Quinonez, Sergiyenko, Gonzalez-Navarro, Basaca-Preciado, & Tyrsa, 2013) and (Rodriguez-Quinonez, et al., 2014). Thereby 3D coordinates of the human spine are measured, in order to classify deformations using the Cobb Angle. More recent research about increasing the scanning speed of the TVS by use of combined variable steps can be found in (Garcia-Cruz, et al., 2014). Also, a power spectrum centroid for measurement improvement got applied in (Flores-Fuentes, et al., 2014).

Positioning Laser

The *Positioning Laser* of the predecessor TVS prototype No.2 is depicted in Figure 12. It can be seen the outer mechanical structure of the PL using two stepping motors (Figure 12a)), as well as the inner optical structure using two 45° staggered mirrors (Figure 12b)). The stepping motor I in Figure 12a) rotates the mirror II over a transmission (worm drive) to position a red laser (635 nm, 20 mW) in the TVS FOV. Stepping motor II in Figure 12a) rotates the TVS around its main axis.

Also, Figure 12b) shows the optical path of the laser beam. Mirror I deflects the laser beam vertically upward to mirror II, which then sends the beam forward onto the object under observation. The used mirrors are made of N-BK7 optical glass with a protected aluminum coating. Due to the 45° construction of the reflecting mirror surface, each incident laser beam is deflected with precisely 90°.

Figure 12. Positioning Laser of TVS No.2 (Rodriguez-Quinonez, Sergiyenko, Gonzalez-Navarro, Basaca-Preciado, & Tyrsa, 2013)

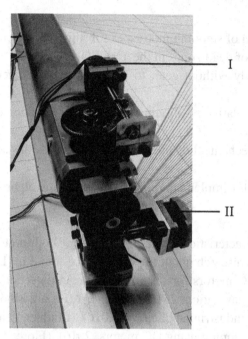

Due to the existing face clearance in every transmission, the successor prototype No.3 has mirror I directly connected to a DC motor, without using a transmission. The new design of the positioning laser is depicted in Figure 13, which shows the optical path of the transmitted laser beam (I), the DC motor Maxon RE-max29 (II), a 45° staggered mirror (III), the pancake motor Printed Motor Works GPM9 (IV), a laser diode module Coherent StingRay-514 (V) and the TVS main axis (VI). Compared to the design of the PL prototype No.2 (Figure 12) first it can be seen, that due to elimination of all transmission parts, a high simplification of the positioning laser is achieved. Systematic errors produced by face clearance in every transmission are eliminated and the PL angle γ (Figure 7) corresponds directly to the actual angular position φ_o f the DC motor (II).

Figure 13. Positioning Laser of TVS No.3 (Lindner L., et al., 2015)

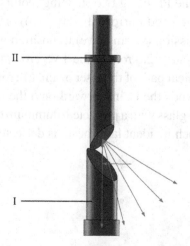

Also, because the TVS is rotated around his horizontal axis without using a transmission, a motor with high torque and flat dimensions is needed. These two conditions are fulfilled by a pancake motor. Also here, the PL angle η (Figure 8) corresponds directly to the actual angular position φo_o f the pancake motor (IV). It must be noted, that now any PL angle η which not corresponds to the TVS balanced state in the earth gravitational field, has to be controlled in closed-loop configuration, since the advantage of a self-locking worm gear is not used anymore. Since the pancake motor constantly has to generate a holding torque, this can lead to higher energy consumption.

Furthermore, it can be seen that this design requires only one 45° staggered mirror to redirect the laser beam in the TVS FOV, due to the vertical installation of the laser source (V). This decreases the systematic error produced by manufacturing tolerances and non-alignment of the staggered mirror and reduces the overall cost of the optical components. The laser source contains a 10mW green laser diode with 514 nm wavelength.

Scanning Aperture

The reflected laser beam from the examined object is received by the *Scanning Aperture* (Figure 14), which basically consists of a 45° staggered mirror, two biconvex lenses, a zero sensor and a stop sensor. The mirror is rotated by a DC motor and redirects the reflected laser beam from the observed object towards the lenses, which concentrate them for the stop sensor (high speed phototransistor). When the received laser beam is located in the orthogonal plane of the 45° staggered mirror, the stop sensor receives the maximum amplitude of the laser beam and converts it to an electrical pulse. The zero sensor produces a reference signal each full revolution and starts two counters to accumulate pulses. The first counter accumulates pulses from a standard reference f_P between two consecutive zero sensor pulses, which defines the pulse number for $N_{2\pi}$ in each full revolution. The standard reference signal represents

a pulse train with fixed period $T_P = \dfrac{1}{f_P}$ and is generated by a microcontroller.

Figure 14. Scanning Aperture of TVS No.2 (Rodriguez-Quinonez, Sergiyenko, Gonzalez-Navarro, Basaca-Preciado, & Tyrsa, 2013)

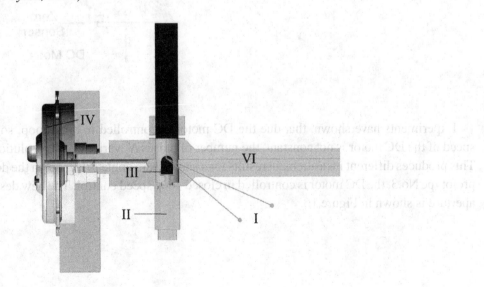

The second counter accumulates pulses of the same standard reference between the zero sensor pulse and the stop sensor pulse, which defines the pulse number for N_A in each revolution. Hence, N_A represents the received beam angle encoded in the pulse number of this variable. The received laser beam angle then is calculated using following equation:

$$\beta = 2\pi \cdot \frac{N_A}{N_{2\pi}}.$$

(12)

Figure 15 depicts the standard reference signal f_p the zero sensor signal $f_{2\pi}$. the period for one revolution $T_{2\pi}$. the full revolution pulse number $N_{2\pi}$. the time phase shift T_A and the SA angle pulse number N_A.

Figure 15. Scanning Aperture Signals

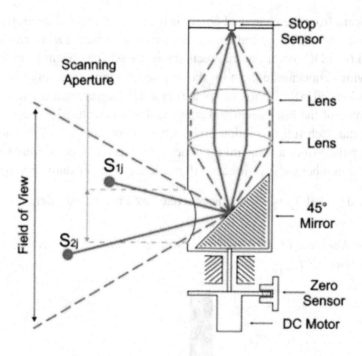

Experiments have shown, that due the DC motor is controlled in open-loop, so the actual angular speed of the DC motor is not constant, the number of pulses N_A varies every revolution of the DC motor. This produces different measurement results for a same beam angle β. Hence, in the developed successor prototype No.3, the DC motor is controlled in closed-loop speed control. The new design of the scanning aperture is shown in Figure 16.

Figure 16. Scanning Aperture of TVS No.3 (Lindner L., et al., 2015)

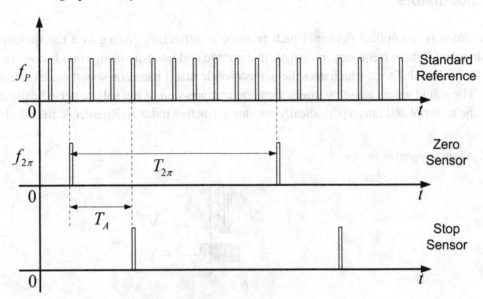

This figure shows the optical path of the reflected laser beam (I), the DC motor (II), a 45° mirror (III), two biconvex lenses (IV), an optical filter (V) and a high-speed photodiode (VI). Like the DC motor has an integrated quadrature encoder with index channel, there is no need for the zero sensor anymore. The used coherent laser signal comes from a $10mW$ laser module with $514nm$ wavelength. The optical filter thereby has a center wavelength of $515nm$ full width-half wavelength of $10nm$ and a transmission of $\geq 45\%$. The photodiode is used with a transimpedance amplifier, to convert the small photocurrent of the diode to a greater voltage, which thereby can be measured using an analog-digital converter (ADC). The DC motor is digitally controlled by a microcontroller, which controls the actual angular speed of mirror III using a PI algorithm.

Laser Beam Positioning

Laser scanning systems, which actively position the laser beam, expand their FOV to a multidimensional scanning area (2D or 3D). This enables the laser scanning systems to high dynamically acquire data from their surroundings and in shortest time. For laser beam positioning in the FOV, essentially four different mechanical principles are used:

- Optical Modulators (Figure 17 and Figure 18)
- Scanning Refractive Optics (Figure 19)
- Galvoscanner (Figure 20 and Figure 21)
- Scanning Mirrors (Figure 12)

These principles are explained in the following, also with examples from the actual optic market and current research references are given, which are using the respective method.

Optical Modulators

Optical modulators are optical devices which produce a diffraction grating in a transparent solid to modulate incident light in frequency, propagation direction and intensity using sound waves or an electric field. Optical modulators, which uses the acousto-optic effect based on sound waves, are shown in Figure 17. The sound waves cause by rarefaction and compression of the solid a periodically changing density of the material and thus a periodically varying refractive index n (Roemer & Bechthold, 2014).

Figure 17. Acousto-optic deflector

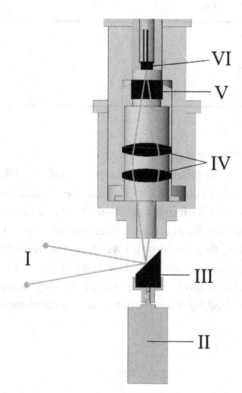

Optical modulators, which use the electro-optic effect based on an electric field, are shown in Figure 18. Atom and molecular structures are disturbed in their position, orientation and shape by the electric field and thereby the refractive index n is changing periodically (Roemer & Bechthold, 2014).

Figure 18. Electro-optic deflector

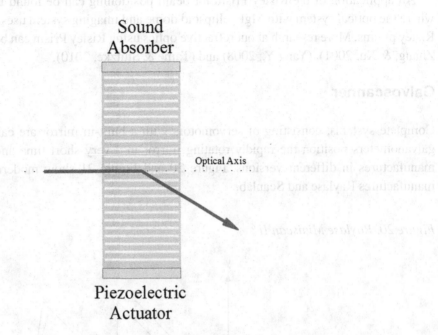

Scanning Refractive Optics

Scanning refractive optics is used with "Risley Prism", where the laser beam is positioned in the FOV only by rotation of the two prisms independently about an axis parallel to the normal of their adjacent faces. With these two prisms a laser beam can be steered in any direction within a ray cone (Wang, y otros, 2010). When the two prisms are oriented as in Figure 19, the laser beam is running parallel to the optical z-axis and got shifted in direction of y-axis. In this orientation, minimum deflection of the laser beam occurs. Maximum deflection is achieved through 180° rotation of the second prism II. In this orientation, both prisms refract the laser beam in the same direction and function as a single prism with double prism angle.

Figure 19. Rotating Risley Prism

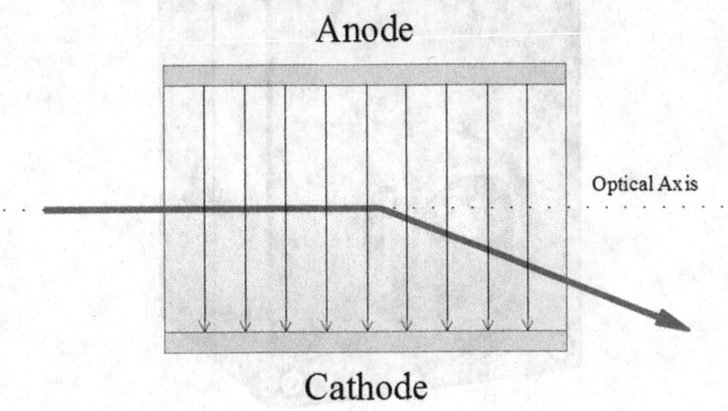

An application of the Risley Prism for beam positioning can be found in (Song & Chang, 2012), where the optical system with MgF_2 ellipsoid dome and imaging system uses a pair of identical rotating Risley prisms. More research about refractive optics using Risley Prism can be found in (Yang, Lee, Xu, Zhang, & Xu, 2001), (Yang Y., 2008) and (Tame & Stutzke, 2010).

Galvoscanner

Complete systems, consisting of servomotors with a built-in mirror are called *Galvoscanner*. These galvanometers position the rapidly rotating mirror in a very short time and are offered by different manufactures in different versions. Figure 20 and Figure 21 show modern galvoscanners from the manufactures Raylase and Scanlab.

Figure 20. Raylase Miniscan II [6]

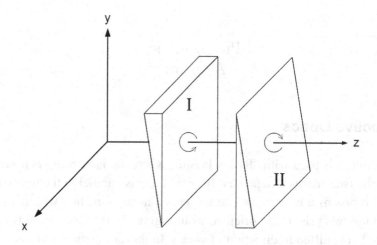

Figure 21. Scanlab dynAXIS S [7]

Galvoscanners shall achieve high speeds and accelerations, thus there are optimized to have less friction in the bearings, commutator rings, etc. and less moment of inertia of all moving masses. An application of a 2D galvoscanner in combination with a fiber laser to provide spatial power modulation can be found for example in (Solchenbach & Plapper, 2013).

Scanning Mirrors

Scanning mirrors are optical devices with a planar surface, to steer a laser beam in a certain direction. The previously introduced TVS use scanning mirrors (Figure 12b)), which reflect and position a laser beam in the FOV of the laser scanner. The principle of operation is explained previously. Also, the laser scanner Faro Focus 3D (Figure 4) uses a rotating mirror in order to position the laser beam in the FOV (Garcia-Talegon, et al., 2015). This laser scanner uses the continuous wave phase shift measuring method to determine 3D spatial coordinates. Another laser scanning device that uses mirrors to position the laser beam can be found in (Hinks, Carr, Gharibi, & Laefer, 2015). Generally, in actual laser scanning systems scanning mirrors are moved using three different mechanical actuators:

- Piezoelectric Actuators
- Stepping Motors
- DC Motors

The advantages and disadvantages of every mechanical actuator are summarized in Table 3:

Table 3. Advantages / Disadvantages of mechanical actuators for scanning mirrors

Method	Advantages	Disadvantages
Piezoelectric Actuators	Continuous working principle, no resolution. Offer a very fast response time. No interaction with magnetic fields.	Very small movement range. Still not widely used in the industry. Currently still high cost.
Stepping Motors	No need for position feedback, while $M_L < M_M$. With modern digital electronic, drive control is very easy. Availability in small sizes and with low cost.	Stalling of motor, when $M_L > M_M$, loss of steps. Torque drop-off at high speed. Due to their discontinuous behavior, a stepping motor can't reach any angle in his FOM.
DC Motors	Can supply about twice their rated torque for short period (acceleration). Can theoretically reach any angle in his FOM.	Require more complex drive circuits and positional feedback for accurate positioning.

Piezoelectric Actuators

Piezoelectric actuators use the piezoelectric effect, which describes the property of certain materials to deform mechanically, when an electrical charge is applied. Classical piezoelectric materials have a very low degree of strain, which is too small for practical applications and thereby has to be amplified. A schematic image of an amplified piezoelectric actuator is shown in Figure 22 (Sixta, Linhart, & Nosek, 2013). It can be seen the piezoelectric element, which oscillations are amplified using an elliptic steel shell structure.

Figure 22. Schematic of an amplified piezoelectric actuator

Another use and more description of these amplified actuators can be found in (Xiang, Chen, Wu, Xiao, & Zheng, 2010). Here an optical tip-tilt actuator based on piezoelectric actuators is used to deflect a laser beam.

Stepping Motors

A *stepping motor* is a special type of a synchronous motor (Figure 23), with a cogged permanent magnet rotor (1) and particular stator poles (2), whose windings (3) are driven by periodic current pulses (Fischer, 2003). An electronic circuit for stepping motors generates these pulses in the correct order to control the desired speed and direction. One pulse moves the stepping motor in one step-angle, which is a typical parameter of each motor. The pulse sequences generate a rotating magnetic field, so that the rotor follows exactly while the load torque is less than the rated torque $M_L < M_M$. Under this condition, the driving of the stepping motor is simpler than that of a DC motor, hence no sensors for position feedback have to be used. Due to the discontinuous operation mode, the stepping motor is suitable as a switching mechanism (printer, quartz clock, etc.) and for positioning tasks. The previously introduced TVS (Figure 10) uses stepping motors, to position an emitted laser line-by-line in the FOV of the system (Sergiyenko O., Optoelectronic System for Mobile Robot Navigation, Optoelectronics, 2010). Stepping motors have a discrete behavior, due to their stepwise rotation of the motor shaft and hence the laser light can only be positioned on discrete points. Despite the fact, that modern stepping motors in certain operating modes have step sizes less than one degree, they will always have unreachable angles in their FOM. Thus, by using stepping motors in the developed TVS, the FOV is discretized and its resolution depends on the distance of the examined scanning object. Targets, which have dimensions less than the FOV resolution in a certain distance, cannot be scanned or captured.

Figure 23. Technology schema of a stepping motor

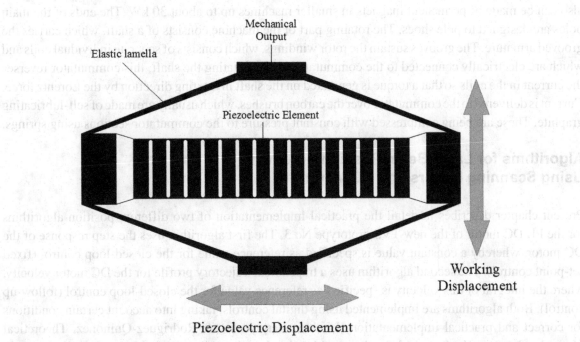

Figure 23. Technology schema of a stepping motor

Direct Current Motors

A *Direct Current* Motor (Figure 24, DC) is a machine, which is energized using direct current and mainly consists of the following parts [27]:

1 Main pole 4 Rotor winding (armature)
2 Stator winding 5 Carbon brush
3 Stator (yoke ring) 6 Commutator

Figure 24. Technology scheme of a direct current motor

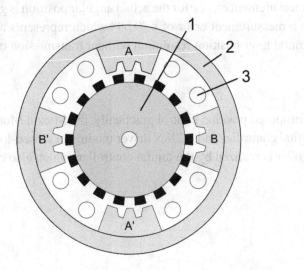

The stator produces the direct magnetic field by the main poles and the stator windings. The stator also can be made of permanent magnets in smaller machines up to about 30 kW. The ends of the main poles are designed to pole shoes. The rotating part of the machine consists of a shaft, which carries the grooved armature. The grooves sustain the rotor windings, which consists of separate individual coils and which are electrically connected to the commutator. When rotating the shaft, this commutator reverses the current in the coils so that a torque is generated on the shaft in rotating direction by the Lorentz force. Current is delivered to the commutator over the carbon brushes, which usually are made of self-lubricating graphite. These are being compressed with constant pressure to the commutator sections using springs.

Algorithms for Laser Beam Positioning When Using Scanning Mirrors and DC Motors

Present chapter describes in detail the practical implementation of two different position algorithms for the PL DC motor of the new TVS prototype No.3. The first algorithm uses the step response of the DC motor, whereby a constant value is specified as reference value for the closed-loop control (fixed set-point control). The second algorithm uses a trapezoidal trajectory profile for the DC motor velocity, where the integral of this velocity is specified as reference value for the closed-loop control (follow-up control). Both algorithms are implemented using digital controls, taking into account certain conditions for correct and practical implementation (Lindner, Sergiyenko, & Rodriguez-Quinonez, Theoretical Method to Increase the Speed of Continuous Mapping in a three-dimensional Laser Scanning System using Servomotors Control, 2017).

Actual Angular Position Measuring

Like it is used the Maxon Motor RE-max29 with encoder MR Type ML (Figure 13), without using a transmission, the actual counted pulses ρ_o (counts) from the encoder represent directly the actual angular position φ_o (degree) and can be converted vice versa by relation:

$$\frac{\varphi_o}{360°} = \frac{\rho_o}{4000}.$$

(13)

An analysis of the measurement error ϵ for the actual angular position is given in section Error Analysis. This section gives a measurement error of $\epsilon \leq 0.09°$ which represents a major improvement, when compared to the horizontal laser position resolution without transmission of the TVS prototype No.2.

Step Response

In order to realize the proposed position control practically, the Maxon Motor RE-max29 with encoder is connected with a digital controller and L298N driver module in closed-loop (CL), depicted in Figure 25. The position controller is realized by the digital controller, which also contains the comparator.

Figure 25. Closed-loop position control

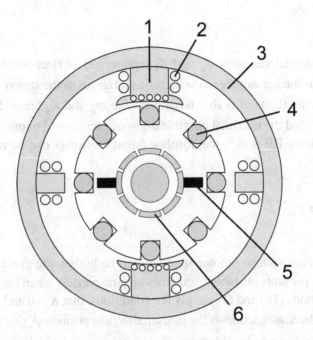

Thereby, φ_r represents the reference angular position from the user, $\varphi_e(k)$ the absolute angular error, $y(k)$ the controller output variable, $u_A(k)$ the motor armature voltage generated by the motor driver, $\varphi_o(k)$ the actual angular position and $\varphi_m(k)$ the measured angular position of the motor shaft. Thereby, k represents discrete time points of the continuous time t. The positioning algorithm is implemented as a software part of the position controller block.

Controller Output Variable

The absolute angular error φ_e is calculated by the difference between the reference angular position φ_r and the measured angular position φm.

$$\varphi_e = \varphi_r - \varphi_m .$$
(14)

To normalize this absolute error to percentage, this value is related to the reference angular position φ_r and thereby converted to the relative angular error φ_e'.

$$\varphi_e' = 100 \cdot \frac{\varphi_r - \varphi_m}{\varphi_r} .$$
(15)

Like the determination of the actual angular position φ_0 has an integral characteristic (accumulating the encoder pulses), only a P-controller (K_R is needed to control the actual angular position of the DC motor shaft. Thus, the controller output variable $y(t)$ is calculated by:

$$y = K_R \cdot 100 \cdot \frac{\varphi_r - \varphi_m}{\varphi_r} = y_{\varphi_r} \,. \tag{16}$$

By multiplying the internal variables φ_e' and K_R represented as float-numbers, with a certain ten-power and storing them as integer-numbers, the significant figures of the stored value is defined. It must be noted that the significant digits of both internal variable φ_e' and K_R are added up by equation (16) and therefore may not exceed the maximal possible digits of precision. To improve the step response of the controlled Maxon Motor RE-max[29], the absolute angular error φ_e can be related to the maximum positioning angle of 360°.

$$y = K_R \cdot 100 \cdot \frac{\varphi_r - \varphi_m}{360°} = y_{360} \,. \tag{17}$$

Thus, the initial armature voltage / motor speed u_A will be higher, the greater is the initial absolute angular error $\varphi_e(0)$. This prevents unstable oscillatory step responses, when the initial absolute error is too small. For both equations (16) and (17) apply the conditions, that $K_R > 0$ and $\varphi_r \geq \varphi_m$. Both controller output variables are considered as functions of the measured angular position φ_m that is $y_{\varphi_r} = y_{360} = f(\varphi_m)$.

Because of real physical constrains, like maximum motor armature voltage u_A the controller output variable must be clamped to the range of [0–100%].

$$y = \max(0, \min(y, 100)) \,. \tag{18}$$

To analyze the influence of both parameters K_R and φ_r to the growth of the controller output functions $y_{\varphi_r} = f(\varphi_m)$ and $y_{360} = f(\varphi_m)$. the zeros, the slope, the intersection with ordinate axis and the intersection with the 100%-line shall be calculated. First, the zeros are calculated by:

$$y_{\varphi_r} = K_R \cdot 100 \cdot \frac{\varphi_r - \varphi_m}{\varphi_r} = 0 \quad y_{360} = K_R \cdot 100 \cdot \frac{\varphi_r - \varphi_m}{360°} = 0 \tag{19}$$

where both functions possess the same zero:

$$\varphi_{m,0} = \varphi_r \,. \tag{20}$$

The slope results by differentiation of equations (19):

$$\frac{dy_{\varphi_r}}{d\varphi_m} = -\frac{100 K_P}{\varphi_r} \quad \frac{dy_{360}}{d\varphi_m} = -\frac{100 K_P}{360°} \tag{21}$$

That means, that the slope of the y_{φ_r} function depends on K_R and φ_r and the slope of the y_{360} function only on K_R. The intersection with the ordinate axis is calculated by:

$$y_{\varphi_r}(0) = 100 K_R \quad y_{360}(0) = \frac{100}{360°} \cdot K_R \varphi_r \qquad (22)$$

The intersection with the 100%-line is calculated by:

In order that $\varphi_m 100 > 0$ and thereby $y_{\varphi_r}(0) > 100$ for $y_{360}(0) > 100$ for the first function must $K_R > 1$ and for the second function must $K_R > \dfrac{360°}{\varphi_r}$.

Figure 26 shows a graphical sketch of both controller output functions $y_{\varphi_r} = f(\varphi_m)$ and $y_{360} = f(\varphi_m)$. For both functions two cases are shown, when y(0)>100 and when y(0)<100. Both controller output variables are not greater than 100%, because they are clamped to the range of [0 – 100%] (18). Both functions have the same zero (20), different slopes (21), different intersection with the ordinate axis (22) and different intersection with the 100%-line (23).

Figure 26. Graph of controller output functions

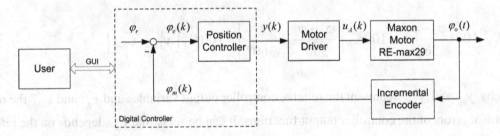

Error Analysis

When calculating the controller output variables, an absolute measurement error ϵ is considered, which is defined by the difference between the actual angular position φ_o and the measured angular position φm:

This error originates from the actual angular position measuring by counting the encoder output pulses, represented by ρ_o (counts). Hereby exists an uncertainty of one pulse $\Delta = 1$, due to the discrete values of the counted output pulses, which are represented by integer numbers, as depicted in Figure 27:

Figure 27. Output pulses of incremental encoder

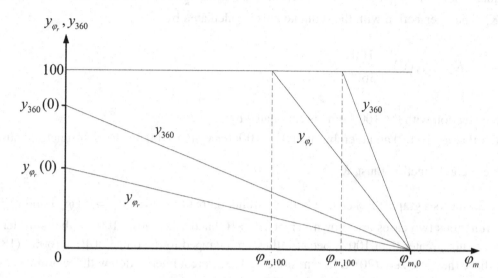

Using equation (13) this one pulse represents 0.09° and thus, the error ϵ is limited by the following range:

$$y_{360} = K_R \cdot 100 \cdot \frac{\varphi_r - \varphi_m}{360°} = 100 K_R \left[\frac{\varphi_r - \varphi_o}{360°} + \frac{\epsilon}{360°} \right] = 100 K_R \left[y'_{360} + \epsilon'_{360} \right]. \qquad (27)$$

Thereby y'_{φ_r} .and y'_{360} .represent the relative controller output variables and ϵ'_{φ_r} and ϵ'_{360} the relative measurement errors of the controller output functions. It can be seen, that ϵ'_{φ_r} .depends on the reference angular position φr while ϵ'_{360} has a fixed value of $\frac{1}{4000}$. Because of (26) and (27), the relative measurement error $\epsilon '$ is producing an armature error voltage, independent if the relative controller output variable y' is zero or not. This means, that the amplification factor K_R must be chosen so small, that this error voltage is small enough so that the static friction force prevents rotation of the DC motor. Equations (26) and (27) also define the range of the controller output functions y_{φ_r} and y_{360}. The resolution of the internal variables K_R, y', $\epsilon '$, and the controller output variable y is defined by the used datatype for storing (Lindner, Sergiyenko, & Rodriguez-Quinonez, Theoretical Method to Increase the Speed of Continuous Mapping in a three-dimensional Laser Scanning System using Servomotors Control, 2017).

Trapezoidal Trajectory Response

With fixed set-point control (step response), a constant reference value for the closed-loop is specified and the controlled system is continuously stabilized around this value. DC motors represent second-order controlled systems, whose step response can overshoot, when certain controller parameters are selected. If the overshoot is undesirable, a tradeoff must be made between speed and overshoot of the step response. One solution to reduce or avoid this overshoot represents the follow-up control of the actual angular position φ_0 of the DC motor shaft. In follow-up control, a time-variable reference value (trajectory) is specified for the closed-loop, which, under certain conditions, guides the actual angular position of the DC motor to the same value as with fixed set-point control.

This new approach to position the PL DC motor in the TVS FOV using the trapezoidal trajectory response was first presented in (Reyes-Garcia, et al., 2018). Using a trapezoidal velocity profile $\omega_r(t)$ as trajectory, which by integrating results in the reference angular position $\varphi_r(t)$ as a ramp, the final angular position $\varphi_o(t)$ of the DC motor shaft $\varphi_o(t)$ is stabilized around this ramp using a PI-controller. This results in a smaller final angular position error and prevents an overshoot of the DC motor shaft. The generation and application of this trapezoidal velocity profile as trajectory was implemented using the digital controller LM_{629}. It could be shown, that using this digital controller, the final angular position error $\varphi_e(t\rightarrow\infty)$ of the PL DC motor shaft can be reduced by at least 82.97% compared to the implementation using a P-algorithm (Reyes-Garcia, et al., 2018). Furthermore, it can be shown, that using a trapezoidal velocity profile as trajectory, the final angular position $\varphi_\infty=\varphi_o(t\rightarrow\infty)$ of the Maxon RE 25 brushed DC motor is defined precisely by integrating this trapezoidal velocity profile and determining the final value of the resulting reference angular position $\varphi_r(t)$. Therefore in (Reyes-Garcia, et al., 2018), the theoretical and practical fundamentals for controlling the final angular position $\varphi_o(t)$ of the Maxon RE 25 brushed DC motor using a trapezoidal trajectory profile are given, as well as the implementation of these profile using the digital controller LM_{629} is presented.

Figure 28. CL position control of RE-max 29 using Step and Trajectory Response

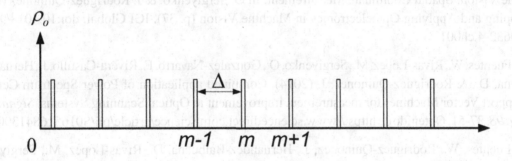

Also, work (Reyes-Garcia, et al., 2018) represents the simulation of the DC Motor Maxon RE-max 29 model in closed-loop configuration, using a constant $\varphi_r=90°$ and an arbitrary defined trajectory $\varphi_r=f(t)$ for the reference angular position. Both reference values are used as input for the same CL algorithm, to position the DC motor shaft on $\varphi_\infty=90°$. Figure 28 shows the comparisons of the CL position control, including the step response (blue) and the trajectory response (red) of the DC motor. This figure shows clearly the overshoot of the step response and the stabilization of the DC motor actual angular position $\varphi_o(t)$ around the trajectory. This figure also emphasizes the actual final angular position for the step response with $\varphi_o(t\rightarrow\infty)=93.13°$ and for the trajectory response with $\varphi_o(t\rightarrow\infty)=89.99°$

REFERENCES

Basaca-Preciado, L., Sergiyenko, O., Rodriguez-Quinonez, J., Garcia, X., Tyrsa, V., Rivas-Lopez, M., Hernandez-Balbuena, D., Mercorelli, P., Podrygalo, M., Gurko, A., Tabakova, I., & Starostenko, O. (2014, March). Optical 3D laser measurement system for navigation of autonomous mobile robot. Optics an*d Lasers in Engineering, 54, 159–169.* doi:10.1016/j.optlaseng.2013.08.005

Benet, G., Blanes, F., Simo, J., & Perez, P. (2002, September 30). Using infrared sensors for distance measurement in mobile robots. Robot*ics and Autonomous Systems, 40(4),* 255–266. doi:10.1016/S0921-8890(02)00271-3

Correal, R., Pajares, G., & Ruz, J. (2014). Automatic expert system for 3D terrain reconstruction based on stereo vision and histogram matching. Exp*ert System with Application, 41,* 2043-2051. Obtenido de https://www.sciencedirect.com/science/article/pii/S0957417413007227

Fang, W., Huang, X., Zhang, F., & Li, D. (2015). Intensity Correction of Terrestrial Laser Scanning Data by Estimating Laser Transmission Function. *IEEE Transactions on Geoscience and Remote Sensing, 53*(2), 942-951. doi:10.1109/TGRS.2014.2330852

Fischer, R. (2003). *Elektrische Maschinen* (12 ed.). Carl Hanser Verlag GmbH & Co. KG. Obtenido de https://www.amazon.de/Elektrische-Maschinen-Rolf-Fischer/dp/3446226931

Flores-Fuentes, W., Rivas-Lopez, M., Hernandez-Balbuena, D., Sergiyenko, O., Rodriguez-Quinonez, J., Rivera-Castillo, J., . . . Basaca-Preciado, L. (2016). Applying Optoelectronic Devices Fusion in Machine Vision: Spatial Coordinate Measurement. In O. Sergiyenko, & J. Rodriguez-Quinonez (Eds.), Developing and Applying Optoelectronics in Machine Vision (p. 37). IGI Global. doi:10.4018/978-1-5225-0632-4.ch001

Flores-Fuentes, W., Rivas-Lopez, M., Sergiyenko, O., Gonzalez-Navarro, F., Rivera-Castillo, J., Hernandez-Balbuena, D., & Rodriguez-Quinonez, J. (2014). Combined application of Power Spectrum Centroid and Support Vector Machines for measurement improvement in Optical Scanning Systems. *Signal Processing, 98,* 37-51. Obtenido de https://www.sciencedirect.com/science/article/pii/S0165168413004337

Flores-Fuentes, W., Rodriguez-Quinonez, J., Hernandez-Balbuena, D., Rivas-Lopez, M., Sergiyenko, O., Rivera-Castillo, J., . . . Mayorga-Ortiz, P. (2015). Photodiode and charge-coupled device fusioned sensors. In *2015 IEEE 24th International Symposium on Industrial Electronics (ISIE)* (pp. 966-971). Buzios: IEEE. doi:10.1109/ISIE.2015.7281602

Flores-Fuentes, W., Sergiyenko, O., Gonzalez-Navarro, F., Rivas-Lopez, M., Rodriguez-Quinonez, J., Hernandez-Balbuena, D., Tyrsa, V., & Lindner, L. (2016, August). Multivariate outlier mining and regression feedback for 3D measurement improvement in opto-mechanical system. *Optical and Quantum Electronics, 48*(8), 21. doi:10.100711082-016-0680-1

Garcia-Cruz, X., Sergiyenko, O., Tyrsa, V., Rivas-Lopez, M., Hernandez-Balbuena, D., Rodriguez-Quinonez, J., ... Mercorelli, P. (2014). Optimization of 3D laser scanning speed by use of combined variable step. *Optics and Lasers in Engineering, 54,* 141-151. Obtenido de https://www.sciencedirect.com/science/article/pii/S0143816613002546

Garcia-Talegon, J., Calabres, S., Fernandez-Lozano, J., Inigo, A., Herrero-Fernandez, H., Arias-Perez, B., & Gonzalez-Aguilera, D. (2015, February 25-27). Assessing Pathologies On Villamayor Stone (Salamanca, Spain) By Terrestrial Laser Scanner Intensity Data. *Remote Sensing and Spatial Information Sciences, XL-5/W4*, 445-451. Retrieved from https://www.int-arch-photogramm-remote-sens-spatial-inf-sci.net/XL-5-W4/445/2015/isprsarchives-XL-5-W4-445-2015.html

Hinks, T., Carr, H., Gharibi, H., & Laefer, D. (2015). Visualisation of urban airborne laser scanning data with occlusion images. *ISPRS Journal of Photogrammetry and Remote Sensing, 104*, 77-87. doi:10.1016/j.isprsjprs.2015.01.014

Hiremath, S., van der Heijden, G., van Evert, F., Stein, A., & ter Braak, C. (2014). Laser range finder model for autonomous navigation of a robot in a maize field using a particle filter. *Computers and Electronics in Agriculture, 100*, 41-50. Obtenido de https://www.sciencedirect.com/science/article/pii/S0168169913002470

Hug, C., & Wehr, A. (1997, September 17-19). Detecting And Identifying Topographic Objects In Imaging Laser Altimeter Data. *IAPRS, 32*, 19-26. Retrieved from https://www.ifp.uni-stuttgart.de/publications/wg34/wg34_hug.pdf

Kumar, P., McElhinney, C., Lewis, P., & McCarthy, T. (2013). An automated algorithm for extracting road edges from terrestrial mobile LiDAR data. *ISPRS Journal of Photogrammetry and Remote Sensing, 85*, 44-55. Obtenido de https://www.sciencedirect.com/science/article/pii/S0924271613001834

Lindner, L. (2016). Laser Scanners. In O. Sergiyenko, J. Rodriguez-Quinonez, O. Sergiyenko, & J. Rodriguez-Quinonez (Eds.), Developing and Applying Optoelectronics in Machine Vision (p. 38). IGI Global. doi:10.4018/978-1-5225-0632-4.ch004

Lindner, L., Sergiyenko, O., Rivas-Lopez, M., Ivanov, M., Rodriguez-Quinonez, J., Hernandez-Balbuena, D., ... Mercorelli, P. (2017). Machine vision system errors for unmanned aerial vehicle navigation. In *Industrial Electronics (ISIE), 2017 IEEE 26th International Symposium on* (pp. 1615-1620). Edinburgh: IEEE. doi:10.1109/ISIE.2017.8001488

Lindner, L., Sergiyenko, O., Rivas-Lopez, M., Rodriguez-Quinonez, J., Hernandez-Balbuena, D., Flores-Fuentes, W., ... Kartashov, V. (2016). Issues of exact laser ray positioning using DC motors for vision-based target detection. In *2016 IEEE 25th International Symposium on Industrial Electronics (ISIE)* (pp. 929-934). Santa Clara: IEEE.

Lindner, L., Sergiyenko, O., Rivas-Lopez, M., Valdez-Salas, B., Rodriguez-Quinonez, J., Hernandez-Balbuena, D., ... Kartashov, V. (2016). Machine vision system for UAV navigation. In *Electrical Systems for Aircraft, Railway, Ship Propulsion and Road Vehicles & International Transportation Electrification Conference (ESARS-ITEC), International Conference on*. Toulouse: IEEE.

Lindner, L., Sergiyenko, O., Rivas-Lopez, M., Valdez-Salas, B., Rodriguez-Quinonez, J., Hernandez-Balbuena, D., & Mercorelli, P. (2016). *UAV remote laser scanner improvement by continuous scanning using DC motors. In Industrial Electronics Society, IECON 2016*. IEEE. doi:10.1109/IECON.2016.7793316

Lindner, L., Sergiyenko, O., Rivas-Lopez, M., Valdez-Salas, B., Rodriguez-Quinonez, J., Hernandez-Balbuena, D., . . . Mercorelli, P. (2016). UAV remote laser scanner improvement by continuous scanning using DC motors. In Industrial Electronics Society, IECON 2016. Florence: IEEE.

Lindner, L., Sergiyenko, O., & Rodriguez-Quinonez, J. (2017). *Theoretical Method to Increase the Speed of Continuous Mapping in a three-dimensional Laser Scanning System using Servomotors Control.* UABC, Editorial Universitario.

Lindner, L., Sergiyenko, O., Rodriguez-Quinonez, J., Rivas-Lopez, M., Hernandez-Balbuena, D., Flores-Fuentes, W., Natanael Murrieta-Rico, F., & Tyrsa, V. (2016). Mobile robot vision system using continuous laser scanning for industrial application. *The Industrial Robot, 43*(4), 360–369. doi:10.1108/IR-01-2016-0048

Lindner, L., Sergiyenko, O., Rodriguez-Quinonez, J., Tyrsa, V., Mercorelli, P., Fuentes-Flores, W., . . . Nieto-Hipolito, J. (2015). Continuous 3D scanning mode using servomotors instead of stepping motors in dynamic laser triangulation. In *Industrial Electronics (ISIE), 2015 IEEE 24th International Symposium on* (pp. 944-949). Buzios: IEEE. doi:10.1109/ISIE.2015.7281598

Lindner, L., Sergiyenko, O., Tyrsa, V., & Mercorelli, P. (2014, June 01-04). An approach for dynamic triangulation using servomotors. In *Industrial Electronics (ISIE), 2014 IEEE 23rd International Symposium on* (pp. 1926-1931). Istanbul: IEEE. doi:10.1109/ISIE.2014.6864910

Lindner, L., Sergiyenko, O., Tyrsa, V., & Mercorelli, P. (2014). An approach for dynamic triangulation using servomotors. In *Industrial Electronics (ISIE), 2014 IEEE 23rd International Symposium on* (págs. 1926-1931). Istanbul: IEEE. doi:10.1109/ISIE.2014.6864910

Murrieta-Rico, F., Hernandez-Balbuena, D., Rodriguez-Quinonez, J., Petranovskii, V., Raymond-Herrera, O., Nieto-Hipolito, J., . . . Melnyk, V. (2015). Instability measurement in time-frequency references used on autonomous navigation systems. In *2015 IEEE 24th International Symposium on Industrial Electronics (ISIE)* (pp. 956-961). Buzios: IEEE. doi:10.1109/ISIE.2015.7281600

Murrieta Rico, F., Petranovskii, V., Raymond-Herrera, O., Sergiyenko, O., Lindner, L., Hernandez-Balbuena, D., . . . Tyrsa, V. (2016). High resolution measurement of physical variables change for INS. In *2016 IEEE 25th International Symposium on Industrial Electronics (ISIE)* (pp. 912-917). Santa Clara: IEEE.

Murrieta-Rico, F., Petranovskii, V., Raymond-Herrera, O., Sergiyenko, O., Lindner, L., Valdez-Salas, B., . . . Tyrsa, V. (2016). Resolution improvement of accelerometers measurement for drones in agricultural applications. In Industrial Electronics Society, IECON 2016. Florence: IEEE.

Murrieta-Rico, F., Sergiyenko, O., Petranovskii, V., Hernandez-Balbuena, D., Lindner, L., Tyrsa, V., Rivas-Lopez, M., Nieto-Hipolito, J. I., & Karthashov, V. (2016, May). Pulse width influence in fast frequency measurements using rational approximations. *Measurement, 86*, 67–78. doi:10.1016/j.measurement.2016.02.032

Ohnishi, N., & Imiya, A. (2013). Appearance-based navigation and homing for autonomous mobile robot. *Image and Vision Computing, 31*, 511-532. Obtenido de https://www.sciencedirect.com/science/article/pii/S0262885612002120

Reyes-Garcia, M., Lindner, L., Rivas-Lopez, M., Ivanov, M., Rodriguez-Quinonez, J., Murrieta-Rico, F., . . . Melnyk, V. (2018). Reduction of Angular Position Error of a Machine Vision System Using the Digital Controller LM629. In *IECON 2018-44th Annual Conference of the IEEE Industrial Electronics Society* (pp. 3200-3205). Washington, DC: IEEE. doi:10.1109/IECON.2018.8592803

Rodriguez-Quinonez, J., Sergiyenko, O., Basaca-Preciado, L., Tyrsa, V., Gurko, A., Podrygalo, M., ... Hernandez-Balbuena, D. (2014). Optical monitoring of scoliosis by 3D medical laser scanner. *Optics and Lasers in Engineering, 54*, 175-186. Obtenido de https://www.sciencedirect.com/science/article/pii/S014381661300242X

Rodriguez-Quinonez, J., Sergiyenko, O., Gonzalez-Navarro, F., Basaca-Preciado, L., & Tyrsa, V. (2013, February). Surface recognition improvement in 3D medical laser scanner using Levenberg–Marquardt method. *Signal Processing, 93*(2), 378–386. doi:10.1016/j.sigpro.2012.07.001

Roemer, G., & Bechthold, P. (2014). Electro-optic and Acousto-optic Laser Beam Scanners. *8th International Conference on Laser Assisted Net Shape Engineering, 56*, 29-39. doi:10.1016/j.phpro.2014.08.092

Sergiyenko, O. (2010). Optoelectronic System for Mobile Robot Navigation. *Optoelectronics, Instrumentation and Data Processing, 46*(5), 414–428. doi:10.3103/S8756699011050037

Sergiyenko, O. (2010). Optoelectronic System for Mobile Robot Navigation, Optoelectronics. *Instrumentation and Data Processing, 46*, 414-428. Obtenido de https://link.springer.com/article/10.3103/S8756699011050037

Sergiyenko, O., Tyrsa, V., Basaca-Preciado, L., Rodriguez-Quinonez, J., Hernandez, W., Nieto-Hipolito, J., . . . Starostenko, O. (2011). Electromechanical 3D Optoelectronic Scanners: Resolution Constraints and Possible Ways of Improvement. In Optoelectronic Devices and Properties. InTech. doi:10.5772/14263

Shan, J., & Toth, C. (2008). *Topographic Laser Ranging And Scanning*. CRC Press. Obtenido de. https://www.amazon.com/Topographic-Laser-Ranging-Scanning-Principles/dp/1420051423/ref=tmm_hrd_title_0?_encoding=UTF8&qid=1441252814&sr=1-1

Sixta, Z., Linhart, J., & Nosek, J. (2013). Experimental Investigation of Electromechanical Properties of Amplified Piezoelectric Actuator. In *Electronics, Control, Measurement, Signals and their application to Mechatronics (ECMSM), 2013 IEEE 11th International Workshop of* (pp. 1-5). Toulouse: IEEE. doi:10.1109/ECMSM.2013.6648957

Solchenbach, T., & Plapper, P. (2013). Mechanical characteristics of laser braze-welded aluminium–copper connections. *Optics & Laser Technology, 54*, 249-256. doi:10.1016/j.optlastec.2013.06.003

Song, D., & Chang, J. (2012, February 29). Super wide field-of-regard conformal optical imaging system using liquid crystal spatial light modulator. *Optik (Stuttgart)*, 2455-2458. Advance online publication. doi:10.1016/j.ijleo.2012.08.013

Tame, B., & Stutzke, N. (2010). Steerable Risley Prism antennas with low side lobes in the Ka band. In *Wireless Information Technology and Systems (ICWITS), 2010 IEEE International Conference on* (págs. 1-4). IEEE. doi:10.1109/ICWITS.2010.5611931

Toth, T., & Zivcak, J. (2014). A Comparison of the Outputs of 3D Scanners. In *24th DAAAM International Symposium on Intelligent Manufacturing and Automation*, (pp. 393-401). Kosice: Elsevier. doi:10.1016/j.proeng.2014.03.004

Volos, C., Kyprianidis, I., & Stouboulos, I. (2013, December). Experimental investigation on coverage performance of a chaotic autonomous mobile robot. *Robotics and Autonomous Systems, 61*(12), 1314–1322. doi:10.1016/j.robot.2013.08.004

Wang, L., Liu, L., Sun, J., Zhou, Y., Luan, Z., & Liu, D. (2010). Large-aperture double-focus laser collimator for PAT performance testing of inter-satellite laser communication terminal. *Optik, 121*(17), 1614-1619. doi:10.1016/j.ijleo.2009.03.004

Xiang, S., Chen, S., Wu, X., Xiao, D., & Zheng, X. (2010). Study on fast linear scanning for a new laser scanner. *Optics & Laser Technology, 42*(1), 42-46. doi:10.1016/j.optlastec.2009.04.019

Yang, S., Lee, K., Xu, Z., Zhang, X., & Xu, X. (2001). An accurate method to calculate the negative dispersion generated by prism pairs. *Optics and Lasers in Engineering, 36*(4), 381-387. doi:10.1016/S0143-8166(01)00055-0

Yang, Y. (2008, November 21). Analytic Solution of Free Space Optical Beam Steering Using Risley Prisms. *Journal of Lightwave Technology, 26*(21), 3576–3583. doi:10.1109/JLT.2008.917323

Zhongdong, Y., Peng, W., Xiaohui, L., & Changku, S. (2014, March). 3D laser scanner system using high dynamic range imaging. *Optics and Lasers in Engineering, 54*, 31–41. doi:10.1016/j.optlaseng.2013.09.003

ENDNOTES

[1] https://portal.motorolasolutions.com
[2] http://www.faro.com
[3] http://www.riegl.com
[4] www.faro.com
[5] http://www.iff.fraunhofer.de
[6] www.raylase.de
[7] www.scanlab.de

Chapter 3
Methods of Reception and Signal Processing in Machine Vision Systems

Tatyana Strelkova
Kharkiv National University of Radio Electronics, Ukraine

Alexander Strelkov
Kharkiv National University of Radio Electronics, Ukraine

Vladimir M. Kartashov
Kharkiv National University of Radio Electronics, Ukraine

Alexander P. Lytyuga
Kharkiv National University of Radio Electronics, Ukraine

Alexander S. Kalmykov
Kharkiv National University of Radio Electronics, Ukraine

ABSTRACT

The chapter covers development of mathematical model of signals in optoelectronic systems. The mathematical model can be used as a base for detection algorithm development for optical signal from objects. Analytical expressions for mean values and signal and noise components dispersion are cited. These expressions can be used for estimating efficiency of the offered algorithm by the criterion of detection probabilistic characteristics and criterion of signal/noise relation value. The possibility of signal detection characteristics improvement with low signal-to-noise ratio is shown. The method is proposed for detection of moving objects and combines correlation and threshold methods, as well as optimization of the interframe processing of the sequence of analyzed frames. This method allows estimating the statistical characteristics of the signal and noise components and calculating the correlation integral when detecting moving low-contrast objects. The proposed algorithm for detecting moving objects in low illuminance conditions allows preventing the manifestation of the blur effect.

DOI: 10.4018/978-1-7998-6522-3.ch003

INTRODUCTION

The dynamic development of engineering and technology over the past decades has led to the possibility of technical implementation of the potentially high capabilities of optoelectronic systems. The main advantages of optoelectronic systems are high accuracy of object location, high range resolution, and high angular resolution. All listed above makes it possible to widely use optoelectronic systems in solving machine vision problems in various fields.

The basis for creating and improving machine vision optoelectronic systems is an understanding of the physical processes of optical radiation generation and propagation, as well as the development and use of the reception and optical signal processing theory taking into account the special features of their space-time structure, wave, corpuscular and statistical properties.

The purpose of this work is to increase the efficiency of the machine vision optoelectronic systems by creating mathematical models, methods and algorithms of signal processing that take into account physical, statistical features of optical signals and the effects of interaction of the received optical radiation with the elements of the optoelectronic system.

The work consists of the following parts:

- Intended use and methods of improvement of optoelectronic systems.
- Physical and mathematical model of signals in optoelectronic systems.
- Imaging in optoelectronic systems.
- Statistical properties of signals in optoelectronic systems.
- Mathematical model of signal detection in optoelectronic systems.
- Image processing in optoelectronic systems.
- Application of accumulation methods for image processing of stationary objects in optoelectronic systems.
- Method for detecting moving objects in optoelectronic systems.

Trends in the improvement of machine vision optoelectronic systems aimed at increasing efficiency and threshold of magnitude require enhancement of time and energy resolution; improvement of signal and image processing methods; expanding the range of observing conditions in which the system can be effectively applied; expanding the range of tasks performed by systems. The development of these directions is based on the in-depth study of physical processes of optical radiation generation and propagation, which determine the wave and corpuscular properties of light; development of optical signal reception and processing methods taking into account the special features of their space-time structure. It is also productive taking into account additional statistical features of input optical signals and output electrical signals.

Spatial resolution and energy detection tasks are performed by signal and image processing based on the threshold and correlation detection methods. The most common model of the output signal used for creating the signal processing methods is based on the description of an uncorrelated Gaussian random process. However, in practice, the signals obey other statistics, and the realization of optimum signal reception assuming that the signals have Gaussian statistics does not always lead to good results. Nowadays, more and more attention is paid to signal processing methods which take into account additional statistical characteristics of received signals.

The main problematic issues in the development of signal processing methods are related to the observation and calculation of the parameters of low-contrast and small-sized objects which signal is characterized by energy value below the sensitivity threshold of optoelectronic systems. There are no universal methods for detecting small-sized, low-contrast objects today. Furthermore, the problems of detection, tracking, and recognition of moving or closely located objects have no unambiguous solution today. For example, correlation methods are good enough to detect signals characterized by small values of signal-to-noise ratio, however, they require a priori information about observed objects. To implement image segmentation techniques, the recognition features are used that are only specific to the objects being observed. Difference methods have proven to be good for recognition of moving objects in the sequence of analyzed frames if no noise component is observed. With relatively homogeneous background, the methods of object recognition using linear and nonlinear spatial filtering and thresholding methods are also efficient.

Since the optical image structure is directly related to the properties of the observed object and the conditions for its observation, in the course of research it turned out to be necessary to clarify and study in more detail the mathematical model of the process of detecting optical radiation from a spatially unsteady object.

The basis of the research was a joint wave-particle description of imaging processes in the plane of the photosensitive element. When setting the time boundaries for signal formation, a transition was made from the conditions of finite amount of time for observation (exposure) to the formation and sharing of a conditionally unlimited sequence of discrete frames.

A description of the optical imaging process was made using the principle of linear superposition. The optical image was represented as a superposition of impulse responses to the functions of independent point sources distributed over the entire plane of objects in a unit frame. The same approach was used to describe the sequence of frames. In other words, this approach established that each point in the plane of the object radiates independently. The resulting image is a weighted sum of these impulse responses.

This chapter describes a method for detecting small-sized, low-contrast moving objects. When such objects are observed in lack of light, the use of these systems becomes ineffective, because of the signal-to-noise ratio reduction under these conditions. This parameter is the main qualitative indicator characterizing the quality of algorithms for imaging of objects, as well as for a system that implements these algorithms. To ensure the proper quality of images obtained in lack of light conditions, it is necessary to increase the exposure time in each frame, or to carry out interframe processing, in particular, post-detector frame accumulation. Due to the high speed of movement of the objects under observation, the "blur" type distortion appears on the object image in lack of light conditions.

The proposed method is used for detection of moving objects and combines correlation and thresholding methods, as well as optimization of the interframe processing of the sequence of analyzed frames. With this method, you can estimate the statistical characteristics of the signal and noise components; calculate the detection probabilities and the correlation integral when detecting moving low-contrast objects.

The developed method allows to reduce the influence of the noise component when using interframe and intraframe signal processing and to improve the energy characteristics of the signal component in proportion to the number of analyzed recognition elements.

Experimental studies of the method for detecting moving objects based on the analysis of the energy and space-time characteristics of simulative and natural objects using an optoelectronic detection system showed that the optimum use of the energy of the spatial distribution of the signal component and the reception time allows increasing the efficiency of optoelectronic system. The statistical characteristics of the signal and background components, which obey the central limit theorem, depending on the accumulation time period. At small exposure time, the effect of averaging of the stochastic signal fluctuations is not observed, but at long exposure time, the effect of space-time averaging of the fluctuations is observed. The use of long exposures allows increasing the system efficiency by the criterion of signal-to-noise ratio by ten folds and by the criterion of conditional probability of false alarm by several orders of magnitude.

The work describes the methods for increasing the efficiency of optoelectronic systems due to the development of signal processing methods and algorithms that take into account the effects of the interaction of the received optical radiation with the elements of the optoelectronic system.

Based on the described approach which provides combined wave and corpuscular signal description and methods of space-time interframe and intraframe signal processing, we can significantly improve the image quality and detection efficiency of small-sized and low-contrast objects including moving objects.

The results of authors' theoretical and experimental research aimed to ensure high accuracy of determination of space-time characteristics of signals and increase probabilistic characteristics of detection are discussed. The coordination of space-energy properties of signals and parameters of optoelectronic systems with a limited dynamic range is taken into account. The calculations of the detectability and detection quality indicators were made on the basis of the developed statistical models; space-time accumulation methods are studied to improve detection parameters of stationary and moving small-sized, low-contrast objects; the detector structure is discussed taking into account the features of the signal and interference structure.

PURPOSE AND METHODS OF IMPROVING OPTOELECTRONIC SYSTEMS

In recent decades the dynamic development of equipment and technologies has provided a way for technical realization of the potentially high possibilities of optoelectronic systems. The main advantages of optoelectronic systems are as follows: a great accuracy of objects coordinates determination, a high resolution in range, angular resolution. All the above allows a widespread use of television systems in many fields of science and technology, for example, in astronomy, vision systems, biology and medicine.

The development of modern optoelectronic systems provides an opportunity to deepen knowledge about the surrounding world and allows you to make discoveries in the field of natural sciences. A deep understanding of physical processes of emergence, propagation of optical radiation and also the theory of receiving and processing of optical signals with consideration of the peculiarities of their spatial-temporal structure, wave, corpuscular and statistical properties lies in creating systems with improving.

There are possible ways to improve and expand the capabilities of optoelectronic systems:

1. Reducing of sensitivity thresholds in the primary information processing devices through the improvement of individual system components, for example, development of new technologies of photodetector elements having higher energy sensitivity.
2. Development of new technologies for creation of elements of an optical link, and also the use of spectral and neutral density filters for matching the dynamic range of the systems during registration of optical signals.
3. Design and development of methods for describing information converted in optoelectronic systems.

4. Improvement of algorithms for data processing, taking into account the statistical characteristics of the received signals in the information processing systems.

The basis of each direction is a rigorous physical and mathematical description of the model of optical signals reception and image processing

Reasoning from the concepts of geometrical optics, image formation in optoelectronic systems is due to employing light rays, which are independent and straightforward in a homogeneous medium and refracted (reflected) on the boundaries of media with different optical properties. The apparatus of geometrical optics makes it possible to describe the process of the image formation, to estimate some parameters of the optical system (the angular field of view, linear magnification, etc.). However, the use of the geometrical optics does not make it possible to consider energy, statistical, spectral characteristics of optical signals.

The formation of signals and images in the systems can be viewed from the position of the wave theory, which is based on the system of differential Maxwell's equations describing electric and magnetic fields strength, electric displacement, magnetic induction and the electric charge density. The system of Maxwell's equations also includes constitutive relations that characterize the behavior of different media in the electromagnetic field. Taking into account the constitutive relations and boundary conditions, the system of Maxwell's equations is a complete one and it allows describing all the properties of the electromagnetic field and many of the processes of interaction of a field with a matter. From the standpoint of the wave theory, it is convenient to describe diffraction phenomena, interference, spectral and polarization properties of optical signals. However, the wave theory application to description of the optical signals of low intensity leads to the results contradictory to the experimental data, when a small number of photons is recorded during the observation time.

Also, the processes of signals and images formation in the systems can be considered from the position of the corpuscle (photon) optics, when light is represented as a stream of discrete particles (photons). Using the photon representation of the optical radiation, it is possible to describe phenomena that have a probabilistic nature (the generation of radiation, absorption in the neutral filters and in the substance of the photocathode). The corpuscular representation enables the analysis of the statistical properties of optical signals and applying methods of the theory of solutions when solving problems of detection, reception of signals and their parameters measurement.

The quality of a particular physical-mathematical model and its applicability limits depend on how fully other factors that characterize the observation conditions and their influence on the parameters of the described signals are taken into account. When analyzing the applicability limits of various models it is advisable to use the criteria for evaluating the efficiency of the optoelectronic systems. These criteria include: the value of the signal/noise ratio, the conditional probability of correct and false detection of objects, etc.

The choice of the basic physical principles and approaches, being the basis for creating mathematical models of signals under different conditions, should be based on the characteristics of the described signals (e.g., the range of intensities) and the background noise and interference condition. So, the concepts of the wave theory of light can be utilized when describing the high intensity signals. In the description of weak signals (for example, weak signals from space objects, signals under low light conditions) the corpuscular structure of the optical radiation should be considered.

Methods of primary and secondary data processing in the optoelectronic systems use the analog and digital methods, which include the spatial-coordinate and the spatial-spectral transformations. Each stage of transformations is aimed at improving the characteristics of signals and images in the optical-electronic systems, for example, spectral redistribution of energy, linear and nonlinear methods of filtering and differentiation. However, the stochastic signal being registered can have different statistical characteristics. This can lead to changes in the accuracy of signals' parameters determination and will not be sufficient to issue reliable results with the best probabilistic characteristics of the observed objects and their parameters. The inclusion of statistical properties of the optical radiation determines the need to develop more complete mathematical models and optimization of algorithms for signal receiving and image processing.

Having reviewed the main approaches to the description of the individual signals, it can be concluded that the mathematical description of signals in optoelectronic systems can be built using a variety of ideas about the structure of the optical radiation and the mechanisms of its interaction with elements of the optical path. So to estimate the spatial position of the signals in the plane of the photodetector the concepts of geometrical optics can be used. The spatial distribution of the signal components from the object is described by the wave representation of the structure and properties of the optical radiation and the theory of electromagnetic waves diffraction. To describe the statistical characteristics of optical signals it is necessary to use elements of the corpuscular (photon) theory of light and a notion of the optical radiation as a random flow of photons. To describe the interaction of light with an absorbent, the process of photocarriers formation, one should use the elements of quantum mechanical representations.

FORMATION OF IMAGES IN OPTOELECTRONIC SYSTEMS

The structure of the optoelectronic system is determined by the range of problems solved with its help. Depending on the purpose and objectives the system can incorporate various technical devices that perform the signals receiving, converting and processing for the purpose of the most effective, justified decision making about the presence of the observation object in the field of view or the value of the signal parameters.

The structural diagram of the optoelectronic system includes: an optical system, an optical receiver, post-detector signal processing shown in Figure 1.

Figure 1. Structural scheme of the generalized optoelectronic system

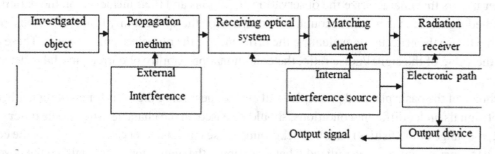

The optical signal in the form of a light field or beam is propagating in the channel, gathered by the system optics and detected by the photodetector. In the general case, the optical field of the useful signal is distorted by the medium of propagation, diffraction and interference effects. The background or noise optical field caused by an external light arrives at the input of the optical receiver. The optical receiver noise is summed to the additive mixture of the useful optical fields and background radiation in the electronic path in the process of converting.

The process of signal formation on the photosensitive elements is not a stationary process. The histograms of the amplitudes of the signals generated by the pixels under the action of optical radiation are presented in Figure 2. Gradual changes of intensity are caused by the uneven lighting, and the abrupt changes in the intensity are stipulated by the quantum structure of optical radiation and dark noise.

Figure 2. Histogram of amplitudes of the light flux intensities

The schematic representation of a fragment of one line in a frame is shown in Figure 3.

Figure 3. Fragment of a row in the frame

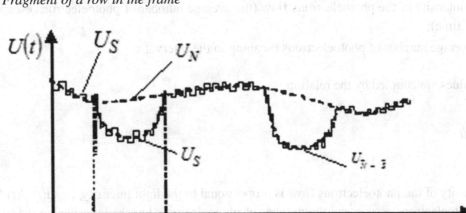

U_N - is the noise component of background, which varies in space (CCD matrix);

U_S - is the signal component (constant in time and varying in space);

$U_{\tilde{N}}$ - is the fluctuation component of noise of the background (varying in time and space);

$U_{\tilde{S}}$ - is the fluctuation component of the signal (varying in time and space).

From the analysis of Figure 3 it follows that the quality of information in images processing depends both on a low-frequency (slowly varying) background noise component, and high-frequency fluctuation components of noise $U_{\tilde{N}}$ and signal $U_{\tilde{S}}$, caused by the corpuscular nature of the light flux and the internal noise of the receiver.

STATISTICAL PROPERTIES OF SIGNALS IN OPTOELECTRONIC SYSTEMS

Normal, log-normal, exponential distributions are used when making statistical models of the received optical radiation in the optoelectronic systems (). Statistical models of the received signals, based on Poisson statistics have gained a large widespread. These models take into account the corpuscular structure of the photon flow, the conditions of the optical radiation formation, the lack of interaction between photons in the stream, as well as the stochastic properties of optical signals with the quantum noise analysis. However, when compiling the statistical signal models based on the Poisson statistics, some assumptions are made, as a rule, and the optical signals properties limitations are accepted, reducing them to three basic properties of the simplest (Poisson) flow: stationarity, ordinary and the lack of the aftereffect. This allows using the invariance property of the Poisson flow when describing its interaction with the elements of the system. The use of such mathematical apparatus makes it possible to obtain analytical expressions for the main performance indicators of the system and to create high-performance algorithms for processing signals in the optoelectronic systems.

Let us introduce the following notation:

$\mu(t)$ - is the intensity of the photoelectrons flow (the average number of photoelectrons escaping per a unit of time);

$\overline{n}$ - is the average number of photoelectrons escaping in the interval τ.

These values are coupled by the relation

$$\overline{n} = \int_0^\tau \mu(t)dt \ .$$

The intensity of the photoelectrons flow is proportional to the light intensity, i.e. $\mu(t) \sim I(t)$. The moments of photoelectrons escape and the number of photoelectrons knocked out during the time τ will be random due to the quantum nature of the signal and quantum nature of the interaction of light with a photodetector.

According to the quantum mechanics statements, the possibility of the photoelectron escape in a small time interval is proportional to the duration of this interval and intensity of the signal $I(t)$. Therefore, the characteristics of the signal at the output of the photodetector are introduced taking into account the temporal dependence of the intensity of the light flux.

If the intensity of the light signal at the input of the photodetector is unchanged (does not fluctuate) in time $I(t)=I_0$, then the probability of the electrons escape in the interval τ depends only on its duration and does not depend on the location of the interval on the time axis. The flow of photoelectrons is stationary. The subsequent escapes of photoelectrons do not depend on the previous ones (absence of an after-effect). In this case, the distribution of the electrons number n in the interval of any duration is described by the Poisson law

$$P(n) = \frac{\overline{n}^n}{n!} e^{-\overline{n}}, \tag{1}$$

where $P(n)$ - is the probability of escape of n photoelectrons in the interval τ.

Mathematical expectation of the photoelectrons number in the interval

$$M(n) = \sum_{n=0}^{\infty} nP_{(n)} = \overline{n}.$$

Dispersion in the number of photoelectrons escapes

$$\sigma^2 = <\left(n-\overline{n}\right)^2> = \sum_{n=0}^{\infty} \left(n-\overline{n}\right)^2 P(n) = \overline{n}.$$

Thus, both dispersion and mathematical expectation are equal to the average number of photoelectrons in the observation interval. Along with stationarity and absence of an after-effect the Poisson flow is characterized by the ordinariness (the emergence of photoelectrons singly and not in pairs or threes).

An important property of the Poisson law is that the sum N of the processes with the Poisson distributions is also the Poisson process with a mean value equal to the sum of average values of N processes.

When $n \gg 10$ the Poisson distribution approaches the normal distribution:

$$P(n) = \frac{1}{\sqrt{2\pi \overline{n}}} e^{-\frac{(n-\overline{n})^2}{2\overline{n}}}. \tag{2}$$

Consider the case where $\mu(t)$ is a random function that occurs in the process of the fluctuations of the light flux received at the input of the photodetector. Thus the probability of the photoelectron emergence in a small interval τ will depend on the value of $\mu(t)$, i.e. it will randomly vary in time.

For easier description, let us introduce the normalized random function $\xi(t)$ such that

$$\mu(t) = \xi(t)\overline{\mu}, \text{ where } \overline{\mu} = \lim_{T\to\infty} \frac{1}{T} \int_0^T \mu(t)dt, \ \overline{\xi} = \lim_{T\to\infty} \frac{1}{T} \int_0^T \xi(t)dt = 1.$$

A random function $\xi(t)$ can be characterized by the correlation interval τC. The statistical characteristics of the signal at the output of the photodetector depend on the ratio of intervals τ and τC.

If the time of observation (duration of implementation) $\tau \ll \tau_C$, then the value $\xi(t)$ can be considered constant and not time-dependent this interval. The average number of photoelectrons in the interval τ is equal to $n_1 = \xi \bar{n}$, where $\bar{n} = \overline{\mu \tau}$.

The distribution of electrons in the interval at a fixed value of ξ corresponds to the Poisson distribution:

$$P(n/\xi) = \frac{\left(\xi \bar{n}\right)^n}{n!} e^{-\xi \bar{n}}. \tag{3}$$

To find $P(n)$ distribution, the joint distribution of n and $\xi P(n,\xi)$ should be averaged over $\xi P(n,\xi)$. In this case $P(n,\xi) = P(n/\xi)P(\xi)$

Usually, the intensity of the signal reflected from the object and the background at the receiver input has normal fluctuations. This is because the signal from the object is created as a result of the interference of statistically independent fields reflected from a large number of "shiny spots" of the scattering surface. Background radiation is also created by a large number of randomly located scatterers (particles). Under normal fluctuations of tension the instantaneous intensity of the signal and, hence, the intensity of the photoelectrons flow, characterized by the value ξ fluctuates according to the exponential law

$$P(\xi) = e^{-\xi}. \tag{4}$$

From (3) and (4) we get

$$P(n) = \int_0^\infty P(n/\xi)P(\xi)d\xi = \int_0^\infty \frac{(\xi \bar{n})^n}{n!} e^{-\xi \bar{n}} e^{-\xi} d\xi = \frac{1}{\bar{n}+1} \left(\frac{\bar{n}}{\bar{n}+1}\right)^n. \tag{5}$$

This distribution is called geometric distribution, or distribution of Bose-Einstein. The mathematical expectation of this distribution is $M(n) = \bar{n}$, and dispersion

$$\sigma^2 = \bar{n}\left(1 + \bar{n}\right) \tag{6}$$

Another practically important case occurs when the observation time is $\tau \gg \tau_C$. The values of the intensity of the flow of electrons $\mu(t)$ through the time intervals $\Delta t \gg \tau_C$ are statistically independent. If we split the observation time τ into $m = \tau/\tau_C$ intervals, then in each interval we will have a distribution $P(n)$ that obeys the law of the Bose-Einstein with a mean value of $\bar{n}_m = \overline{\mu \tau_C} = \bar{n}/m$. Over the interval τ the addition m of statistically independent values occurs, each having the distribution of Bose-Einstein. Thus we turn to the negative binomial distribution

$$P(n) = \frac{(m+n-1)!}{(m-1)!n!}\left(\frac{\bar{n}_m}{\bar{n}_m+1}\right)^n \cdot \left(\frac{1}{\bar{n}_m+1}\right)^m. \tag{7}$$

The mathematical expectation of this distribution is equal to the average number of photoelectrons escaping per the interval τ:

$$M(n) = \bar{n} = \bar{n}_m\, m,$$

And dispersion

$$\sigma^2 = \bar{n}\left(1+\frac{\bar{n}}{m}\right). \tag{8}$$

The number of photons in the light flux is proportional to the modulus square of the electric component of the electromagnetic wave. The photocathode converts the signal photons flow in the flow of signal carriers of N_S charges, the number of which is proportional to the number of photons in the light flux.

In addition to the signal flow of N_S charges the interfering flow of the N_N carriers is formed caused by the action of the background optical radiation quanta and internal noise of the photodetector. Resolution elements of the photodetector produce an accumulation of flows N_S and N_N for the accumulation time T_a.

Let's break the plane of the photodetector into the elementary areas of the size $\Delta S_{ij} = (u_i - u_{i-1})(v_j - v_{j-1})$, where $i=1,\ldots,k$; $j=1,\ldots,m$ (Figure 4).

Figure 4. Plane of photodetector

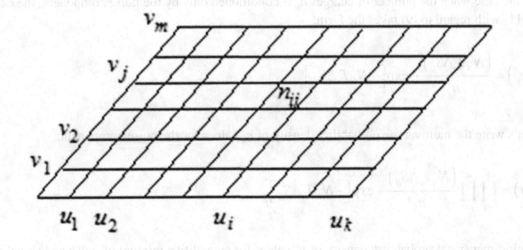

Each ΔS_{ij} elementary area has n_{ij} charges for the accumulation time T_a. Let's write the average number of charges in the cases, when only the interference or only the signal components are present in the implementation, as

$$\bar{n}_{ij} = N_N(u,v) = N_N T_a \Delta S_{ij},$$
$$\bar{n}_{ij} = N_S(u,v) = N_S T_a \Delta S_{ij},$$

(9)

where N_S and N_N - are the average count rate, respectively, of the signal and interference charges per unit time on a unit square.

The spatial distribution of the number of charges N_S over the photocathode:

$$N_S(u,v) \approx N_{S_0} \left[\frac{\sin\left(c(u-u_0)\frac{a}{2}\right)}{c(u-u_0)\frac{a}{2}} \right]^2 \left[\frac{\sin\left(d(v-v_0)\frac{b}{2}\right)}{d(v-v_0)\frac{b}{2}} \right]^2,$$

(10)

where N_{S_0} - is the number of charges, formed under the action of light radiation of the intensity $|E_0|^2$.

As the flows of charge carriers N_S and N_N have the Poisson statistics, the probability of the n_{ij} charges appear on the ΔS_{ij} site for the accumulation time T_a

$$P(n_{ij}) = \frac{(\bar{n}_{ij})^{n_{ij}}}{n_{ij}!} \exp\left[-\bar{n}_{ij}\right].$$

(11)

In the case when the number of charges n_{ij} is conditioned only by the noise component, the expression (11) with regard to (9) takes the form:

$$P_N(n_{ij}) = \frac{(N_N T_a \Delta S_{ij})^{n_{ij}}}{n_{ij}!} \exp\left[-N_N T_a \Delta S_{ij}\right].$$

(12)

Let's write the multivariate probability density of n_{ij} values for the noise component:

$$P_N(\hat{n}) = \prod_{i=1}^{k} \prod_{j=1}^{m} \frac{(N_N T_a \Delta S_{ij})^{n_{ij}}}{n_{ij}!} \exp\left[-N_N T_a \Delta S_{ij}\right].$$

(13)

Multidimensional probability density of n_{ij} values for an additive mixture of noise and signal components is determined by the expression:

$$P_{S+N}\left(\hat{n}\right) = \prod_{i=1}^{k}\prod_{j=1}^{m}\frac{\left(T_a\Delta S_{ij}\left(N_S+N_N\right)\right)^{n_{ij}}}{n_{ij}!}\exp\left[-N_N T_a\Delta S_{ij}\right]. \tag{14}$$

Using expressions (13) and (14) we write the logarithm of the likelihood ratio for the case when

$$\frac{N_S}{N_N} \ll 1. \tag{15}$$

Under the condition (15) it can be assumed that $\ln\left(1+\dfrac{N_S}{N_N}\right) \sim \dfrac{N_S}{N_N}$. Then

$$\ln L = \sum_{i=1}^{k}\sum_{j=1}^{m}\left[n_{ij}\frac{N_S}{N_N} - N_S T_a\Delta S_{ij}\right]. \tag{16}$$

Substituting in (16) N_S value we will receive:

$$\ln L = \frac{N_{C_0}}{N_{\Pi}}Y\left(\hat{n}\right) - W_C. \tag{17}$$

For brevity sake, the following expressions are used in the expression (17):

$$Y\left(\hat{n}\right) = \sum_{i=1}^{k}\sum_{j=1}^{m}n_{ij}\left[\frac{\sin\left(c\left(u_i-u_0\right)\frac{a}{2}\right)}{c\left(u_i-u_0\right)\frac{a}{2}}\right]^2\left[\frac{\sin\left(d\left(v_j-v_0\right)\frac{b}{2}\right)}{d\left(v_j-v_0\right)\frac{b}{2}}\right]^2, \tag{18}$$

where $\hat{n} = n_{1j}, n_{2j}, \ldots, n_{kj}, n_{i1}, n_{i2}, \ldots, n_{km}$ - is the assumed implementation.

It is obvious that the value W_S is proportional to the energy of the received signal.

$$W_S = \sum_{i=1}^{k}\sum_{j=1}^{m}N_{C_0}\left[\frac{\sin\left(c\left(u_i-u_0\right)\frac{a}{2}\right)}{c\left(u_i-u_0\right)\frac{a}{2}}\right]^2\left[\frac{\sin\left(d\left(v_j-v_0\right)\frac{b}{2}\right)}{d\left(v_j-v_0\right)\frac{b}{2}}\right]^2 T_a\Delta S_{ij}. \tag{19}$$

From the expression (16) it follows that the logarithm of the likelihood ratio with a precision of constants is determined by the value of the function $Y\left(\hat{n}\right)$. Therefore, the algorithm of detection of a signal of low intensity using an optoelectronic system can be represented in the form shown in Figure 5.

Figure 5. The algorithm of detection of a signal

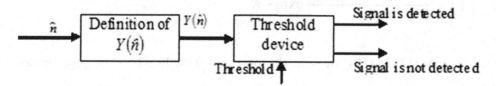

The $Y\left(\hat{n}\right)$ function values are a random variable and depend on the adopted implementation of $\hat{n}$.

The distribution law, which $Y\left(\hat{n}\right)$ value from the expression (17) is subject to, is not obvious. Although n_{ij} values involved in (17) have the Poisson statistics, the sum of $k \bullet m$ terms, each of which is the product of independent Poisson variables by different factors, in a certain way dependent on the spatial coordinates, obviously, will not obey the Poisson law.

However, if k and m are large enough to satisfy the condition $\Delta S_{ij} \to 0$, for large n_{ij} values, due to the law of large numbers, the law of $Y\left(\hat{n}\right)$ value distribution will tend to the normal law.

Therefore, in the first approximation $Y\left(\hat{n}\right)$ value will be assumed to be distributed by the normal (Gaussian) law.

To estimate the effectiveness of the proposed algorithm it is possible to use the criterion of the value of the signal/noise ratio. As it is known from (Levin, 1968), this criterion is the universal one in evaluating the effectiveness of the known signal detection on the background of Gaussian noise. Let us write down the expression for the signal-to-noise ratio ϕ in the form similar to that given in (Gal'jardi & Karp, 1978; Strelkova, 1999):

$$\phi = \frac{\bar{\alpha}}{\sqrt{D(\alpha)}}, \tag{20}$$

where $\bar{\alpha}$ - is the average value of the measured value α; $D(\alpha)$- is the dispersion α.

In our case the measured value is the value of the function $Y\left(\hat{n}\right)$ defined by expression (18). To calculate the value of the signal/noise ratio (20) it is necessary to determine the mean value $\overline{Y\left(\hat{n}\right)}$ and the dispersion $D\left(Y\left(\hat{n}\right)\right)$. Since the information about the signal component is contained in its additive mixture with noise, it is necessary to center the noise component. Then the expression (15) can be written in the form:

$$\phi = \frac{\bar{Y}_{S+N}\left(\hat{n}\right) - \bar{Y}_{N}\left(\hat{n}\right)}{\sqrt{D\left(Y_{S+N}\left(\hat{n}\right)\right) + D\left(Y_{N}\left(\hat{n}\right)\right)}}. \tag{21}$$

Let us calculate the values appearing in (21). Let us write the expression for $\overline{Y_{N}\left(\hat{n}\right)}$ average value:

$$\overline{Y_N(\hat{n})} = \sum_{i=1}^{k}\sum_{j=1}^{m}\overline{n_{ij}}\left[\frac{\sin\left(c(u_i-u_0)\dfrac{a}{2}\right)}{c(u_i-u_0)\dfrac{a}{2}}\right]^2\left[\frac{\sin\left(d(v_j-v_0)\dfrac{b}{2}\right)}{d(v_j-v_0)\dfrac{b}{2}}\right]^2 \cdot \tag{22}$$

Given that the linear size of the main lobe of the intensity distribution is considerably less than the linear size of the photodetector, and replacing the sum by the integral in infinite limits, we get an expression for the average value of noise:

$$\overline{Y_N(\hat{n})} = N_N T_a \int_{-\infty}^{\infty}\left[\frac{\sin\left(c(u_i-u_0)\dfrac{a}{2}\right)}{c(u_i-u_0)\dfrac{a}{2}}\right]^2 du \int_{-\infty}^{\infty}\left[\frac{\sin\left(d(v_j-v_0)\dfrac{b}{2}\right)}{d(v_j-v_0)\dfrac{b}{2}}\right]^2 dv \tag{23}$$

or after integration:

$$\overline{Y_N(\hat{n})} = \frac{4\pi^2}{cadb}N_N T_a \ . \tag{24}$$

The expression for the noise component dispersion:

$$D\left(Y_N(\hat{n})\right) = N_N T_a \int_{-\infty}^{\infty}\left[\frac{\sin\left(c(u_i-u_0)\dfrac{a}{2}\right)}{c(u_i-u_0)\dfrac{a}{2}}\right]^2 du \int_{-\infty}^{\infty}\left[\frac{\sin\left(d(v_j-v_0)\dfrac{b}{2}\right)}{d(v_j-v_0)\dfrac{b}{2}}\right]^2 dv \tag{25}$$

or after integration:

$$D\left(Y_N(\hat{n})\right) = \frac{16\pi^2}{9cadb}N_N T_a \ . \tag{26}$$

Similar arguments will be held for the determination of the mean value and dispersion of the additive mixture of signal and noise:

$$\overline{Y_{S+N}(\hat{n})} = N_N T_a \int_{-\infty}^{\infty}\left[\frac{\sin\left(c(u_i-u_0)\dfrac{a}{2}\right)}{c(u_i-u_0)\dfrac{a}{2}}\right]^2 du \int_{-\infty}^{\infty}\left[\frac{\sin\left(d(v_j-v_0)\dfrac{b}{2}\right)}{d(v_j-v_0)\dfrac{b}{2}}\right]^2 dv +$$

$$+N_{S_0}T_a\int\limits_{-\infty}^{\infty}\left[\frac{sin\left(c(u_i-u_0)\dfrac{a}{2}\right)}{c(u_i-u_0)\dfrac{a}{2}}\right]^4 du\int\limits_{-\infty}^{\infty}\left[\frac{sin\left(d(v_j-v_0)\dfrac{b}{2}\right)}{d(v_j-v_0)\dfrac{b}{2}}\right]^4 dv=\frac{4\pi^2}{cdab}N_NT_a+\frac{16\pi^2}{9cdab}N_{S_0}T_a, \quad (27)$$

$$D\left(Y_{S+N}(\hat{n})\right)=N_NT_a\int\limits_{-\infty}^{\infty}\left[\frac{sin\left(c(u_i-u_0)\dfrac{a}{2}\right)}{c(u_i-u_0)\dfrac{a}{2}}\right]^2 du\int\limits_{-\infty}^{\infty}\left[\frac{sin\left(d(v_j-v_0)\dfrac{b}{2}\right)}{d(v_j-v_0)\dfrac{b}{2}}\right]^2 dv+$$

$$+N_{S_0}T_=\int\limits_{-\infty}^{\infty}\left[\frac{sin\left(c(u_i-u_0)\dfrac{a}{2}\right)}{c(u_i-u_0)\dfrac{a}{2}}\right]^6 du\int\limits_{-\infty}^{\infty}\left[\frac{sin\left(d(v_j-v_0)\dfrac{b}{2}\right)}{d(v_j-v_0)\dfrac{b}{2}}\right]^6 dv=\frac{16\pi^2}{9cdab}N_NT_a+\frac{121\pi^2}{100cdab}N_{S_0}T_a.$$

$$(28)$$

The expression for the signal-to-noise relation value will be:

$$\phi=\frac{16\pi^2 T_a N_{S_0}}{9cdab\sqrt{\dfrac{121\pi^2}{100cdab}T_aN_{S_0}+2\dfrac{16\pi^2}{9cdab}T_aN_N}}. \quad (29)$$

IMAGE PROCESSING IN OPTOELECTRONIC SYSTEMS

The problem of signal detection on the background of additive noise in the image field formed by the optical system is the primary objective for most optoelectronic systems. The results of digital processing of images depend strongly on the quality of information contained in the resulting image, which is determined by the magnitude of the contrast ratio K and the value of the signal/noise ratio ϕ. In practice, the post-detector accumulation method is the most widely used when recording low-contrast objects characterized by small values of the signal/noise ratio. The registration time has a significant impact on the quality of the signals detection in the optoelectronic systems, and it is one of the parameters determining the energy ratio between the signal and noise components of the response of the system's photosensitive element to the optical exposure. With the increase in the accumulation time Ta the radiation energy of the object and the background detected by the photodetector increases, average values and dispersion of the signal and noise components are increasing in proportion to Ta According to expression (29) the process of energy accumulation use will result in the increase in the magnitude of the signal/noise ratio at the output of the detector in proportion to $\sqrt{T_a}$, and, therefore, in the increase in the penetrating ability of the optoelectronic system.

Traditional methods of detecting signals from the objects in optoelectronic systems are based on the threshold signal processing, that is, on comparing the magnitude of the photosensitive elements response to the impact of the additive mixture of signal and background components with a fixed threshold, the value of which is conditioned by the selected criterion of the decision making quality. The decision about the signal detection is making, when the amplitude of the electrical signal generated by one resolution element of the matrix of photodetectors, exceeds the threshold value (single count method).

In (Gal'jardi & Karp, 1978; Levin, 1968; Lytyuga, 2009) it is shown that the quasi-optimal solution for the problem of detecting optical signals on the background of additive noise can be implemented by comparing the received implementation of the "signal + noise" additive mixture with the threshold. The decision about the presence of the signal component in the received implementation (TV frame) is made when exceeding a certain threshold valuably the response amplitude of the element of the photodetector matrix resolution. The so-called energy detection of a signal is realized.

Experimental studies have been devoted to the research on the effect of registration time on the quality of the signals detection in optoelectronic systems. A video clip was used that contained an image of the Beehive (the Manger) constellation recorded under conditions of compensation of the Earth's daily rotation. Video material was provided by the staff of the Nikolaev Astronomical Observatory (A.V. Shulga, etc.). The method of successive accumulation, based on the logical summation of a sequence of images, was implemented in the course of the experiment.

The original image and the result of processing of the video are shown in Figure 6.

Figure 6. Result of experimental verification of signal processing.

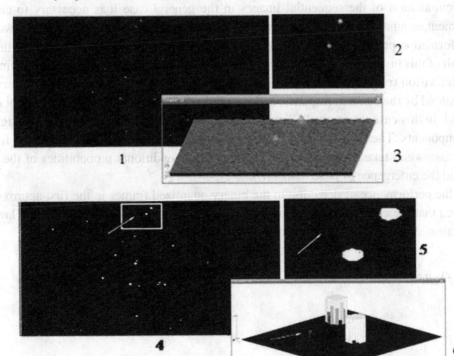

1) the image formed for the accumulation time of 40 MS; 4) the resulting image; 2), (5) the parts of the image containing the detected point objects; 3), 6) histograms of the intensity of the photodetector output signal. Arrows show a point object for which the value of the signal/noise ratio was calculated.

The star magnitude of the investigated object in the catalog Redshift – $9\,{}^{m}_{,}6$

The dependence of the signal-to-noise ratio value in the resulting image on the number of original frames is shown in Figure 7. The calculations were performed for the object indicated by arrows in the Figure 6. The penetrating power of optical-electronic systems is increased by $3\,{}^{m}_{,}0$.

Figure 7. Dependence of signal/noise ratio on the number of realizations chosen for processing

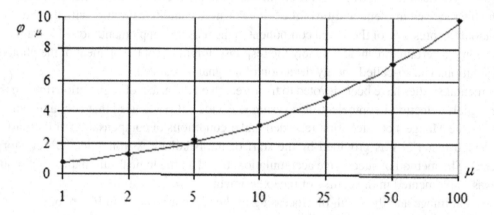

With the accumulation of the sequential images in the general case it is necessary to carry out element-by-element summation of the digital images. For the convenience of the digital processing and reducing the information processing time, it may be best to perform a binary quantization of the video signal. The result of this processing is a sequence of numeric binary arrays corresponding in dimension to the size of a television frame in which the marks "1" correspond to the case of exceeding permission of a certain threshold by the element response, and the marks "0" correspond to the case of not exceeding the threshold. In this case the marks "1" can be formed under the influence of both the signal and background components. Therefore, at a co-processing of the sequence of binary quantized frames it is necessary to assess and take into account the change of the conditional probabilities of the correct detection (C) and the emergence of false marks (F).

To evaluate the performance of detection in the binary quantized frames in the first approximation it can be assumed that the centered noise component is distributed according to the normal law. Then detection performance in a single binary quantized frame will be equal to:

$$C_1 = \int_{U_0}^{\infty} w_{N+S}(U)\, dU\,;$$

(30)

$$F_1 = \int\limits_{U_0}^{\infty} w_N\left(U\right) dU, \tag{31}$$

where C_1 - is the probability of correct detection of the signal;

F_1 - is the probability of the false marks emergence;
$w_{N+C}\left(U\right)$ - is the probability density of the total fluctuation component of the signal and background;
$w_N(U)$ - is the probability density of the fluctuation component of the background.

To obtain the resulting image the logical summation of the binary quantized frames will be held, for example, for two frames, according to the logical rule "And":

$$\begin{cases} 1+1=1 \\ 1+0=0 \\ 0+1=0 \\ 0+0=0 \end{cases} \tag{32}$$

To summarize let's select a finite number n of binary frames. Let's calculate the detection characteristics for the resulting image (C_R and F_R). The probability of occurrence of the marker in the resulting frame can be defined as the probability k of favorable outcomes in a series of n tests with the likelihood of success in a single trial p. The probability of a correct detection and the probability of a false marks emergence can be described by the binomial distribution:

$$C_R = C_n^k p^k q^{n-k}, \tag{33}$$

$$F_R = C_n^k p^k q^{n-k}, \tag{34}$$

where k – is the number of frames to be marked;

n- is the number of frames in the series;
$P=C_1$ and – is determined using (30) to calculate the probability of correct detection;
$p=F_1$ and – is determined by the expression (31) for the probability of occurrence of false marks;

$q=1-p$; $0 \leq k \leq n$.

Figure 8 shows the dependence C_R and F_R calculated in accordance with (33) and (34), from k for $n=15$ for the value of the signal-to-noise ratio $\phi=1$; $C1\_0.5$; $F1\_0.16$.

Figure 8. Probabilities of correct detection and false alarm

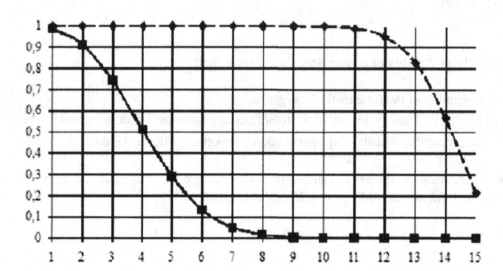

From the analysis of the curves shown in Figure 8, it follows that in the summation of *n* frames the detection characteristics in the total image greatly depend on *k*. It is advisable to choose *k* values within certain limits to ensure the optimum ratio of probabilities of the correct detection of the signal and an emergence of the false marks.

The possibility emerges to stabilize one of the detection characteristics when optimizing the other at the co-processing of a sequence of binary quantized images. For example, the stabilization of the probability of a correct detection and reduction of the probability of false marks in the resulting frame. This requires the quantization of the original video frames with an adaptive threshold.

The adaptive choice of the threshold should be conducted in such a way as to provide the desired stabilized value of the probability in the resulting image.

Taking into account the above, it is proposed a method of joint processing of the frames to ensure a given probability of correct detection and minimization of a probability of false marks occurrence in the resulting image.

The method consists in the sequential execution of the following operations. Upon reception of the first frame, the noise component should be centered and dispersion should be measured. The probability of correct detection and the probability of false marks should be calculated for a specific value of the signal component (signal/noise ratio). Based on the values of the detection characteristics, the number of frames for summing *n* should be selected. It is possible to use the values of the integral distribution function for the normalized random variable, distributed according to the normal law for computing. The value of an adaptive threshold, providing a specified value of C_R and F_R, should be calculated according to the expressions (33) and (34).

A binary quantization of the summed frames should be performed taking into account the selected threshold. The binary quantized frames should be summed up according to the logical rule (32). The dependence of the detection characteristics on the number of the summed frames for the case of stabilization of the correct detection probability at the level $C_R=0.9$ and the value of the signal/noise relation in a single frame $\phi1\_2$ are shown in Figure 9.

Figure 9. Value of signal/noise relation in a single frame

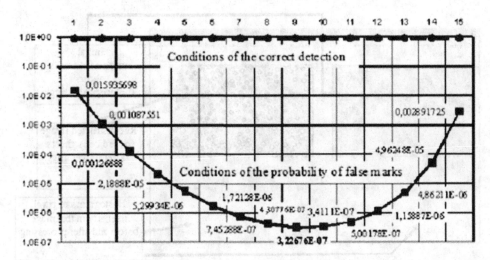

From the analysis of the curves in the Figure 9 it is seen that the probability of the false alarm in the resulting image will be of a minimal value when $k=9$ for the given initial conditions. This procedure makes it possible to detect the objects with conditional probabilities $C_R=0.9$ and $F_R=3.23 \cdot 10^{-7}$ in the resulting image.

The method for selecting an adaptive threshold is implemented in the optical-electronic system, which is used for morphological studies in microscopy. Experimental investigations were devoted to the research on the action exerting by the process of optimization of time accumulation on the quality of signals detection in optoelectronic systems. A video clip, containing an image of erythrocytes in the process of hemolysis, was used. In the process of lysis, hemolytic destroys erythrocytes and the hemoglobin releases into the blood plasma, the cells become low-contrast, complicating greatly the process of determining the shape of erythrocytes in microscopy in Figure 10.

Figure 10. The image of erythrocytes before and after lysis

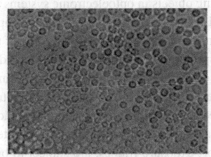 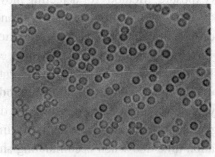

The first television frame, the red blood cells prior to the hemolysis beginning

Last TV frame, the end-stage of hemolysis

Figure 11. Results of experimental verification of image enhancement

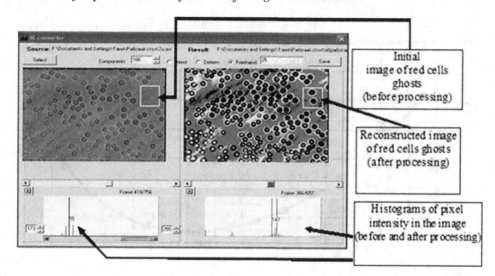

The results of processing are presented in Figure 11. The application of this method has made it possible to estimate accurately the time intervals for which the morphological changes of cells were taking place, to recognize automatically the shape of erythrocytes.

SIMULATION OF THE METHOD FOR DETECTING IMAGES OF MOVING OBJECTS WITH KNOWN SHAPE IN TELEVISION FRAMES IN LOW ILLUMINANCE CONDITIONS

Expansion of the dynamic range at the lower limit is achieved by lowering the sensitivity thresholds. Detection and tracking of low-contrast moving objects is much more difficult when observing moving objects, i. e. when the displacement of the object image during exposure and in one frame significantly exceeds the size of one image resolution element. As a result, the image has distortions caused by blurring during its exposure. Methods for detecting moving objects in optoelectronic systems are described in (Prokopenko, 2011), (Bondarev et al., 1999) and (Forsyth & Pons, 2004), for example, first- and second-order differential methods, tensor, correlation, and phase methods. Only some of the described methods can be used in low illuminance conditions, especially at high speeds. These methods are difficult to implement because the required speed of calculations limits their application in real time.

Description of the method and simulation model. The signal at the input of optoelectronic systems is described by its statistical parameters, which are determined taking into account the space-time structure of this signal. In low illuminance conditions, images of fast-moving objects, both after a relatively long accumulation time and when accumulating short-exposure frames, are subject to blur effect. To demonstrate this effect, a simulation model is built, which forms an object image in the frame in low illuminance conditions and at fast movement of the object. The simulation result is shown in Figure 12.

Analyzing Figure 12, we can conclude that it is impossible to determine the main features of the object image: its shape, size and location (according to the speed and trajectory).

To reduce the blur effect, it is necessary to reduce the accumulation time (exposure), but this reduces the signal contrast and, accordingly, reduces the signal-to-noise ratio. As an example, let us take the signal model shown in Figure 12 as a basis, and reduce the accumulation time by 20 times. As a result, we shall obtain the image of the frame shown in Figure 13. You can see that the blur effect is practically invisible, but the object image against the background of interference is difficult to distinguish visually, indicating that the signal-to-noise ratio is low.

Figure 12. Blur effect for a long time of signal accumulation

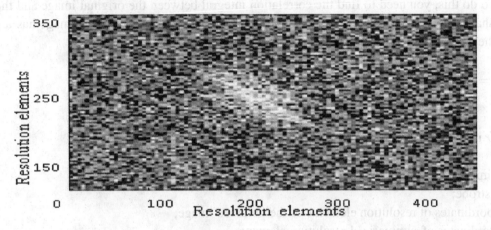

Figure 13. Location of the object image: a) initial; b) final

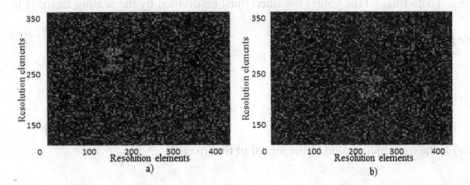

The value of signal-to-noise ratio in one separation (resolution) element is determined according to expression (20):

$$\phi = \frac{\overline{n}_S}{\sigma_{S+N}}.$$

(35)

where: σ_{S+N} is a standard deviation of an additive mixture of signal and noise; and $\bar{n}_A$ is a mean value of the signal.

For the image shown in Figure 13, the following parameters are selected: the variance of the additive mixture of signal and noise $\sigma_{S+N}{}^2 = 20$ and the mean value of the signal $= 5$; respectively, the signal-to-noise ratio in one frame will be equal to . However, the object image has a large size, more than one separation (resolution) element; respectively, given the total signal energy, you can accurately determine the location of the object image in the frame. Since the shape and size of the object image are known, it is advisable to use the correlation detection method to detect it and determine the coordinates of its center. To do this, you need to find the correlation integral between the original image and the strobe (which shape and size correspond to the object image). Because the simulated image has a discrete nature, the expression for cross-correlation has the following form:

$$Kr(k,l) = \sum_i \sum_j St(i,j) U(i+k, j+l). \tag{36}$$

where Kr is the obtained image of the correlation integral function;

U is the initial image of the frame;
St is the strobe;
k,l are coordinates of resolution elements of the obtained image;
i,j are coordinates of summarized resolution elements.

Let us present the correlation integral as an image and give it a convenient form for observation. First, perform scaling. To do this, all the values obtained must be divided by the scaling factor (Figure 14a):

$$k_m = \max(Kr)/255 \tag{37}$$

$$Kr_m = Kr / k_m \tag{38}$$

where $\max(Kr)$ is the maximum value of the signal in the image;

k_m is the scaling factor;
Kr_m is the image obtained after scaling.

To increase the contrast of this image for all values obtained after scaling, subtract the minimum value, and multiply by a factor that increases the brightness level (0-255) to the full dynamic range of the monitor (Figure 14 b):

$$k_p = 255/(255 - \min(Kr_m)), \quad Kr_{mp} = (Kr_m - \min(Kr_m)) * k_p. \tag{39}$$

where k_p is the contrast enhancement factor;

Figure 14. Image: a) after scaling; b) after increasing the contrast

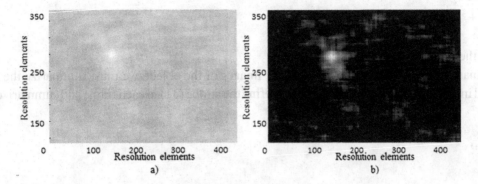

a)   b)

Kr_{np} is the image obtained by increasing the contrast.

Find the coordinates of the object location in the frame image. To do this, determine the coordinates of the maximum value of the correlation integral K_r in the frame:

$$U_{max} = \max(K_r),(i_k,j_k) => U_{max} = K_r(i_k,j_k). \tag{40}$$

where U_{max} is the maximum value of the signal in the image K_r;

i_k,j_k are coordinates of the center of the object image in the TV frame.

After determining the coordinates (i_k,j_k), select the image area that corresponds to the object location in the frame image, so that it completely covers the object. Taking into account the errors in determining the coordinates of the object, we find the image of this area (strobe) from the expression:

$$Q = U(x_1:x_2,y_1:y_2), \tag{41}$$

where

$$x_1 = i_k - V/2 - k_s, x_2 = i_k + V/2 + k_s, y_1 = j_k - S/2 - k_s, y_2 = j_k + S/2 + k_s,$$

Q   is the strobe, i. e. the frame area that includes the object image;

k_s is the correction that takes into account the error of determining the coordinates $k_s = 1,2$;
V,S are the height and width of the object, respectively.

Since the reduction of the accumulation time by 20 times has resulted in reduction of the signal-to-noise ratio (by the square root of 20), to obtain an acceptable image and achieve a high signal-to-noise ratio, it is necessary to accumulate a series of frames. The expression for calculating the signal-to-noise ratio after accumulating over time, provided that the location of the object image in each frame is unchanged, will be as follows:

$$\phi = \sqrt{M} \, \frac{\bar{n}_A}{\sigma_{A+?}},$$

(42)

where M is the number of summarized frames.

To eliminate distortions caused by mixing the images of the object, we must summarize the frames with aligned images of the object. To do this, we define the strobe (41) in each frame and summarize them:

$$Us = \sum_{n=1}^{N} U_n,$$

(43)

$$Qs = \sum_{n=1}^{N} Q_n,$$

(44)

where N is the number of summarized frames;

Us is the sum of the sequence of images of frames;
Qs is the sum of strobes.

Then in the image of the sum of frames US, the area corresponding to the object location in the last frame is replaced by the sum of strobes Qs:

$$Us(x_1 : x_2, y_1 : y_2) = Qs.$$

(45)

where

$$x_1 = i_N - V/2 - k_s, x_2 = i_N + V/2 + k_s, y_1 = j_N - S/2 - k_s, y_2 = j_N + S/2 + k_s,$$

The obtained images transformed according to the expression (39) are shown in Figure 15. In these images, the object is clearly highlighted against the background of noise, indicating an increase in the value of the signal-to-noise ratio.

Figure 15. Image of the sum of frames: a) after scaling; b) after increasing the contrast

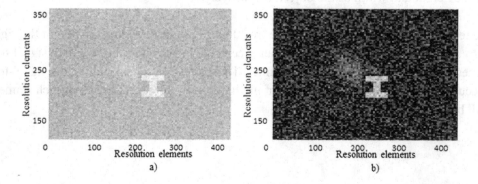

The location of the object image corresponds to the end point of its trajectory.

The proposed algorithm for detecting moving objects in low illuminance conditions allows preventing the manifestation of the blur effect. The simulation model of this algorithm has proved its efficiency even at a small signal-to-noise ratio. However, this method requires a priori knowledge of the object size and shape. Eliminating the blur effect and detecting moving objects in low illuminance conditions with unknown signal parameters requires a more complex algorithm using a number of reference strobes.

Experimental studies of the proposed algorithm for detecting moving object in low illuminance conditions were carried out for the optoelectronic detection system.

Goal. The analysis of the possibility to increase the efficiency of optoelectronic detection system by the criterion of the signal-to-noise ratio.

Description of the Experimental Installation

The experimental installation of the optoelectronic system consists of the following components: optical system – the input aperture diameter $D=150mm$, the focal distance $f=500mm$; photo detector – the photosensitive area $L=1cm^2$, the number of pixels $N=1000\times1000$. Observed object is a test chart No. 1 with an area $S=1mm^2$, see Figure 16.

Figure 16. Scheme of the experimental installation. 1 – photo detector; 2 – optical system; 3 – collimator; 4 – divider; 5 – test chart No. 1; 6 – neutral filter; 7, 8 – light sources; 9 – computer.

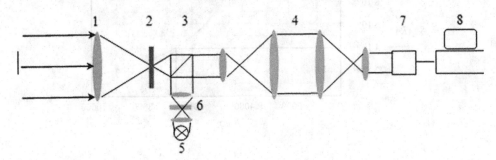

Let us analyze registration an optical radiation and making decision of the detection of the object under study on the basis of the stochastic-determined methods of signal processing in optoelectronic systems. The illuminance created by the object under study on the photodetector can be determined for point and extended objects accordingly as:

$$E_{im} = \frac{E_0 \sigma S_{in} \cos \omega}{4\pi R^2 S_{im}} \tau_{pm}\tau_{os} ; \tag{46}$$

$$E_{im} = \frac{E_0 \sigma S_{in} \cos \omega}{4\pi S_{ob} f_{ekv}^2} \tau_{pm}\tau_{os} , \tag{47}$$

where E_0 is the illuminance created on the object by the radiation source;

$S_{in} = \pi D^2/4$; D is the input aperture diameter;

$\sigma = 4\pi R_g r_k^2 \sin\alpha \cos\varphi$,

$\tau_{pm}\tau_{os}$ are the attenuation coefficients of the propagation medium and optical system;

S_{ob} is the object area.

Determine the distance r^* at which the optoelectronic system distinguishes an object with given parameters $S_{im} > S_{pix.} = 4\cdot10^{-10}\,m^2$. An analysis shows that the system will recognize the object under study at a distance of up to 6.6 km.

The illuminance in one resolution element (pixel) of the photo detector depends on the brightness of the background radiation, the diameter of the input aperture, the equivalent distance of the optical system, the attenuation coefficients of the optical system and the propagation medium. Taking into account the values of the brightness of the background radiation with given characteristics, for example, it is possible to estimate the illuminance created on the photocathode, see Figure 17, 18 and 19.

Figure 17. Relationship between the object image area and its distance

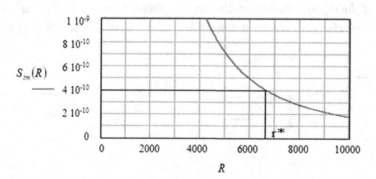

Table 1. Photo detector illuminance lux

	Clear sky	Clouds	Twilight	Full moon
Photo detector illuminance, lux	10^3	76	0.36	$1.6\cdot10^{-3}$

The value of the signal-to-noise ratio in the resolution element is determined according to expression (20):

$$\phi = \frac{\bar{n}_s}{\sqrt{\sigma_s^2 + \sigma_n^2 + \sigma_d^2}}, \tag{48}$$

Figure 18. a) the relationship between the signal-to-noise ratio and the distance when discriminating the target signal from the signal of the clear sky (2) and clouds (1); b) the relationship between the signal-to-noise ratio and the distance when discriminating the target signal from the signal of the sky under full moon illumination; c) analytical calculations of the relationship between the signal-to-noise ratio and the target distance under conditions of full moon illumination (1 – the relationship between the signal-to-noise ratio and the distance when detecting in one resolution element); 2 – the relationship between the signal-to-noise ratio and the distance under the optimum accumulating the energy of spatial coordinates; 3 – the relationship between the signal-to-noise ratio and the distance under the optimum accumulating the energy of spatial and 10 time modes (frames))

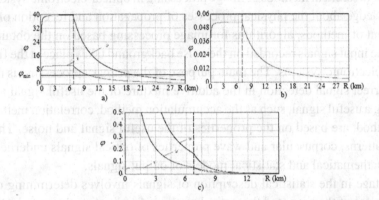

Figure 19. Method of detecting moving objects: a) the histogram of the photo detector illuminance values in the illuminance conditions characterized by the signal-to-noise ratio of 20; c) the histogram of values of the photo detector illuminance in the illuminance conditions characterized by the signal-to-noise ratio of 1.6; b), d) the image of the observed object on the monitor of the optoelectronic system; e) the histogram of the correlation integral function according (36) obtained with using of rectangular strobe; f) the image of the correlation integral on the monitor of the optoelectronic system.

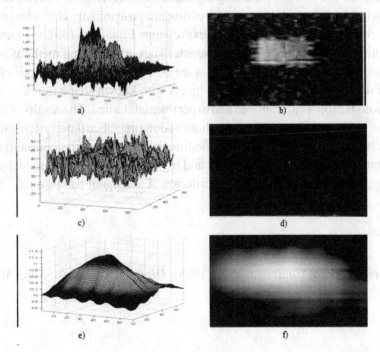

where $\overline{n}_s$ is the average value of the signal component; and $\sigma_s, \sigma_n, \sigma_d$ are standard deviations of signal and noise components and the standard deviation of the photo detector internal noise level (dark signal).

It is clear that the object detection in conditions characterized by the low value of the signal-to-noise ratio requires additional space-time processing.

CONCLUSION

The development of theoretical methods of image processing in optical-electronic systems includes, both deepening of knowledge about the physical processes of propagation and reception of optical radiation and the development of methods, algorithms for image processing based on the obtained knowledge.

Processing of the input signals recorded on the noise background takes place at the first stage of image processing in optoelectronic systems. The main purpose of the primary processing is the increase in the probability of a correct signal detection in the additive mixture of the useful signal and noise. Known methods of filtering a useful signal, such as the accumulation method, correlation method, nonlinear and linear filtering method, are based on the properties of the useful signal and noise. The peculiarities of spatial-temporal patterns, corpuscular and wave properties of optical signals underlie the development of more detailed mathematical and statistical models of output signals.

The principle stage in the statistical description of signals involves determining of their statistical characteristics. The use of the proposed theoretical method of estimating the mean value and dispersion of the received useful signal and noise, based on the joint use of the corpuscular and wave descriptions of signals and noise, makes it possible to describe the parameters of the received signals more fully, the expression Equations (22) – (28).

The aim of the second stage of image processing in optoelectronic systems is the provision of increasing the magnitude of the signal/noise ratio Equation (29). The basis for these methods is the statistical analysis of the system's output signals. Taking into account Poisson statistics with Gaussian approximation of the output signals the image processing algorithms are developed. Criteria of efficiency of the proposed algorithms and methods for image processing are probabilistic characteristics of signals and noise calculated according to combine using the expressions Equations (30), (31) and (32) – (34) (for binary images). Optimization of the described characteristics allows using methods of frame-by-frame and inter-frame processing with the parameters that are best for different conditions of the experiments.

The method for detecting images of moving objects with known shape in television frames in low illuminance conditions is proposed, emulated and experimental studied. The results of experimental studies aimed at increasing the efficiency of the machine vision optoelectronic systems using the criterion of signal-to-noise ratio in difficult observation conditions showed that the optimum use of the energy of the spatial distribution of the signal component and registration time allows to increase the efficiency of the optoelectronic detection system using the criterion of the signal-to-noise ratio by 5 to 10 times.

REFERENCES

Bondarev, V. N., Trester, G., & Chernega, V. S. (1999). *Digital signal processing: Methods and tools.* Academic Press.

Cokes, D., & Lewis, P. (1969). *Stochastic analysis of chains of events*. Academic Press.

Cunninghom, I. A., & Shaw, R. (1999). Signal-to-noise optimizatiom of medical imaging systems. *Journal of the Optical Society of America. A, Optics, Image Science, and Vision, 19*(3), 621–632. doi:10.1364/JOSAA.16.000621

Don, H. (2013). *Johnson*. Statistical Signal Processing.

Fedoseev, V. I. (2011). *Priem prostranstvenno-vremennyh signalov v optiko-jelektronnyh sistemah* (puassonovskaja model'). Sp-b.: Universitetskaja kniga.

Forsyth, D., & Pons, J. (2004). *Computer vision. The modern approach*. Moscow, Russia: "Williams" Publishing House.

Gal'jardi, R., & Karp, Sh. (1978). Opticheskaja svjaz. Moscow: Svjaz.

Gonzalez, R. (2008). *Richard Woods*. Digital Image Processing.

Levin, B.R. (1968). *Theoretical foundations of statistical radio engineering*. Sov. radio.

Lytyuga, A. (2009). Mathematical model of signals in television systems with low-orbit space objects observation in daytime. *Collected Works of Kharkiv University of Air Force, 4*(22), 41–46.

Miranda-Vega, J. E., Rivas-López, M., Flores-Fuentes, W., Sergiyenko, O., Lindner, L., & Rodríguez-Quiñonez, J. C. (2019). Implementación digital de filtros FIR para la minimización del ruido óptico y optoelectrónico de un sistema de barrido óptico. *Revista Iberoamericana de Automática e Informática., 16*(3), 344–357. doi:10.4995/riai.2019.10210

Nikitin, V. M., Fomin, V. N., Borisenkov, A. I., Nikolaev, A. I., & Borisenkov, I. L. (2008). Adaptive noise protection of optical-electronic information systems. Academic Press.

Prokopenko, I. G. (2011). Statistical signal processing. Academic Press.

Rivas-Lopez, M., Sergiyenko, O., Flores-Fuentes, W., & Rodríguez-Quiñonez, J. C. (Eds.). (2018). *Optoelectronics in Machine Vision-Based Theories and Applications*. IGI Global.

Seitz, P. (2011). *Fundamentals of Noise in Optoelectronics*. Single-Photon Imaging. *Springer Series in Optical Sciences, 160*, 1–25. doi:10.1007/978-3-642-18443-7_1

Sergiyenko, O., Flores-Fuentes, W., & Mercorelli, P. (Eds.). (2019). *Machine Vision and Navigation*. Springer Nature.

Strelkov, A., Zhilin, Ye., Lytyuga, A., & Lisovenko, S. (2007). Signal Detection in Technical Vision Systems. *Telecom. and Radio Engineering, 66*(4), 283–293. doi:10.1615/TelecomRadEng.v66.i4.10

Strelkov, A. I., Stadnik, A. M., Lytyuga, A. P., & Strelkova, T. A. (1998). Comparative Analysis of Probabilistic and Determinate Methods for Attenuating Light Flux. *Telecommunications and Radio Engineering, 52*(8), 54–57. doi:10.1615/TelecomRadEng.v52.i8.110

Strelkova, T. A. (1999). The Potentialities of Optical and Electronic Devices for Biological Objects Investigations. Telecommunications and Radio Engineering, 53, 190-194.

Strelkova, T. (2014a). Influence of Video Stream Compression on Image Microstructure in Medical Systems. *Biomedical Engineering*, 47(6), 307–311. doi:10.100710527-014-9398-1

Strelkova, T. (2014b). Statistical properties of output signals in optical-television systems with limited dynamic range. *Eastern-European Journal of Enterprise Technologies*, 9(68), 38–44. doi:10.15587/1729-4061.2014.23361

Strelkova, T.A. (2014c). Studies on the Optical Fluxes Attenuation Process in Optical-electronic Systems. *Semiconductor physics, quantum electronics & optoelectronics (SPQEO)*, 4, 421-424.

Yang, F., Lu, Y. M., Sbaiz, L., & Vetterli, M. (2012). Oversampled Image Acquistion Using Binary Poisson Statistics. *IEEE Transactions on Image Processing*, 21(4), 1421–1436. doi:10.1109/TIP.2011.2179306 PMID:22180507

Chapter 4
Reducing the Optical Noise of Machine Vision Optical Scanners for Landslide Monitoring

Jesús Elias Miranda-Vega

iD https://orcid.org/0000-0003-0618-0455

Autonomous University of Baja California, Mexico

Oleg Sergiyenko

iD https://orcid.org/0000-0003-4270-6872

Engineering Institute, Autonomous University of Baja California, Mexico

Javier Rivera-Castillo

Engineering Institute, Autonomous University of Baja California, Mexico

Julio C. Rodríguez-Quiñonez

Faculty of Engineering, Autonomous University of Baja California, Mexico

Moisés Rivas-López

Engineering Institute, Autonomous University of Baja California, Mexico

Daniel Hernández-Balbuena

Faculty of Engineering, Autonomous University of Baja California, Mexico

Wendy Flores-Fuentes

iD https://orcid.org/0000-0002-1477-7449

Faculty of Engineering, Autonomous University of Baja California, Mexico

ABSTRACT

An application of landslide monitoring using optical scanner as vision system is presented. The method involves finding the position of non-coherent light sources located at strategic points susceptible to landslides. The position of the light source is monitored by measuring its coordinates using a scanner based on a 45° sloping surface cylindrical mirror. This chapter shows experiments of position light source monitoring in laboratory environment. This work also provides improvements for the optical scanner by using digital filter to smooth the opto-electronic signal captured from a real environment. The results of these experiments were satisfactory by implementing the moving average filter and median filter.

DOI: 10.4018/978-1-7998-6522-3.ch004

INTRODUCTION

Landslide is a term associated to downslope movements of rock, debris or earth under the influence of gravity, which may cover a wide range of spatial and temporal scales. Physical factors, human activities and natural environmental changes like earthquakes, volcanic activity, heavy rainfalls and changes in ground water are typical natural triggering mechanisms for landslides, which amplifies the inherent weakness in rock or soil (Savvaidis, 2003).

Landslides are a major hazard worldwide causing fatal disasters, property losses and degradation of the environment, so it is necessary to understand the mechanisms that trigger them. To understand the dynamics of the landslides it is necessary to monitor de surface displacements including accurate inventories of past slides, their location, extent, type and triggering mechanism (Kerle, Stump, & Malet, 2010). The magnitude, velocity and acceleration of displacements can provide an indication of the stability of the slope. These movements, if detected early enough, can indicate impending catastrophic failure of a sloping mass. (Savvaidis, 2003).

The long term monitoring of instability phenomena has become relatively inexpensive and affordable due to the recent technological growth in the field of data storage and transmitting. Measurements of landslide movements that integrate hydrological and geological data have greatly improved the knowledge of landslide mechanics and the integration of different techniques allows for a better understanding of this kind of phenomenon and thus better protecting human settlements and infrastructures. (Anderson & Tallapally, 1996), (Rentschler & Moser, 1996).

There are many landslide types depending on the slope movement and the type of terrain. Slope movement may be fall, topple, lateral spreading, slide and flow and the earth material could be rock or soil. In past decades, photograph sensors, including extensometers, in-place inclinometers, tiltmeters, pressure transducers and rain gauges, were used to landslides monitoring. However, researchers had several difficulties for accurate assessments and evaluations, monitoring times were long and the measurement process very expensive.

In this chapter, a machine vision application to landslide monitoring tasks is presented. Machine vision is a new alternative that uses different optoelectronics devices as vision systems, nowadays the most used devices include photographs and video cameras. However, camera images are detailed and therefore contain a large amount of information.

Our application uses an optical scanner with incoherent light emitter as a vision system.

The scanner used in this application was also proposed for structural health monitoring tasks (Rentschler & Moser, 1996), (Rivas López, Flores Fuentes, Rivera Castillo, Sergiyenko, & Hernández Balbuena, 2013), but in this chapter, we will show the advantages in landslide monitoring using incoherent light. Several position measurements were carried out with a laser emitter and incoherence light emitter. Position measurements were carried out during the scanning process moving the position of light emitter source from the nearest distance to the optical aperture sensor to the farthest position from the optical aperture sensor along the same angle line. Experiments show that with laser emitter, alignment difficulties were presented in all positions. When the laser emitter was moved to 10mm up, down, right or left, the light beam was not detected in the scanner aperture.

On the other hand, an incoherence light emitter was positioned to distances of 2m, 4m, 6m, and 8m. In each position, the emitter was moved 10cm, 20cm, 50cm, to up, down, right and left. The light was detected in the scanner aperture without problem in all positions. In this way, we opted to use optical scanning with incoherent light for measurements of coordinates and calculate displacements.

The Optical Scanner System (OSS) of this study consists of the following elements:

a) An incoherent light source emitter (non-rotating) mounted on the structure under monitoring.
b) A passive rotating optical aperture sensor designed with a 45°-sloping mirror, and embedded into a cylindrical micro rod. The beam of light is deviated by the 45°-sloping mirror to a double convex lens and filtered, in order to remove any interference and enhance the focus.
c) A photodiode to capture the beam of light while the cylindrical micro rod mounted on a dc electrical motor shaft is rotating.

This last element generates the targeted signals to be analyzed by the proposed method.

The light emitter source is set at a distance from the receiver; the receiver includes a mirror, which spins with an angular velocity ω. The beam emitted arrives with an incident angle β with respect to the perpendicular mirror and is reflected with the same angle β, according to the reflecting principle to pass through the lens that concentrates the beam to be captured by the photodiode, which generates a signal with a shape similar to the Gaussian function. When the mirror starts to spin, a sensor is synchronized with the origin generating a pulse that indicates the starting of measurement that finishes when the sensor releases the next start signal. This signal is captured by a data acquisition module to start the signal processing method (Rivas López, Flores Fuentes, Rivera Castillo, Sergiyenko, & Hernández Balbuena, 2013).

One important thing to consider after the data acquisition process is filtering the optical and electrical noise from OSS. The performance and accuracy can be affected because of the optical noise captured in a real environment. The spectral components dominating of the sun's solar radiation are the infrared radiation and visible light. Designers of the OSS must carefully consider the longer wavelengths like infrared radiation the reason is that photosensors of the system can be influenced by this radiation that is present *in situ* during the day.

Researches for filtering the effects caused to the OSS in a real environment have been examined in (Wendy Flores-Fuentes, 2018), in which it is found that an LED can be used as a natural filter of the sunlight. In this work, the performance and accuracy incremented when it was used digital filters. The main characteristics of digital signal processing are that signal captured by photosensors of the OSS can be used in a proper manner because unwanted signals can be removed by this process; another relevant characteristic is that in case of any change of the filter specifications these can be updated immediately in comparison of analog filters.

A complete description of the application is explained in the present chapter, experimental results are also described.

1. BACKGROUND

Sohn and Farrar established that structural health monitoring problem is fundamentally one of a statistical pattern recognition paradigm consisting of four processes: (1) operational evaluation, (2) data acquisition, fusion and cleansing, (3) feature extraction and information condensation, and (4) statistical model development for feature discrimination.

During the operational evaluation, four questions must be answered: (1) how damage is defined for the system to be monitored? (2) What are the operational and environmental conditions under which functions? (3) What are the limitations of acquiring data in the operational environment? And (4) what are the economic and/or life safety motives for performing the monitoring?

In the data acquisition, fusion end cleansing step it is necessary to establish the type, number and placement of sensors, the data acquisition system, the storage and transmission type and the data cleansing methods.

In the feature extraction and information condensation process it is possible to distinguish between undamaged and damaged structure by means of the analysis of the measured data and to do some kind of compression data.

Statistical model development involves the selection of the algorithms to be used. Usually, they fall into two categories: (1) supervised learning when there is available data from undamaged and damaged structure. (2) Unsupervised learning refers to algorithms applied to data that do not contain data from the damaged structure. (Hoon Sohn, 2002)

According to USGS in 2001 the economic cost of landslides in the United States was $ 3.5 billion dollars and 20 to 50 people per year are killed during these catastrophes. Rock falls, rock slides and debris flows are the most important events involving people's deaths.

On December 29th, 2013, a collapse of 30 centimeters in the Tijuana-Ensenada Scenic Road became the collapse of up to 40 meters along a stretch of 300 meters at kilometer 93. This damage is attributable to the San Andreas fault, and although there were no deaths it will take at least one year to be repaired at a high economic and social cost (see **Figure 1** and **Figure 2**).

Figure 1. Collapse in the Scenic Road, Ensenada, Baja California, Mexico

Figure 2. Another view of the Collapse in the Scenic Road, Ensenada, Baja California, Mexico

There are various types of landslides depending on the earth's material and the mode of movement where the Rotational and Translational are the two major types. The next list is a brief description of such a landslide taken from the USGS paper "Landslide Types and Processes".

1. **Rotational slide**: This is a slide in which the surface of rupture is curved concavely upward and the slide movement is roughly rotational about an axis that is parallel to the ground surface and transverse across the slide.
2. **Translational slide**: In this type of slide, the landslide mass moves along a roughly planar surface with little rotation or backward tilting. A block slide is a translational slide in which the moving mass consists of a single unit or a few closely related units that move downslope as a relatively coherent mass.
3. **Falls**: Falls are abrupt movements of masses of geologic materials, such as rocks and boulders that become detached from steep slopes or cliffs. Separation occurs along discontinuities such as fractures, joints, and bedding planes, and movement occurs by free-fall, bouncing, and rolling. Falls are strongly influenced by gravity, mechanical weathering, and the presence of interstitial water.
4. **Topples**: Toppling failures are distinguished by the forward rotation of a unit or units about some pivotal point, below or low in the unit, under the actions of gravity and forces exerted by adjacent units or by fluids in cracks.
5. **Flows**: There are five basic categories of flows that differ from one another in fundamental ways.
   a. Debris flow: A debris flow is a form of a rapid mass movement in which a combination of loose soil, rock, organic matter, air, and water mobilize as a slurry that flows downslope. Debris

flows include <50% fines. Debris flows are commonly caused by intense surface-water flow, due to heavy precipitation or rapid snowmelt, that erodes and mobilizes loose soil or rock on steep slopes. Debris flows also commonly mobilize from other types of landslides that occur on steep slopes, are nearly saturated, and consist of a large proportion of silt- and sand-sized material. Debris-flow source areas are often associated with steep gullies, and debris-flow deposits are usually indicated by the presence of debris fans at the mouths of gullies. Fires that denude slopes of vegetation intensify the susceptibility of slopes to debris flows.

b. Debris avalanche: This is a variety of very rapid to extremely rapid debris flow.

c. Earthflow: Earthflows have a characteristic "hourglass" shape. The slope material liquefies and runs out, forming a bowl or depression at the head. The flow itself is elongate and usually occurs in fine-grained materials or clay-bearing rocks on moderate slopes and under saturated conditions. However, dry flows of granular material are also possible.

d. Mudflow: A mudflow is an earthflow consisting of material that is wet enough to flow rapidly and that contains at least 50 percent sand-, silt-, and clay-sized particles. In some instances, for example in many newspaper reports, mudflows and debris flows are commonly referred to as "mudslides."

e. Creep: Creep is the imperceptibly slow, steady, downward movement of slope-forming soil or rock. Movement is caused by shear stress sufficient to produce permanent deformation, but too small to produce shear failure. There are generally three types of creep: (1) seasonal, where movement is within the depth of soil affected by seasonal changes in soil moisture and soil temperature; (2) continuous, where shear stress continuously exceeds the strength of the material; and (3) progressive, where slopes are reaching the point of failure as other types of mass movements. Creep is indicated by curved tree trunks, bent fences or retaining walls, tilted poles or fences, and small soil ripples or ridges.

6. **Lateral Spreads:** Lateral spreads are distinctive because they usually occur on very gentle slopes or flat terrain. The dominant mode of movement is lateral extension accompanied by shear or tensile fractures. The failure is caused by liquefaction, the process whereby saturated, loose, cohesionless sediments (usually sands and silts) are transformed from a solid into a liquefied state. Failure is usually triggered by rapid ground motion, such as that experienced during an earthquake, but can also be artificially induced. When coherent material, either bedrock or soil, rests on materials that liquefy, the upper units may undergo fracturing and extension and may then subside, translate, rotate, disintegrate, or liquefy and flow. Lateral spreading in fine-grained materials on shallow slopes is usually progressive. The failure starts suddenly in a small area and spreads rapidly. Often the initial failure is a slump, but in some materials, movement occurs for no apparent reason. A combination of two or more of the above types is known as a complex landslide.

There are a lot of Landslide causes. It can be geological causes like weak or sensitive materials, weathered materials, sheared, jointed, or fissured materials, adversely oriented discontinuity (bedding, schistosity, fault, unconformity, contact, and so forth), contrast in permeability and/or stiffness of materials. Also, it can be caused by morphological causes like tectonic or volcanic uplift, glacial rebound, fluvial, wave, or glacial erosion of slope toe or lateral margins, subterranean erosion (solution, piping), deposition loading slope or its crest, vegetation removal (by fire, drought), thawing, freeze-and-thaw weathering, shrink-and-swell weathering.

Human activities can be a cause for landslides, for example: excavation of slope or its toe, loading of slope or its crest, drawdown (of reservoirs), deforestation, irrigation, mining, artificial vibration, water leakage from utilities.

2. OPTOELECTRONIC DEVICES FOR LANDSLIDE MONITORING

P. D. Savvaidis made a summary of the systems and techniques for landslide monitoring including remote sensing or satellite techniques, photogrammetric techniques, geodetic techniques or observational techniques, and geotechnical or instrumentation or physical techniques **(Savvaidis, 2003)**

I. Baron and R. Supper applied a Questionnaire on National State on Landslide Site Investigation and monitoring which was disseminated among European institutes and representatives within the frame of the Safeland project with the following objectives: (1) Assessing general state of the slope instability investigation and monitoring in different European countries; (2) Assessing effectiveness/reliability of each method for slope instability investigation and monitoring; and (3) Applicability of the monitoring techniques for early warning.

They found that active rotational and translational slides with recent moving rates less than 10 mm/month are the most observed and the most frequently applied investigation methods are geological, geomorphic and engineering-geological and core drilling, testing of strength properties / deformability and clay mineralogy, studying of aerial photographs, LiDAR airborne Laser scans (ALS), radar interferometry, resistivity measurements and refraction seismic.

Aerial photographs, satellite optical very high resolution (VHR) imagery, LiDAR ALS, radar interferometry and measurement of resistivity, reflection and refraction seismic, time-domain electromagnetic, passive acoustic emission, geophysical logging were the most reliable investigation methods.

They also found that the monitoring of movement and deformation was most frequently done by repeated orthophotos, radar interferometry, differential LiDAR ALS, webcam, dGPS, total station, inclinometer and wire extensometers. The most frequently monitored hydro-meteorological factors were precipitation amount, pore water pressure and air temperature; the most frequently monitored geophysical parameters were passive seismic/acoustic emissions, electromagnetic emissions, and direct current resistivity (I. Baron, 2010)

In a more recent study, Michoud in "New classification of landslide-inducing anthropogenic activities" introduces a table with the techniques available for landslide monitoring. Such a table includes passive optical sensors (ground based imaging, aerial imagery and satellite imaging, active optical sensors (distance meters, terrestrial Lidar (TLS), and airborne Lidar (ALS)), active microwave sensors (interferometric radar distance meter, differential InSAR, Advanced InSAR, ground based InSAR and Polsar and Polinsar, ground-based geophysics sensor (seismic, electricity, electromagnetic (low frequency), ground penetrating radar, gravimetry and borehole geophysics), offshore sensors (2D and 3D seismic, sonar and multi-beam), geotechnical sensors (extensometers, inclinometers, piezometers, contact earth pressure cells and multiparametric in place systems), and other sensors like global navigation satellite systems (GNSS) and core logging. (C. Michoud, 2012).

As can be seen in the above section, there are many techniques to monitor areas with the probability of a catastrophic landslide. In terms of optoelectronic devices for landslide monitoring are the next ones: Passive and active optical sensors. As an example of the first one is ground based and aerial imaging working on the basis of photogrammetric techniques where the combination of aerial photography and infrared imagery results in a more accurate and complete portrayal of terrain conditions. Infrared imagery provides information that can lead to landslide susceptible terrain, like conditions in drainage, moisture in surface and near surface, the presence of bedrocks at shallow deeps, and the distinction between loose materials. At this time, aerial photographs can be georeferenced and taking photos at different times and comparing them with the use of a computer can show changes in the terrain. One problem with this technique occurs when there are heavy changes in vegetation between epochs.

As an example of active optical sensors are the Electronic distance meters instruments widely used in landslide monitoring due to the ability to measure distances with high accuracy. They are electro-optical instruments with visible or near infrared continuous radiation with an accuracy in the range of mm. Electronic theodolites and electronic distance meters are widely used in some countries for angle and distance measurements in high accuracy deformation surveying so triangulation and trilateration horizontal networks are employed. For height determination, differential leveling is the traditional technique. Vertical position can be determined with these instruments with very high accuracy ranging in the ± 0.1 – 1 mm over distances of 10 – 100 m. High accuracy electronic theodolites and electronic distance meters replace geodetic leveling instruments in a more economical trigonometric height measurement way, obtaining accuracy better than 1 mm in the determination of height between two targets 200m apart. This combination equipment is now called "total surveying station and as with all the optical instruments working over the earth's surface, the refraction error is still the major problem. CCDs are now installed in accuracy theodolites so the centroid can be located in relation to the cross-hairs as well as robotic systems to automate the process. The main advantage of using total station instruments is that they provide 3D coordinate information of the points measured and some disadvantages are that it is required an unobstructed line of sight between the instrument and the targeting prisms and the price.

LIDAR technologies are used to create a point cloud with latitude, longitude and height information by means of a laser beam, a scanner, and a GPS. Two types of LIDAR are topographic and bathymetric. Topographic LIDAR typically uses a near-infrared laser to map the land, while bathymetric lidar uses water-penetrating green light to also measure seafloor and riverbed elevations (NOAA, 2015). The operating principle of the LIDAR is by means of emitting a laser beam and measuring the reflection from the earth's surface to get the distance measurement and combining that information with position and orientation data generated by GPS and Inertial measurement units integrated. Basically, are two types of LIDAR: airborne and terrestrial. Airborne has the advantage of covering a large area but it needs a plane or satellite to correct operation, so the time between scanning is large. On the other hand, Terrestrial LIDAR has the advantage of continuous monitoring versus small area coverage.

Digital cameras and video cameras are now used to measure slow motion landslides creating a lot of data to process by way of pattern recognition software to recognize topography and landform changing.

Ming-Chih Lu proposed an image based landslide monitoring system using a laser projector placed in a non-landslide area, a grid plate, and a wireless camera to measure land displacements (Ming-Chih, Su-Chin, Yu-Shen, & Cheng-Pei).

3. TYPICAL OPTOELECTRONIC SCANNERS

Nowadays optoelectronic scanners are widely used for multiple applications; most of the position or geometry measuring scanners use the triangulation principle or a variant of this measurement method. There are two kinds of scanners for position measuring tasks: scanners with static sensors and scanners with rotating mirrors. Optical triangulation sensors with CCD or PSD are typically used to measure manufactured goods, such as tire treads, coins, printed circuit boards and ships, principally for monitoring the target distance of small, fragile parts or soft surfaces likely to be deformed if touched by a contact probe.

3.1. Scanners with Position Triangulation Sensors Using CCD or PSD

A triangulation scanner sensor can be formed by three subsystems: emitter, receiver, and electronic processor as shown in Figure 3. A spotlight is projected onto the work target; a portion of the light reflected by the target is collected through the lens by the detector which can be a CCD, CMOS or PSD array. The angle (α) is calculated, depending on the position of the beam on the detectors CCD or PSD array, hence the distance from the sensor to the target is computed by the electronic processor. As stated by Kennedy William P. in (Kennedy, 2012), the size of the spot is determined by the optical design, and influences the overall system design by setting a target feature size detection limit. For instance, if the spot diameter is 30 μm, it will be difficult to resolve a lateral feature < 30 μm.

Figure 3. Principle of Triangulation

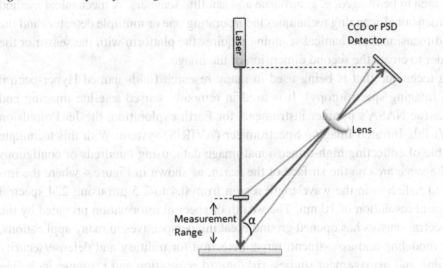

Techniques are then used to determine the location of the object. For situations requiring the location of a light source on a plane, a position sensitive detector (PSD) offers the potential for better resolution at several devices are commonly utilized in different types of optical triangulation position scanners and have been built or considered in the past for measuring the position of light spot more efficiently. One method of position detection uses a video camera to electronically capture an image of an object. Image processing lower system cost **(Vahelal, 2010)**. However, there are other kinds of scanners used commonly in large distances measurement or in structural health monitoring tasks, these scanners will be explained in the next section.

3.2. Scanners with Rotating Mirrors and Remote Sensing

In the previous section, we described the operational principle of scanners for monitoring the distance of small objects, now we will describe the operational principle of scanners with rotating mirrors for large distances measurement or in structural health monitoring tasks.

There are two main classifications of optical scanning: remote sensing and input/output scanning. Remote sensing detects objects from a distance, as by a space-borne observation platform. For example, infrared imaging of terrain. Sensing is usually passive and the radiation incoherent and often multispectral. Input / output scanning, on the other hand, is local. A familiar example is the document reading (input) or writing (output). The intensive use of the laser makes the scanning active and the radiation coherent. The scanned point is focused via finite-conjugate optics from a local fixed source; see (Bass, 1995).

In remote sensing, there are a variety of scanning methods for capturing the data needed for image formation. These methods may be classified into framing, push broom, and mechanical. In the first one, there is no need for physical scan motion since it uses electronic scanning and implies that the sensor has a two-dimensional array of detectors. At present, the most used sensor is the CCD and such array requires an optical system with 2-D wide-angle capability. In push broom methods linear array of detectors are moved along the area to be imaged, e. g. airborne and satellite scanners. A mechanical method includes one- and two-dimensional scanning techniques incorporating one or multiple detectors and the image formation by one dimensional mechanical scanning requires the platform with the sensor or the object to be moved in order to create the second dimension of the image.

These days there is a technique that is being used in many research fields named Hyperspectral imaging (also known as imaging spectroscopy). It is used in remotely sensed satellite imaging and aerial reconnaissance like the NASA's premier instruments for Earth exploration, the Jet Propulsion Laboratory's Airborne Visible-Infrared Imaging Spectrometer (AVIRIS) system. With this technique the instruments are capable of collecting high-dimensional image data, using hundreds of contiguous spectral channels, over the same area on the surface of the Earth, as shown in Figure 4 where the image measures the reflected radiation in the wavelength region from 0.4 to 2.5 μm using 224 spectral channels, at nominal spectral resolution of 10 nm. The wealth of spectral information provided by the latest generation hyperspectral sensors has opened ground breaking perspectives in many applications, including environmental modeling and assessment; target detection for military and defense/security deployment; urban planning and management studies, risk/hazard prevention and response including wild-land fire tracking; biological threat detection, monitoring of oil spills and other types of chemical contamination (Plaza, 2009).

Figure 4. The concept of hyperspectral imaging illustrated using NASA's AVIRIS (Plaza, 2009)

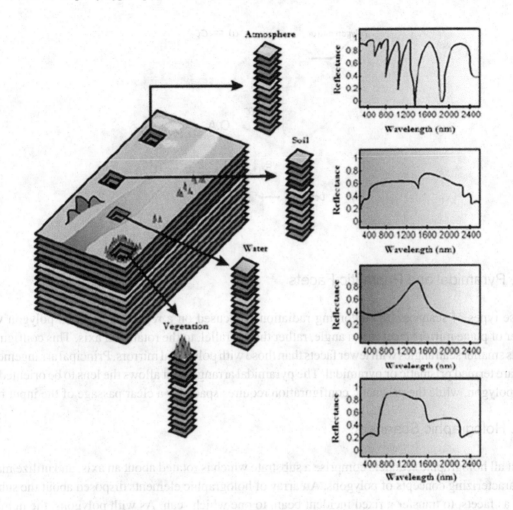

While remote sensing requires capturing passive radiation for image formation, active input/output scanning needs to illuminate an object or medium with a "flying spot" derived typically from a laser source. Therefore, it can be said that in input scanning the radiation is modulated by the target to form a signal and in the output scanning it is modulated by a signal.

3.3. Examples of Input/Output Scanning

3.3.1. Polygonal Scanners

These scanners have a polygonal mirror rotating at a constant speed by way of an electric motor and the radiation received by the lens is reflected on a detector. The primary advantages of polygonal scanners are speed, the availability of wide scan angles, and velocity stability. They are usually rotated continuously in one direction at a fixed speed to provide repetitive unidirectional scans that are superimposed in the scan field, or plane, as the case may be. When the number of facets reduces to one, it is identified as a monogon scanner, Figure 5 illustrates a hexagonal rotating mirror scanner.

Figure 5. Polygon scanner

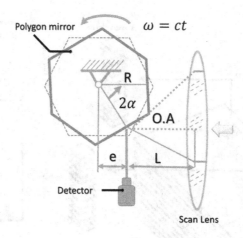

3.3.2. Pyramidal and Prismatic Facets

In these types of scanners, the incoming radiation is focused on a regular pyramidal polygon with a number of plane mirrors facets at an angle, rather than parallel, to the rotational axis. This configuration permits smaller scan angles with fewer facets than those with polygonal mirrors. Principal arrangements of facets are termed prismatic or pyramidal. The pyramidal arrangement allows the lens to be oriented close to the polygon, while the prismatic configuration requires space for a clear passage of the input beam.

3.3.3. Holographic Scanners.

Almost all holographic scanners comprise a substrate which is rotated about an axis, and utilize many of the characterizing concepts of polygons. An array of holographic elements disposed about the substrate serves as facets, to transfer a fixed incident beam to one which scan. As with polygons, the number of facets is determined by the optical scan angle and duty cycle, and the elemental resolution is determined by the incident beam width and the scan angle. In radially symmetric systems, scan functions can be identical to those of the pyramidal polygon. Meanwhile, there are many similarities to polygons, there are significant advantages and limitations, see Figure 6.

Figure 6. Polygonal scanner (From https://www.globalspec.com/reference/34369/160210/chapter-4-3-5-4-scanner-devices-and-techniques-postobjective-configurations).

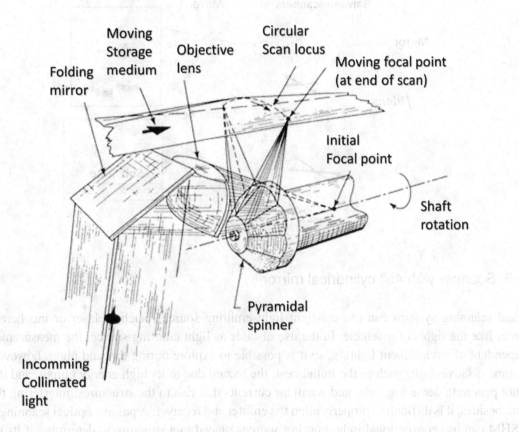

3.3.4. Galvanometer and Resonant Scanners.

To avoid the scan nonuniformities which can arise from facet variations of polygons or holographic deflectors, one might avoid multi-facets. Reducing the number to one, the polygon becomes a monogon. This adapts well to the internal drum scanner, which achieves a high duty cycle, executing a very large angular scan within a cylindrical image surface. Flat-field scanning, however, as projected through a flat-field lens, allows limited optical scan angle, resulting in a limited duty cycle from a rotating monogon. If the mirror is vibrated rather than rotated completely, the wasted scan interval may be reduced. Such components must, however, satisfy system speed, resolution, and linearity.

Vibrational scanners include the familiar galvanometer and resonant devices and the least commonly encountered piezoelectrically driven mirror transducer as shown in Figure 7.

Figure 7. Galvanometer scanner (From http://www.yedata.com).

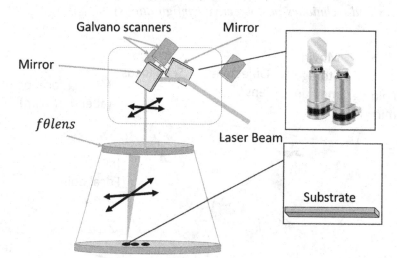

3.3.5. Scanner with 45° cylindrical mirror

Optical scanning systems can use coherent light emitting sources, such as laser or incoherent light sources like the lights of a vehicle. In the use of laser as light emitting source, the measurements are independent of environment lighting, so it is possible to explore during day and night, however, there are some disadvantages such as the initial cost, the hazard due to its high energy output, and that they cannot penetrate dense fog, rain, and warm air currents that rise to the structures, interfering the laser beam, besides, it is difficult to properly align the emitter and receiver. A passive optical scanning system for SHM can use conventional light emitting sources placed in a structure to determine if its position changes due to deteriorating. Figure 8 illustrates a general schematic diagram with the main elements of the optical scanning aperture used to generate the signals to test the proposed method.

Figure 8. Scanner with 45° Cylindrical Mirror

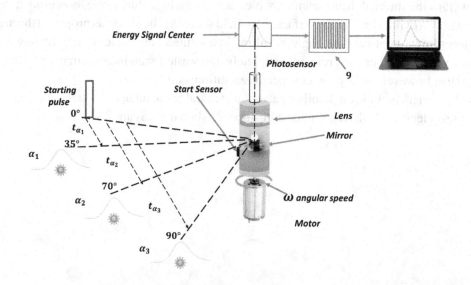

As can be seen in Figure 8, the optical system is integrated by the light emitter source set at a distance from the receiver; the receiver is compound by the mirror E, which spins with an angular velocity ω. The beam emitted arrives with an incident angle β with respect to the perpendicular mirror, and is reflected with the same angle β, according to the reflecting principle to pass through a lens that concentrates the beam to be captured by the photodiode, which generates a signal "f" with a shape similar to the Gaussian function. When the mirror starts to spin, the sensor "s" is synchronized with the origin generating a pulse that indicates the starting of measurement that finishes when the photodiode releases the stop signal. This signal is released when the Gaussian signal energetic centre has been detected.

Figure 9 shows that light intensity increments in the centre of the signal generated by the scanner. The sensor "s" generates a starting signal when tα =0, then the stop signal is activated when the Gaussian function geometric centre has been detected.

Figure 9. Signal generated by a 45° cylindrical mirror scanner

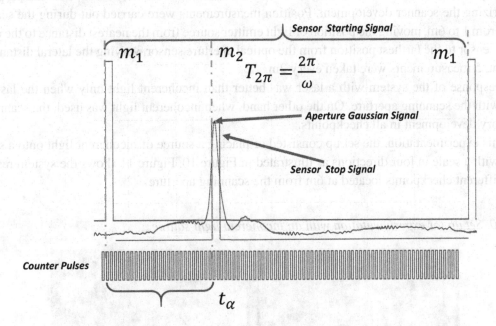

The distance $T_{2\pi}$ is equal to the time between m_1 and m_1, that is expressed by the code $N_{2\pi}$ as defined in equation 1.

$$N_{2\pi} = T_{2\pi} f_0.$$ (1)

On the other hand, the time tα is equal to the distance between m_1 and m_2, which could be expressed by the code defined in equation 2.

$$N_{\alpha} = t_{\alpha} f_0.$$ (2)

Where f_0 is a standard frequency reference. With this consideration, the time variable could be eliminated from equation 2 obtaining equation 3.

$$\alpha = \frac{2\pi N_\alpha}{N_{2\pi}}.$$
(3)

4. INCOHERENT VS COHERENT LIGHT IN SCANNERS FOR LANDSLIDE MONITORING

The objective of this study was to compare the effectiveness of laser and incoherent light used in a scanning system for landslide monitoring. A total of 63 measurements were taken, with the purpose of characterizing the scanner development. Position measurements were carried out during the scanning process from 1 to 6m, moving the position of light emitter source from the nearest distance to the optical aperture sensor to the farthest position from the optical aperture sensor, varying the lateral distance and the height, 5 measurements were taken every 2m.

The response of the system with a laser was better than incoherent light only when the laser was aligned with the scanning aperture. On the other hand, when incoherent light was used, the scanner has satisfactory development in all checkpoints.

For this experimentation, the set up consisted of placing a source of incoherent light onto a surface marked with a scale in four directions as illustrated in Figure 10. Figure 11 shows the system response in four different checkpoints located at 6m from the scanning aperture.

Figure 10. Set up for experimentation with an incoherent light source

Figure 11. System response in four different checkpoints located at 6m from the scanning aperture

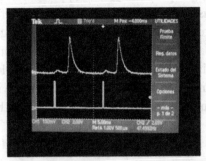

a) Response of scanner with incoherent source light located 30cm above of center.

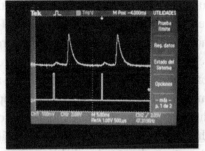

b) Response of scanner with incoherent source light located 30cm below of center.

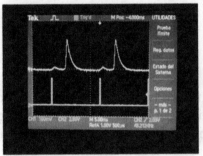

c) Response of scanner with incoherent source light located 30cm to rigth of center.

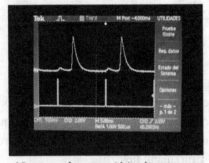

a) Response of scanner with incoherent source light located 30cm to left of center

As can be seen in Figure 11, the response to each checkpoint did not differ significantly between them.

These results show the advantage of using incoherence light with optical scanners. However, the application of such scanners in landslides monitoring involves measurements containing the coordinates of the light-emitting source to determine the amount of displacement.

Experimental readings showed that, in order to find the position of a light source, the targeted signal resemblances a Gaussian shaped signal. This is mainly observed when the light source searched by the optoelectronic scanning is a punctual light source. This last assertion corresponds to the fact that when the punctual light source expands its radius, a cone-like or an even more complex shape is formed depending on the properties of the medium through which the light is traveling. To reduce errors in position measurements, the best solution is taking the measurement in the energy center of the signal generated by the scanner. The Energy Centre of the signal concept employs different mathematical methods as a way to assess which one yields the most precise measurement. The following section presents an application of an optical scanner in landslides monitoring, with experimentation in a laboratory environment.

5. A NEW APPROACH BASED IN OPTICAL SCANNERS

Typically, a landslide-prone area is a remote one and the sensors need to be rugged, weather resistant, portable, with low power consumption, with a low cost and easy interfacing with the data acquisition system. The novelty of the proposed method is the combination of the optical scanner system with incoherent light detection and machine learning techniques, specifically the Support Vector Machine method to reduce error in measurements. This method was previously validated with successful results, taking measurements at distances less than 1.5m. The prototype used to test the method was significantly improved to increase the distance range of measurements. The new experiment (Figure 10) and results, taking measurements at a distance of 6m, are showed in this section. The experiment was developed with low cost sensors and incoherent light emitter. An additional advantage of using inexpensive sensors is that it allows more units to be deployed. For example, Tungsten Quartz Halogen Lamps may be used as a beacon for the 45 ° mirror scanner explained above.

Consider the emitted power of a light bulb filament as a function of wavelength λ and given by Planck's radiation formula: (Corke, 2013)

$$E(\lambda) = \frac{2hc^2}{\lambda^5 \left(e^{\frac{hc}{k\lambda T}} - 1 \right)}. \tag{4}$$

where T is the absolute temperature (K) of the source, h is the Planck's constant, k is the Boltzmann's constant end c is the speed of light. This is the power emitted per unit area per unit wavelength.

The total amount of power radiated (per unit area) is the area under the blackbody curve and is given by the Stefan-Boltzmann law:

$$E = \frac{2\pi^5 k^4}{15c^2 h^3} T^4. \tag{5}$$

if

$$\sigma = \frac{2\pi^5 k^4}{15c^2 h^3}. \tag{6}$$

Then

$$E = \sigma T^4. \tag{7}$$

For lighting applications, a typical filament temperature is between 2800°K to 3400°K. On the other hand, for infrared heating applications, tungsten temperature could be anywhere from 2000°K to 3200°K, the most commonly used temperature being 2500°K. As the filament temperature is raised, the radiation spectrum changes. For a given filament temperature T(°K), the wavelength at which the radiant energy is maximum can be estimated by the Wien's Law:

$$\lambda_{max} = \frac{2.8978 \times 10^{-3}}{T}. \tag{8}$$

The filament temperature can be estimated from the ratio of the hot resistance (RH) and cold resistance (RC @ 25°C) of the filament. Since the data of resistance of tungsten as a function of the temperature is accurately known, the ratio RH / RC gives a fairly accurate estimate of filament temperature. The emissivity of tungsten at blue end is higher than at red end and therefore, the apparent colour temperature of a tungsten filament is higher than the actual filament temperature by 50 to 100°K around 3000°K (Litex Electricals Pvt. Ltd., 2015).

As can be seen in Figure 12, Halogen Spectrum, the greatest Energy of a Tungsten Quartz Halogen Lamp is in the infrared section with a peak between 700-800 nm, leaving a small part in the visible spectrum. If this type of lamp is coupled with an optical sensor like the OPT 301 with peak sensitivity between 700-800 nm it is possible to get a very good system response (see Figure 13). With this in mind, the system shown in Figure 8 was constructed and tested.

Figure 12. Tungsten Quartz Halogen Lamp Spectrum

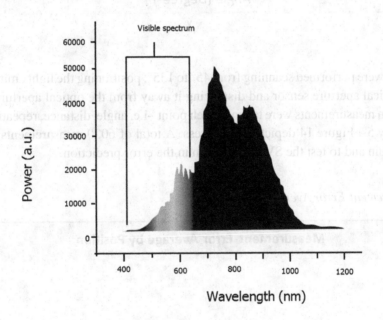

Figure 13. Spectral responsivity of OPT 301

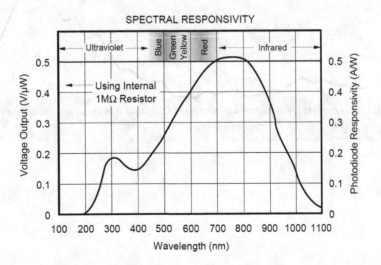

Figure 14. Measurements angle and distance representation

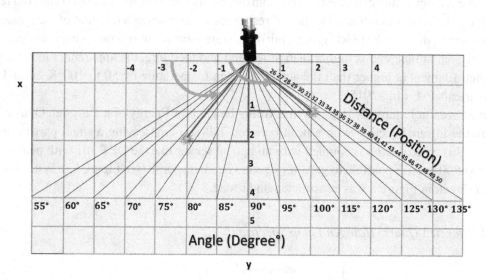

Measurements were performed scanning from 45° to 135°, positioning the light emitter source at near distance to the optical aperture sensor and displacing it away from the optical aperture sensor through the same angle. Ten measurements were taken at each point –i.e. angle, distance, repeating this measurement process every 5°. Figure 14 depicts this process. A total of 6020 measurements built the dataset that was used to train and to test the SVM algorithm in the error prediction.

Figure 15. Measurement Error Average by Position

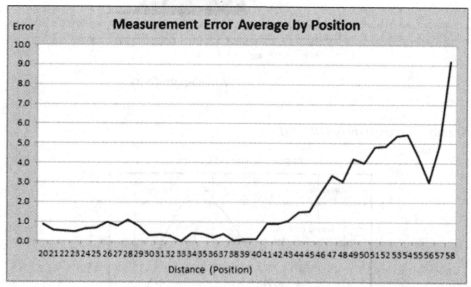

Figure 16. Measurement Error Average by Angle°

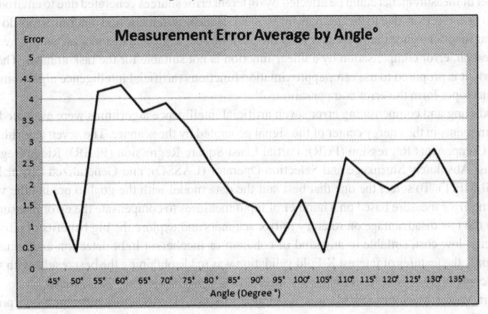

Figure 15 and Figure 16, gives an error profile by position and angle. It is seen that the error does not follow a well-defined behavior or any known function.

As far as the light emitter source is moved away from the optical aperture sensor, the error increments. Also, it was noticed that the best scanning window angle was close the 90°.

Measurement errors have been calculated by (9) without error adjustment and by (10) with error adjustment using the most appropriate method to be implemented on the optical scanning system for better measurement accuracy.

$$E = \alpha_m - \alpha_t \tag{9}$$

$$E_P = \alpha_{mc} - \alpha_t . \tag{10}$$

Where:

E   is the real Error

E_p is the predicted Error

α_m is the angle measured by the system

α_t is the target angle measured

α_{mc} is the angle measured by the system corrected (digital rectified).

Before correcting the measurements, 53.39% of measurement had an error of less than 2.99°.

Each set of measurements could be affected by different error sources generated due to environmental conditions or even errors due to the mechanism by itself. Hence, systematic and random errors do not follow a linear function, since their behavior is by the position -i.e. angle and distance, scanning frequency. For this reason, error compensation by a linear function is not suitable for the task at hand. Therefore, in this work, it is proposed to use error approximation functions (artificial intelligence algorithms under assessment) toper form the error compensation.

To predicting and compensating error seven artificial intelligence algorithms were evaluated taking the measurements in the energy center of the signal generated by the scanner. The seven algorithms are: Principal Components Regression (PCR); Partial Least Square Regression (PLSR); Ridge Regression (RR); Least Absolutely Shrinkage and Selection Operator (LASSO); and Generalized Linear Models Fitting (GLMFIT). To select the one that best suit the data model with the goal to predict the value of the outcome error measure based on a number of input measures to compensate the error measurement, caused by the few disadvantage on rotatory mirror scanners and explore the big benefits of advantages as simplicity, low cost, suitable to any kind of coherent or incoherent light detection and focus in the search of only the features of interest K-Fold validation was made obtaining the best results with support vector machine regression (Flores Fuentes, y otros, 2014; Gascón-Moreno, 2011, June).

Support Vector Machines (SVM) was developed to solve classification and regression problems in machine learning or pattern recognition area. The statistical learning theory goal in modeling is to choose a model from the hypothesis space, which is closest (with respect to some error measure) to the underlying function in the target space as described on (Gunn, 1998; Gascón-Moreno, 2011, June). According to Qingsong Xu, SVM is capable of modeling nonlinear systems by transforming the regression problem into a convex quadratic programming (QP) problem and solving it with a QP solver (Xu, 2014), basically, support vector machines regression can be applied to: a) Linear regression by the introduction of an alternative loss function (as Quadratic, Laplace, Huber and ε-insensitive); and b) Non-linear regression by a non-linear mapping of the data into a high dimensional feature space where linear regression is performed. In this chapter we show an application of prediction measurement error by means of an SVM regression to perform the digital rectification by measurement correction, adding the predicted error to each measurement.

Figure 17 shows measurement improvement through a rotatory mirror scanner applying Support Vector Machine as an algorithm for error compensation. Darker points indicate error bigger than 6.00°. White points indicate error bigger than 3.00° and less than 5.99°. Gray points indicate an error less than 2.99°. So, error bigger than 6.00° are reduced to 0, error bigger than 3.00° and less than 5.99° are reduced from 28.81% to 1.70%, and error less than 2.99° are increased from 53.39% to 98.30%

The selection of the compensation method was published by the authors in (Flores Fuentes, et al., 2014), (Flores Fuentes, y otros, 2014).

As shown in Figure 17, the error is reduced to less than 3° in 98% of measurements. However, it is necessary to reduce this error more. The next paragraph is focused on digital signal processing in order to improve the performance of the OSS and reduce the optical noise.

Figure 17. Results after the compensating error

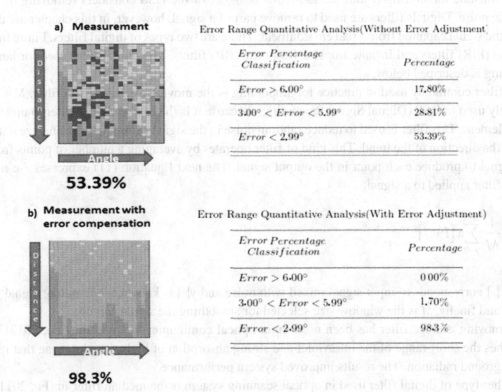

a) Measurement

53.39%

Error Range Quantitative Analysis(Without Error Adjustment)

Error Percentage Classification	Percentage
Error > 6.00°	17.80%
3.00° < Error < 5.99°	28.81%
Error < 2.99°	53.39%

b) Measurement with error compensation

98.3%

Error Range Quantitative Analysis(With Error Adjustment)

Error Percentage Classification	Percentage
Error > 6.00°	0.00%
3.00° < Error < 5.99°	1.70%
Error < 2.99°	98.3%

6. DATA ACQUISITION AND DIGITAL FILTERS FOR OPTICAL SCANNING SYSTEM

The acquisition of the signal is the main process that ensures the accuracy of the OSS system because as a result of this process, the SHM system can determine if damage is present or not. The current system implemented in this chapter has two sensors to capture the opto-electrical signal from the reference source that is mounted on a specific point of the structure under monitoring.

The first sensor of the OSS consists basically of a photodiode to convert an optical signal reference to a voltage signal, this signal is acquired in one digital channel of Data acquisition system (DAQ). Another photosensor measures the pulses of the revolution of the dc motor where each revolution of this motor is counted. With this kind of sensors, two pulses per revolution are acquired, processed by a DAQ as a digital input. The sensor mentioned before consist of a packaging system as an infrared emitting diode coupled with a silicon phototransistor in a plastic housing.

Data acquisition was carried out by using DAQ USB-6003 this device provides eight single-ended analog input (AI) channels with a resolution of 16-Bit and allows 100 kS/s, two analog output (AO) channels at 5 kS/s/ch, 13 DIO channels, and a 32-bit counter with a full-speed USB interface. The sampling rate of 50k/samples per channel was used for the experiments only using two analog inputs. The first analog input was used to measure the angular frequency of motor DC. The second was used to acquire the reference signal.

For landslide monitoring is indispensable that designer of the OSS considers removing the opto-electronic noise. Digitals filters are used to remove part of a signal, however, in this chapter are used to smooth the signal captured from a real environment. There are two types of digital filters: Finite Impulse Response (FIR) filters and Infinite Impulse Response (IIR) filters. The digital filter used for landslide monitoring is described below.

FIR filter commonly used in practice for smoothing is the moving average (MA) filter. MA filters are widely used in DSP (Digital Signal Processing), because it is the easiest digital filter to understand and implement. This filter is used to reduce random noise in the signal (Smith, 1997), analyze data, and identify the direction of the trend. This kind of filter operates by averaging a number of points from the input signal to produce each point in the output signal. The next Equation (11) expresses the moving average filter applied to a signal:

$$y[n] = \frac{1}{M} \sum_{j=0}^{M-1} x[i+j].$$

(11)

Where $x[\ ]$ corresponds to input signal mixed with noise and $y[\ ]$ is known as the output signal that is filtered, and finally, M is the window size selected for smoothing the signal captured.

The moving average filter has been used in an optical communication system (Luo, 2012) where approaches the short-range of the ultraviolet and strong absorption of high altitude ozone that reduces the background radiation. The results improved system performance.

Another type of digital filter used in optical scanning system is the median filter in (Fu, 2011), this work is based on laser technology in order to get information like coordinates of 3D measurement where the radiation of the laser is projected on the detection of objects, and use the CCD of sensitive, different positions and then the analysis image algorithms.

The median filter replaces every data by the median of the windows sizes, it is important to mention that this process is similar to MA filter. (MIČEK & KAPITULÍK, 2003)

Median filter represents a nonlinear dynamic system derived from the vector of values.

$$x(n) = \left\{ x(n), x(n-1), \ldots, x(n-M) \right\}^T.$$

(12)

Output smoothed of the signal $y(n)$ can be defined as the middle sample from a classified collection of elements of vector $x(n)$.

$$y(n) = med\left\{ x(i) \right\}. \text{ where } i = n, n-1, n-M$$

(13)

7. IMPROVEMENTS IN THE OSS FOR LANDSLIDE MONITORING

When the system OSS is applied to a real-environment the possible issues that can arise are optical and electrical noise. There are several methods for reducing the optical and electrical noise, however, the cost of the system can increment when a new element is used. An alternative way to reduce the cost of implementation and minimize the effects caused by environmental influences is using devices like an LED as a transductor of the optical system (Wendy Flores-Fuentes, 2018). As indicated in that research, devices, such as a phototransistor, LED as a photosensor, photodiode (PD), and light-dependent resistor (LDR) can be implemented as a natural filter of the sunlight because of their narrow electromagnetic spectrum. In (Brooks, 2001; Mims, 2000), the authors examined that it was possible of using an LED as a photosensor with relevant and convenient results that increment the accuracy of the OSS. One advantage of using this optoelectronic device is that can be available on the commercial market at low cost in comparison with other already commercial instrumental devices.

In the present chapter, an optoelectronic circuit is proposed for landslide monitoring by using an LED as a transductor of an OSS with the constraint that analog signal captured from the sensor it should be conditioned before processing.

On the other hand, devices such as operational amplifiers (op-amps) are commonly implemented to amplify the output of the photodiode and the most widely used circuit are a transimpedance amplifiers (TA). The disadvantage of using this configuration is that electrical noise appears in the output of the signal. Another disadvantage of using op-amp is that it requires a symmetrical power-supply voltage. One alternative to solve this inconvenience is using the configuration common-mode rejection. This configuration has the property of canceling out any signals that are common (the same potential on both inputs).

By using an LED as a photosensor this circuit it will be employed for the experiments.

On the other hand, a transistor field-effect transistor (FET) is used in the experiments for the acquisition process and illustrated and detailed in this work as a signal conditioner. In the case of using transistor FET, the cost of design can be reduced considerably and at the same time, this device eliminates the need for using asymmetrical power supply by comparing it with some configurations of the op-amp. In Figure 18 is illustrated the circuit used for optical signal conditioning.

Figure 18. Optoelectronic circuit by using an FET as a conditioner of the OSS.

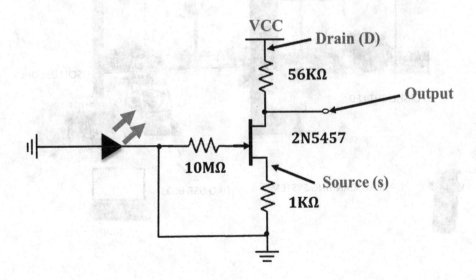

One of the main objectives of an OSS is to reduce the optical and optoelectronic noise. In order to ensure the accuracy and precision of an OSS, it's very important to minimize electro-optical noise. The electro-optical noise is an undesired signal that can affect and reduce the resolution of the optical system. The electro-optical noise is composed of two parts, such as electrical and optical noise. The electrical part of an optical system allows converting a light signal into an electrical signal. In all electronics systems, the noise is an inherent problem that the designer has to consider it.

There are several types of electrical noise that affects all photosensors, however, highlight three main types of noise such as thermal noise, shot noise, and flicker noise. Due to the blue LED working as a photosensor involves small current signals, a signal undesired can be imposed on the signal of interest causing serious accuracy problems, and in order to minimize its effects it is necessary a suitable opto-electronic circuit design.

Another part of the OSS is based on optical devices such as lens, mirrors, and optical filters. These devices play an important role in order to collect and gather light. The principal environmental condition factors to produce optical noise in an optical system is the bright sunlight that affects the accuracy and resolution of the optical system. In real operation, an OSS is exposed to the shorter-wavelength radiant energy and these can saturate the optoelectronic sensors. Depending on the photosensor and reference source used, it should take in account into the wavelength of these, because in outdoors the incident radiations caused by the sun contain several wavelengths that can cause interference with the sensing stage. For the purpose of this chapter, the optical noise is considered as sources that cause interference between the optical system and the reference source. For example, the reference source and photodetector system are based on blue light (470nm) while the rest of the wavelength is considered optical noise. The next Figure 19 shows the experimental setup and measurement equipment used in a real environment.

Figure 19. Experimental Set up in real-environment

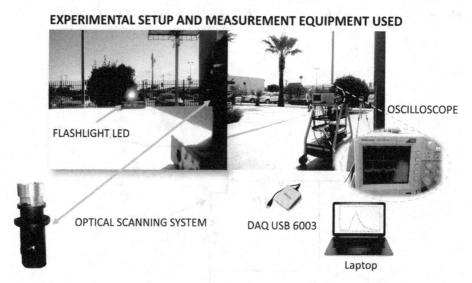

EXPERIMENTAL SETUP AND MEASUREMENT EQUIPMENT USED

FLASHLIGHT LED

OSCILLOSCOPE

OPTICAL SCANNING SYSTEM

DAQ USB 6003

Laptop

The Figure 20 above, illustrates a signal that was captured by OSS mixed with noise that it was captured in a real-world environment. The potential problem that OSS system can be exposed is that several peaks appear. When more than one peak is detected by the optical system, this might cause a problem for calculating the angle measurement because only one peak is related to the energy center. The peaks that appeared in Figure 20 can be smoothed by applying a digital filter.

Figure 20. Signal captured from reference source mixed with electrical noise

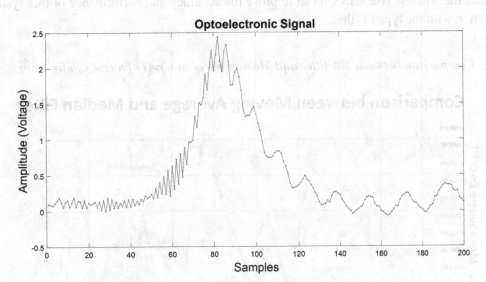

The signal smoothed by the digital filter is shown in the next Figure 21. This capture was taken from a real environment according to the previous Figure 20.

Figure 21.Signal smoothed by using digital filters.

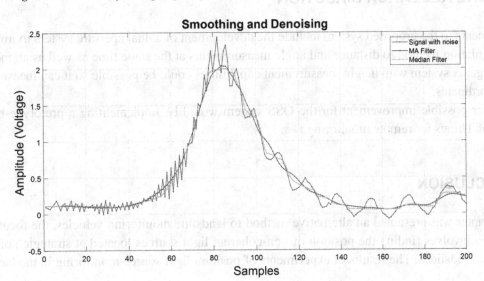

Figure 22 illustrates the comparison between the moving average and median filter applied to signal with noise (showed in figures 20 and 21). The experiments were based on 20 configurations where each one corresponds to the windows sizes of the digital filter (MA and Median Filter). The behavior in terms of mean square error (MSE) of the median filter by using a window size of 8 elements is better than MA, for example, the MSE calculated for Median and MA filter were 0.3687°, MA 1629° respectively. For this configuration, the difference of the median filter was 0.5° according to the real position of aperture and reference source. The difference calculated when it was used MA it was 8.2°. It is important to mention that the window size selected can improve the accuracy and performance of that system OSS in conjunction with the type of filter.

Figure 22. Comparison between MA filter and Median Filter in terms of mean square error

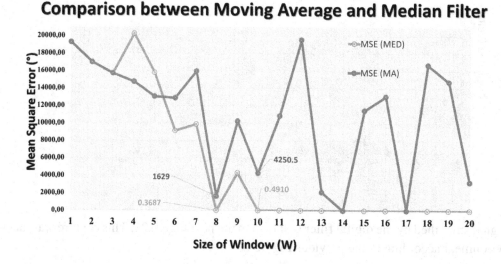

8. FUTURE RESEARCH DIRECTION

Improvements to the proposed system include the development of a dual aperture system to implement the triangulation method to distance and angle measurements at the same time as well as an increment in the range. A system with height measurement capabilities could be possible so it can measure x and y axis coordinates.

Another possible improvement for the OSS system would be implementing a prototype-based of Internet of Things for remote monitoring.

9. CONCLUSION

In this chapter was presented an alternative method to landslide monitoring vehicles, the focus of this application involves finding the position of non-coherent light sources located at strategic points susceptible to landslides. The results of experiments of position light source monitoring in the laboratory environment showed that the incoherent light is better than laser scanners for in such applications because of no alignment problems.

The experiments also showed that the error is significantly reduced when using artificial intelligence techniques like support vector machine to compensate it.

Thus, the proposed method has been proved to be successful and was shown to be an inexpensive way of detecting damages. However, the scope of the features explored in this research is limited, therefore the authors have been visualized other improvements for further research, on the mechanical-electrical optical system, as in the energy signal center, digital processing and its rectification by the analysis of the motor rotation frequency effects.

For removing and smoothing the optoelectronic noise a FET circuit and an LED as a photosensor has been presented. According to the comparison of the results illustrated in Figure 22, we concluded that Median filter requires fewer elements of the window size than MA filter. For this reason, the Median filter can be used instead of MA filter, for landslide monitoring. One advantage of using fewer elements of the windows sizes is that the system requires less computational.

REFERENCES

Anderson, S. A., & Tallapally, L. K. (1996). Hydrologic response of a steep tropical slope to heavy rainfall. In 7th International Syposium on Landslides (pp. 1489-1498). Trondheim, Norway: Brookfield.

Baron, I. R. S. (2010). State of the Art of Landslide Monitoring in Europe: Preliminary results of the Safeland Questionnaire. In Landslide Monitoring Technologies & Early Warning Systems (pp. 15-21). Vienna: Geological Survey of Austria.

Bass, M. (1995). Handbook of Optics, Vol. II- Devices, Measurements and Properties. New York: McGraw-Hill, Inc.

Brooks, D. R., & Mims, F. M. III. (2001). Development of an inexpensive handheld LED-based Sun photometer for the GLOBE program. *Journal of Geophysical Research, D, Atmospheres*, *106*(D5), 4733–4740. doi:10.1029/2000JD900545

Corke, P. (2013). *Robotics, Vision and Control*. Springer.

Flores Fuentes, W., Rivas López, M., Sergiyenko, O., González Navarro, F. F., Rivera Castillo, J., Hernández Balbuena, D., & Rodríguez Quiñonez, J. (2014). Combined Application of Power Spectrum Centroid and Support Vector Machines for Measurement improvement in Optical Scanning Systems. *Signal Processing*, *98*, 37–51. doi:10.1016/j.sigpro.2013.11.008

Flores Fuentes, W., Rivas López, M., Srgiyenko, O., González Navarro, F. F., Rivera Castillo, J., & Hernández Balbuena, D. (2014). Machine Vision Supported by Artificial Intelligence Applied to Rotary Mirror Scanners. In *Proceedings of the 2014 IEEE 23th International Symposium on Industrial Electronics* (pp. 1949-1954). Istambul, Turkey: ISIE.

Fu, H. L. (2011). Home-Made 3-D image measuring instrument data process and analysis. In *2011 International Conference on Multimedia Technology* (págs. 6386-6389). Hangzhou, China: IEEE.

Gascón-Moreno, J. O.-G.-S.-T.-M.-F. (2011, June). Multi-parametric gaussian kernel function optimization for ε-SVMr using a genetic algorithm. In *International Work-Conference on Artificial Neural Networks* (pp. 113-120). Berlin: Springer.

Gunn, S. R. (1998). Support vector machines for classification and regression. *ISIS Technical Report*, 5-16.

Hoon Sohn, C. F. (2002). *A Review of Structural Health Monitoring Literature 1996-2001*. www.lanl.gov.projects/damage

Kennedy, W. P. (2012). *The Basics of triangulation sensors*. Obtenido de http://archives.sensorsmag.com/articles/0598/tri0598/main.shtml

Kerle, N., Stump, A., & Malet, J.-P. (2010). Object oriented and cognitive methods for multidata event-based landslide detection and monitoring. *Landslide Monitoring Technologies & Early Warning Systems*.

Litex Electricals Pvt. Ltd. (2015). Obtenido de http://www.litexelectricals.com/halogen_lamp_characteristics.asp

Luo, P. a. (2012). A moving average filter based method of performance improvement for ultraviolet communication system. In *2012 8th International Symposium on Communication Systems, Networks & Digital Signal Processing (CSNDSP)* (pp. 1- 4). Poznan, Poland: IEEE.

Miček, J., & Kapitulík, J. (2003). Median filter. *Journal of Information, Control and Management Systems*, 51-56.

Michoud, M. J.-H. (2012). New classification of landslide-inducing anthropogenic activities. *Geophysical Research Abstracts*.

Mims, F. M. (2000). Solar radiometer with light-emitting diodes as spectrally-selective detectors. *Applied Optics*, *39*(34), 6517–6518. doi:10.1364/AO.39.006517

Ming-Chih, L., Su-Chin, C., Yu-Shen, L., & Cheng-Pei, T. (n.d.). *Machine Vision in the realization of Landslide monitoring applications*. Academic Press.

NOAA. (2015). Obtenido de http://oceanservice.noaa.gov: https://oceanservice.noaa.gov/facts/lidar.html

Plaza, J. P., Plaza, A., & Barra, C. (2009). Multi-Channel Morphological Profiles for Classification of Hyperspectral Images Using Support Vector Machines. *Sensors (Basel)*, *9*(1), 197. doi:10.339090100196 PMID:22389595

Rentschler, K., & Moser, M. (1996). Geotechniques of claystone hillslopes-properties of weathered claystone and formation of sliding surfaces. In *7th International Symposium on Landslides* (pp. 1571-1578). Tronheim, Norway: Brookfield.

Rivas López, M., Flores Fuentes, W., Rivera Castillo, J., Sergiyenko, O., & Hernández Balbuena, D. (2013). A Method and Electronic Device to Detect the Optoelectronic Scanning Signal Energy Centre. In S. L. Pyshkin, & J. M. Ballato (Eds.), Optoelectronics (pp. 389-417). Crathia: Intech.

Savvaidis, P. D. (2003). Existing Landslide Monitoring Systems aand Techniques. School of Rural and Surveying Engineering, The Aristotle University of Thessaloniki, 242-258.

Smith, S. W. (1997). Digital Signal Processors. In The scientist and engineer's guide to digital signal processing (pp. 503-534). San Diego, CA: California Technical Pub.

Vahelal, A. (2010). *Sensors on 3D Digitization*. HasmukhGoswami College of Engineering.

Wendy Flores-Fuentes, J. E.-V.-L.-Q. (2018). Comparison between different types of sensors used in the real operational environment based on optical scanning system. *Sensors (Basel)*, *18*(6), 1684. doi:10.339018061684 PMID:29882912

Xu, Q. (2014). Impact detection and location for a plate structure using least squares support vector machines. *Structural Health Monitoring*, *13*(1), 5–18. doi:10.1177/1475921713495083

Chapter 5
Statistical Characteristics of Optical Signals and Images in Machine Vision Systems

Tatyana A. Strelkova
Kharkiv National University of Radio Electronics, Ukraine

Alexander P. Lytyuga
Kharkiv National University of Radio Electronics, Ukraine

Alexander S. Kalmykov
Kharkiv National University of Radio Electronics, Ukraine

ABSTRACT

The chapter is devoted to the creation of a comprehensive approach to the physical and mathematical description of signals in optoelectronics in machine vision, taking into account the phenomena of interaction of optical radiation with system elements. A new methodology for the study of the statistical properties of input and output signals in optoelectronic systems is proposed, taking into account the availability of grouped statistical properties that do not obey the Poisson statistics. The basis is the joint use of wave and corpuscular description of signals in systems, stochastic flow theories, and elements of statistical detection theory. Information and energetic technology have been developed that integrates the theoretical justification of signal description under various observation conditions and decision-making methods.

INTRODUCTION

The monograph is devoted to the creation of a comprehensive approach to the physical and mathematical description of signals in machine vision optoelectronic, take into account the phenomena of interaction of optical radiation with system elements. A new methodology for the study of the statistical properties of input and output signals in optoelectronic systems is proposed, taking into consideration the availability

DOI: 10.4018/978-1-7998-6522-3.ch005

of grouped statistical properties that do not obey the Poisson statistics. The basis is the joint use some aspects of stochastic flow theory, and elements of statistical detection theory. Information and energetic technology have been developed that integrates the theoretical justification of signal description under various observation conditions and decision-making methods. We have considered the application of the developed stochastic models in applied problems: formation and analysis of signals and images; detection of signals against the interference in various observation conditions, and properties of photosensitive and optical elements.

The problematic issues that arise when registering objects are related to changes in statistical behavior of both signals and interference. The authors propose an approach to the description of the statistical characteristics of the output signals, taking into account the presence of grouped statistical properties of optical radiation. Stochastic models of output signals of optoelectronic systems are based on the analysis of Poisson fluxes with modified dispersion. The analysis of such stochastic processes is based on the process of generating a random variable under the conditions of applicable generalized limit theorems, which use the family of stable distribution laws as limit distributions. When random processes describing the operation of a stochastic system are non-Markovian.

The first section focuses on determining the directions for improving and developing optoelectronic systems. The analysis of methods aimed at increasing the efficiency of optoelectronic systems has been carried out, and the factors constraining the development and improvement of systems have been determined. The constraining factors include physical processes that affect the formation of output signals of systems; statistical instability of signals received by systems and ambiguity of methods of their description (for example, presence of groupings, correlations of photons); uncertain state of the input optical radiation; technical limitations of elements of optoelectronic systems in machine vision. The scientific and technical problem aimed at eliminating contradictions in the existing theoretical methods of receiving and processing signals and experimental data has been posed. Research problems are formulated.

The second section contains the systematization of theoretical methods underlying the description of the input and output signals of optoelectronic systems.

The analysis of theoretical methods for describing input signals with allowance for corpuscular and wave properties has been carried out. The wave theory makes it possible to describe the processes of diffraction and interaction of the field with matter; the corpuscular theory takes into account the absorption processes, since the optical range quantum energy is sufficiently large.

An analysis of the methods for describing the output signals has been made with the consideration of the statistical properties of the signals. The main attention has been paid to statistical models based on the Poisson, Gaussian, log-normal, negatively binomial, Bose-Einstein distributions. Fluctuations in the background and signal components are characterized by distribution laws with finite dispersion, and, according to the theory of errors, are united by a central limit theorem.

The principle of signal formation in Optoelectronics in Machine Vision has been considered as a process having the stochastic character of electromagnetic radiation. Spatial-temporal changes in signals, from the position of the corpuscular theory, take into account the conditions of linearity, invariance, physical feasibility, and stability. The relationship between the input and output streams has been defined as stochastic (at the corpuscular description) dependence.

The third section is aimed at carrying out theoretical and experimental studies of the statistical properties of the output signals of optoelectronic systems with a limited dynamic range. Based on the Poisson and Gaussian statistics, theoretical models of interaction of optical radiation with the optical link of the system on the basis of corpuscular descriptions have been compiled. An approach to the evaluation of the efficiency of optoelectronic systems with a limited dynamic range has been developed by the criterion of the magnitude of the signal-to-noise ratio. The dependence of the change in the value of the signal-to-noise ratio on the absorption coefficient of the optical link as *a deterministic parameter* $1/\sqrt{k}$ (*k* is the attenuation coefficient) has been determined.

Experimental studies of the statistical characteristics of the output signals of systems with a limited dynamic range have shown that the quality and composition of the optical glass affect the characteristics of the output signals. The assumption has been made that the optical link must be regarded as a *stochastic element*, which introduces distortions into the output signal of the system. It has been proposed to consider optical glasses as *stochastically inhomogeneous* media that can influence the path of photons passing through the optical link, violating the principle of linear superposition and the *conditions of linearity, invariance, physical feasibility, and stability*.

Statistical models for the interaction of an optical signal with an optical link in optoelectronic systems in machine vision have been developed on the basis of the distribution laws having finite and infinite variances. The model based on the central limit theorem has been presented as a process of thinning a stochastic flow, in which individual points disappear independently of other points. The experimental results have shown that this model was applied in approximating the probability density of the output signals of optical-electronic systems that do not use attenuators. The model based on a family of stable laws has been presented as a stochastic process characterized by the presence of the probability of a large but rare event that cannot be neglected and which makes a significant contribution to the description of the results obtained. The regions of attraction of the limiting distributions of the output signals of optoelectronic systems in machine vision have been determined experimentally. Investigations of the asymptotic behavior of the tailings of the distribution density of the output signals for optoelectronic systems have shown the possibility of using stable laws for the distribution of the description of the output signals of optoelectronic systems in machine vision. The dependence of the attenuation coefficient of the optical link *k* and the characteristic factor ($0<\alpha<2$) has been established. It has been suggested that the efficiency of the optoelectronic system can be limited in different energy modes of operation by the composition and quality of the *optical link*.

The limits of applicability of statistical models on the basis of central and generalized limit theorems have been determined. The consideration of additional factors affecting the formation of the output signal and the description of the stochastic behavior of the signal on the basis of generalized limit theorems have made it possible to determine a unified approach to assessing the effect of attenuators on the statistical properties of signals and the efficiency of the optoelectronic systems in machine vision as a whole.

THE FIRST SECTION. STATISTICAL PROPERTIES OF OUTPUT SIGNALS OF OPTOELECTRONIC SYSTEMS WITH LIMITED DYNAMIC RANGE

To analyze the statistical characteristics of the optical field, it is necessary to adopt a certain physical-mathematical statistical model. When developing mathematical models of output signals, we use normal, lognormal, Bose-Einstein, and exponential distribution (Cox & Lewis, 1969; Nikitin et al., 2008; Strelkov et al., 2010). Widespread mathematical models of received signals are developed on the basis of Poisson statistics (Strelkova & Sautkin, 2013; Fedoseev, 2011; Lytyuga, 2009; Yang at all, 2012; Bedard, 1967). These models take into account the corpuscular structure of the photon flux, the conditions for the formation of optical radiation, the lack of interaction between photons in the flux, as well as the stochastic properties of optical signals in the analysis of quantum noise. However, when developing a mathematical model of signals based on Poisson statistics, as a rule, some assumptions are made and restrictions are imposed on the properties of optical signals, which are reduced to the three main properties of the simplest (Poisson) flow – stationarity, ordinariness, and absence of aftereffect. This allows us to use the invariance property of Poisson flow in describing its interaction with the elements of the system. The use of such a mathematical apparatus allows us to obtain analytical expressions for the main indicators of system efficiency and to create highly efficient algorithms of signal processing in optoelectronic systems (Lytyuga, 2010; Flores-Fuentes at all, 2013; Strelkova, 2016).

The analysis results presented in (Nikitin et al., 2008; Strelkova at all, 2012; Strelkov at all, 2013; Strelkova & Sautkin, 2013), and the experience of applied research indicate that in the case of using attenuators as part of the optical link, the statistical characteristics of the output signals do not obey conventional Poisson and Gaussian statistics. This leads to significant errors in assessing the system efficiency by known criteria. Accordingly, there is a need to revise the mathematical description and take into account the action of additional factors that affect the formation of output signals in systems, and improve the statistical model of output signals.

Based on the theoretical principles of the corpuscular theory, we describe the process of generating the output signal of the optoelectronic system (Figure 1).

The input of the optical link (lens, attenuator) receives an additive mixture of useful signal and background radiation, which is collected by the lens and registered by the photodetector. The photodetector converts the received radiation into a space-time distribution of charges inside the photosensitive layer. During the conversion process in the electronic path, the noise of the photodetector is added to the additive mixture of useful and background radiation.

The data array, which is a response of photosensitive elements to the intensity of optical radiation, is supplied to the input of the post-detector processing device. This array is supplied as the sum of independent random variables that characterize the signal and background components.

Figure 1. Block diagram of the optoelectronic system

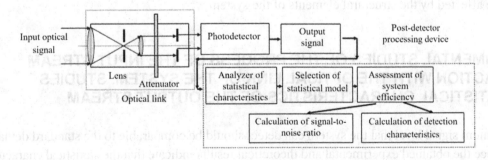

Photodetector The conversion of optical radiation that has passed the optical link into an electrical signal by a photosensitive element is described from the standpoint of corpuscular theory. When developing statistical models of output signals, it is assumed that the physical and statistical characteristics of the received optical signal interacting with the photosensitive element obey the Poisson, sub-Poisson, and Gaussian statistics.

Optical link. The conversion of optical radiation by an optical link can be described from the standpoint of geometric optics, wave, or corpuscular theory of light. When describing the interaction of optical radiation with the optical link from the standpoint of the wave theory of light, attenuators are considered as a linear element that does not affect the statistical characteristics of the output signal. The attenuation coefficient of the neutral filter does not affect the criteria for assessing the system quality, in particular the value of signal-to-noise ratio.

When describing the interaction from the standpoint of the corpuscular theory, light is described as a stream of discrete particles (photons), herewith optical radiation makes it possible to describe phenomena of a probabilistic nature, such as absorption in neutral filters, and, therefore, take into account changes in statistical characteristics of output signals based on Poisson and Gaussian statistics. The probabilistic nature of the process of attenuation of the input signal by optical elements is taken into account when calculating the value of signal-to-noise ratio ϕ. In this case, the attenuation coefficient was taken into account as a deterministic parameter. Analogously to expressions used in (Lytyuga, 2009), (Strelkova et al., 2016), we write the expression for the signal-to-noise ratio for systems with a limited dynamic range as:

$$\phi = \frac{1}{\sqrt{k}} \frac{\overline{n}_{s+i} - \overline{n}_i}{\sqrt{\sigma^2_{(n_{s+i})} + \sigma^2_{(n_i)}}}, \qquad (1)$$

where k is the attenuation coefficient;

$\overline{n}_{s+i}$ is a mean of an additive mixture of signal and interference;

$\overline{n}_i$ is a mean of interfering component;

$\sigma_{(n_{s+i})}$ and $\sigma_{(n_i)}$ are standard deviations of an additive mixture of signal and interference, and an interfering component, respectively.

When calculating one of the most important criteria of system efficiency – the value of signal-to-noise ratio (expression 1), it is necessary to take into account statistical characteristics of output signals, which are affected by the structural elements of the system.

EXPERIMENTAL STUDIES OF THE PROCESS OF THE INPUT STREAM INTERACTION WITH THE OPTICAL LINK OF THE SYSTEM. STUDIES OF STATISTICAL CHARACTERISTICS OF THE OUTPUT STREAM

The minimum signal level that the system can detect should be comparable to the standard deviation of noise. Since the obtained experimental and theoretical results indicate that the statistical characteristics deviate from the accepted theoretical models, additional studies were carried out to study the stochastic characteristics of output signals of optoelectronic systems with a limited dynamic range versus the composition and quality of the optical link (Strelkov et al., 2010), (Khoroshun et al., 2020; Strelkov et al., 2013; Strelkova, 2014a; Strelkova, 2014b; Strelkova, 2015; Strelkova & Sautkin, 2013; Strelkova et al., 2012; Strilkova et al., 2019; Strilkova & Lytyuga, 2018).

The purpose of the experiment. Studying the effect of attenuators on the statistical characteristics of the output signals of optoelectronic systems for various applications.

The research was carried out with the help of installations, which block diagrams are shown in Figure 2.

Figure 2. Schemes for experimental research

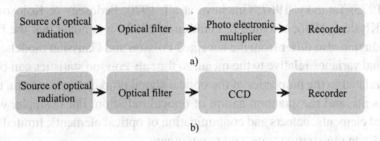

A low-intensity optical signal was applied to the system input, while the photomultiplier tube was operating in the photon-counting mode. The photomultiplier tube was previously kept in the dark to reduce the effect of internal noise. The probabilistic attenuation of the flux was carried out using neutral filters with a given (GOST 9411-91) optical density (Figure 2a) (Strelkov et al., 2010).

We measured the values of ϕ_m and ϕ_{ou} taking into account (1):

$$\phi_{in} = \frac{\bar{u}_{in}}{\sigma_{in}}, \phi_{ou} = \frac{\bar{u}_{ou}}{\sigma_{op}}, \tag{2}$$

where $\bar{u}_{in}$ and $\bar{u}_{ou}$ are mean values of input and output streams; and σ_{in} and σ_{ou} re standard deviations of input and output streams.

The results of experimental measurements of ϕ_{in} nd ϕ_{ou} values were obtained by averaging over more than 10,000 realizations. Figure 3 shows the results for the normalized value of $\phi = \phi_{ou}/\phi_{in}$.

Figure 3. The results of experimental and theoretical dependences of the normalized value of ϕ on the attenuation coefficient. We have accepted the Poisson model of the stochastic signal

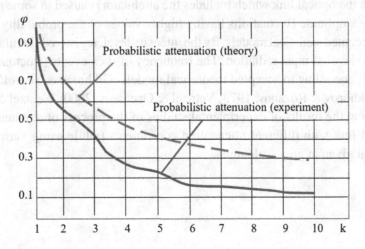

T$_o$ calculate the theoretical relationship for the normalized value of $\phi=\phi$ou/ϕin, $_w$e h$_{av}$e used the Poisson flux model.

Data analysis of Figure 3 indicates that the results of experimental studies show deviations from the accepted theoretical Poisson model with an increase in the attenuation coefficient. Such changes can be explained by an increase in the standard deviation of the recorded flux compared to its mean value. In the study of photon statistics described in (Nikitin et al., 2008), (Cox & Lewis, 1969), and (Kupchenko, 2009; Klyatskin, 1975; Tatarsky, 1967; Gracheva at all, 1970; Fedoseev & Kolosov, 2007; Parfenov & Kirillov, 2012; Klyshko, 1996; Milyutin & Gumbinas, 2002). it is noted that the fluctuations of the received optical radiation also differ from the accepted Poisson and Gaussian models. The studies show a change in the signal variance relative to the mean, so the sub-Poisson statistics can be used. These differences are often related to the properties of the optical signal propagation medium, the characteristics of the system elements, and the quantum nature of optical radiation. For example, with the scattering properties of optical elements, defects and contamination of optical elements, limited dynamic range of the photodetector, finite registration time, and many others.

To find out the reasons for the deviation of experimental results from theoretical dependences, the following statistical characteristics of signals were additionally studied and analyzed at different levels of attenuation – probability density of output signals, asymptotic behavior of tails, density distribution; as well, limit theorems of the distribution of summarized independent random variables are determined that characterize the output signal. The installation shown in Figure 2b was used to conduct an experiment, which allows studying both the time and space behavior of the output signal (Strelkov et al., 1998).

An optical signal was applied to the input of the optoelectronic system, and its intensity was registered by a photodetector in the range from 0 to 255 gradations of brightness. An optical radiation was supplied from incoherent source; neutral filters were used to match the dynamic range. Frame by frame study of video fragments with a record frame rate of 25 frames per second and resolution of 1000×1000 resolution elements allows drawing conclusions about sufficient space-time statistics and assuming the number of tests $N\rightarrow\infty$.

The interaction of optical radiation with an optical link can be characterized as a stochastic attenuation of input radiation. Downsampling of the input stream, taking into account the accepted Poisson statistics, should lead to a natural change in the statistical parameters of the signal proportionally to the attenuation coefficient of the neutral filter, namely, reduce the mean and reduce the variance of the input signal while maintaining statistics.

However, when studying the signals registration process in systems that use neutral filters, the radiation (passed through the optical link which includes the attenuator) caused in some resolution elements of photodetector the amplitude fluctuations having higher values of the probability of occurrence of a large event than in accepted statistical models. By the intensity level, registered fluctuations are exceeding or comparable to the level of input radiation. The frequency of such events (fluctuations) exceeded the expected probability, according to accepted theoretical models (Gikhman & Skorokhod, 1977; Mosyagin et al., 1990; Prokhorov & Rozanov, 1973; Vetnzel & Ovcharov, 2000; Wentzel & Ovcharov, 1988).

Figures 4 – 7 show the results of experimental studies of the process of attenuation of a stochastic signal by an optical link with different attenuation coefficients (while using various neutral filters). Amplitudes of signal given in grey scale units.

Figure 4. Stochastic signal without attenuation

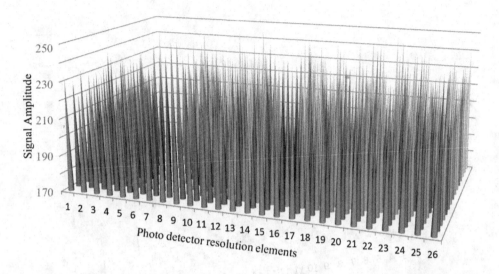

Figure 5. Stochastic signal attenuated by a neutral filter. Attenuation coefficient $k = 2$

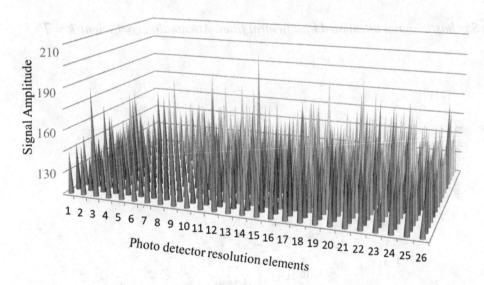

In order to approximate the probability density of output signals, a statistical analysis of the obtained data was carried out. A comparative analysis of the probabilities of occurrence for normalized and centered values obtained during experimental studies and the normal distribution law (at the level of 2, 3, 4, 5 standard deviations of the normal distribution) is given in Table 1. The histogram of the probability density of the output signal and the normal distribution law is shown in Figure 8.

Analysis of the data in Table 1 makes it possible to assert that in the range from 2 to 4 SD (Standard Deviation) the statistical characteristics of radiation at the output of the optical system that does not use attenuators are well approximated by Poisson and Gaussian statistics (Figure 8). When using attenuators,

Figure 6. Stochastic signal attenuated by a neutral filter. Attenuation coefficient $k = 4$

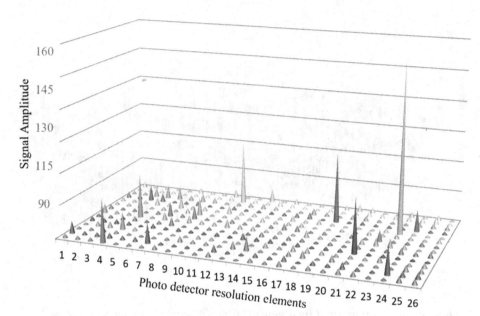

Figure 7. Stochastic signal attenuated by a neutral filter. Attenuation coefficient $k = 7$

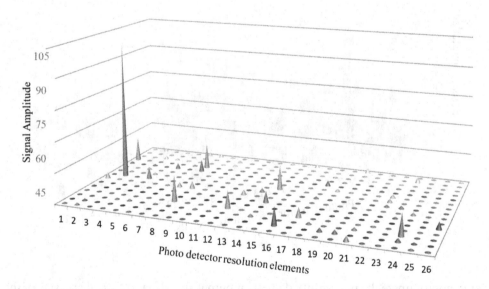

the asymptotic behavior of the "tail" of the distribution density of experimental results in the range of 2–5 SD indicates the impossibility of approximation by Gaussian statistics (Figure 9).

The process of evaluating the quality of detection algorithms for random signals against the background of random interference can be carry out with using of limit distributions that characterize the output signal. The output signal of optoelectronic systems is described as the sum of a large number of

Table 1. Probability of occurrence for normalized and centered values obtained during experimental studies of the attenuation process

SD (σ)	Without attenuator	Attenuator $k=2$	Attenuator $k=4$	Normal distribution
	Probability of occurrence			
2σ	$4\cdot10^{-2}$	$8\cdot10^{-2}$	$8\cdot10^{-2}$	$5\cdot10^{-2}$
3σ	$3\cdot10^{-3}$	$7\cdot10^{-2}$	$1\cdot10^{-2}$	$3\cdot10^{-3}$
4σ	$5\cdot10^{-4}$	$1\cdot10^{-2}$	$8\cdot10^{-3}$	$1\cdot10^{-4}$
5σ	$1.6\cdot10^{-4}$	$7\cdot10^{-3}$	$5\cdot10^{-3}$	$1.4\cdot10^{-6}$

*Figure 8. Approximation of experimental data for output signals of the optoelectronic system that **does not** use attenuators, by normal distribution law*

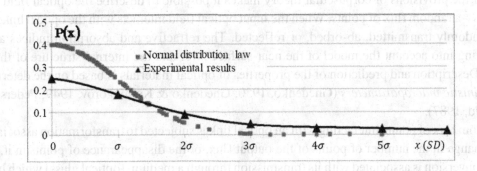

Figure 9. Approximation of experimental data for output signals of the optoelectronic system that use attenuators, by normal distribution law

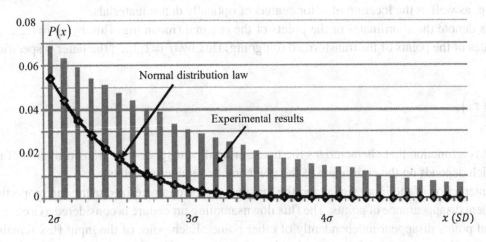

independent random variables, and according to the central limit theorem tends to the normal distribution law as a limit.

As a rule, when implementing threshold detection methods in optoelectronic systems, it is necessary to ensure the probability level of false alarm about <10^{-3}. Conditional probabilities of correct detection and false alarm are optimized based on the choice of threshold values. However, the asymptotic behavior of the probability density of the experimental data indicates that the choice of threshold values based on Poisson and Gaussian statistics will lead to errors in calculating the operating characteristics of the system.

THE SECOND SECTION. A STATISTICAL MODEL OF THE INTERACTION OF AN OPTICAL SIGNAL WITH AN OPTICAL LINK IN OPTOELECTRONIC SYSTEMS. CENTRAL LIMIT THEOREM

The use of the provisions of corpuscular theory makes it possible to describe the optical field as a flux of photons – a random flux of points. When the input optical field interacts with the optical link, photons can be randomly transmitted, absorbed, or reflected. The refractive and absorption indices are determined taking into account the model of the near-surface layer and the internal structure of the optical material. Description and prediction of the properties of optical materials is based on the determination of *deterministic macroparameters* (Gnedenko, 1950; Gnedenko & Kolmogorov, 1949; Pedersen at all, 2005; Field, 1987).

A random flux of points interacting with an optical link is subjected to transformation associated with either a change in the number of points of the output flux, or the disappearance of points in it.

Flux conversion is associated with its transmission through a medium (optical glass) which has linear or nonlinear characteristics. To analyze the process of interaction of optical radiation with glass, it is necessary to adopt an ordered or disordered model of the medium. Optical glass is mainly considered in terms of order and symmetry. When describing the interactions of light quanta with glass material, it must be taken into account the fact that amorphous materials disorder can occur due to spatial asymmetry of nodes in volume, in the presence of chaotically located defects, under atmospheric influences and aging, as well as the location of color centers of optically dense materials.

Let us denote the coordinates of the points of the original (incoming) flux by τj j=1,2,..., and the coordinates of the points of the transformed (outgoing) flux by tj j=1,2,...; the latter is specified by the formula:

$$t_j = D_j\left(\tau_j\right),\tag{3}$$

where D_j is a function that characterizes the linear or nonlinear space-time transformation of the input flux, which depends on the coordinate of the point and its sequential number.

The interaction of the input signal and the medium can be characterized using the properties of the independent disappearance of points. The flux downsampling procedure is considered as a case in which individual points disappear independently of other points. Each point of the input flux remains in the flux with probability $p(\tau_j$, and disappears with probability $1-p(\tau j)$. In this event, the probability p(τj) $_d$epends on the coordinate of the point in question and does not depend on the number and coordinates of other points. This case is a case of a degenerate process of reproduction, in which each point can either transform into a single point with the same coordinate as the original point, or disappear.

The input flux can be subjected to linear or nonlinear transformation with the simultaneous extraction of a certain number of points. The transformation involves overlay and screening procedures in which the overlay and screening procedures are independent of each other. In this case, it is convenient to use the method of successive transformations, assuming that the overlay of fluxes is carried out first, followed by the screening of the flux obtained after the first transformation. After the procedure of independent downsampling, the conditional flux density decreases by v times, keeping the statistics.

According to the accepted theoretical statistical models of output signals in optoelectronic systems, the flux of discrete particles is described by Poisson statistics. According to Theorem (Cox & Lewis, 1969): if in the sequence of coordinates of points xi that form a Poisson flux with intensity f, a downsampling operation is used with the point extraction probability γ, then the flux of remaining points is a Poisson flux with intensity $(1-\gamma)f$. These statements, according to (), allows us to presuppose that the central limit theorem can be applied to the analyzed sequence of independent random events, and the fluctuations of the output flux have Gaussian statistics.

These experimental results indicate that this model can be used to approximate the probability density of the output signals of systems that do not use attenuators in the range from 2 to 4 SD.

When using an attenuator, it is necessary to take into account additional factors, such as the aging of optical glass that affect the formation of output signals in systems operating in a limited dynamic range, as well as improve the statistical model of output signals.

A STATISTICAL MODEL OF THE INTERACTION OF AN OPTICAL SIGNAL WITH AN OPTICAL LINK IN OPTOELECTRONIC SYSTEMS. GENERALIZED LIMIT THEOREMS

Disordered materials, such as neutral filters, at the microscopic level are fundamentally inhomogeneous and differ in structure in different parts of the macroscopically homogeneous optical glass. All processes of photon energy transformation are probabilistic. Taking into account microparameters, such as quantum properties of radiation, i.e. the processes that occur after the absorption of a quantum of light, groups and antigroups of photons, fluctuations and photon-phonon interaction, allows more adequately interpret the obtained results.

The statistics of the original radiation can be empirically described by asymmetric distribution laws that have "heavy" tails, i.e. there is a probability of a large but rare event that cannot be ignored, which significantly affects the description of the obtained results. In explaining these processes, we can use the theory of anomalous diffusion processes ().

The mathematical basis is the theory of stable probability distributions, which allows the study of anomalous diffusion processes from a unified position (Gnedenko, 1950), ().

Stable Laws

In the process of registration of optical radiation in many empirical studies, the fluctuations of the received optical radiation differ from the accepted models. Studies show a change in the variance of the received signals relative to the mean, while using sub-Poisson statistics.

Differences are often associated with the properties of the optical signal propagation medium; with the characteristics of the system elements, such as scattering properties, defects and contamination of optical elements; with limited dynamic range of the photodetector; finiteness of the registration time, as well as with the quantum nature of optical radiation. However, to evaluate the quality of signal processing algorithms, the normal distribution law is used as a limit.

Recently, research on stochastic processes that not obey Gaussian statistics and the classical central limit theorem are of considerable interest. A statistical feature of such processes is a significantly higher probability of large fluctuations. In such cases, generalized limit theorems are used. The distribution of the sum of independent random variables is described by a family of stable distributions (Lifshits, 2007).

The concept of stable distributions was introduced in 1925 by P. Levy as a result of studying the properties of sums of equally distributed random variables. This class of distributions includes heavy-tailed distribution and distributions with asymmetric densities. Often in the analysis of statistical dependencies, the possibility of great events located on the distribution "tail" is neglected. Heavy-tailed distribution is a distribution whose tail cannot be "cut off", i.e. the influence of great but rare events cannot be neglected. Great events mean the phenomena which impact can outweigh the damage from all other events in this class.

Stable distribution laws are characterized by four parameters $0<\alpha\leq2$, $-1<\beta\leq1$, $\lambda>0$, $-\infty<\gamma<\infty$. For fixed α, the characteristic coefficient β and parameters γ and λ play the role of centering and normalizing constants. The values $\beta=0$ correspond to the symmetric standard stable law. The greatest distortion of symmetry occurs after reaching $\beta=1$ ($\beta=-1$). Characteristic index α is the main indicator of a stable law, because its value mainly determines the analytical structure of the distribution density.

Multidimensional stable laws have distribution densities P= $(x,\alpha,\beta,\gamma,\lambda)$ *(Figure 10)*:

a) $\alpha=2$, other parameters are arbitrary. This case corresponds to the family of normal distributions:

$$P = \left(x, 2, \beta, \gamma, \lambda\right) = \left(4\pi\lambda\right)^{-1/2} \exp\left\{-\frac{\left(x-\gamma\right)^2}{4\lambda}\right\}; \tag{4}$$

b) $\alpha=1$, $\beta=0$, parameters γ and λ are arbitrary. This case corresponds to the family of Cauchy distributions:

$$P = \left(x, 1, 0, \gamma, \lambda\right) = \frac{\lambda}{\pi}\left(\lambda^2 + \left(x-\gamma\right)^2\right)^{-1}; \tag{5}$$

c) $\alpha=1/2$, $\beta=1$, parameters γ and λ are arbitrary. The corresponding countless set of stable laws is called the family of Levy distributions:

$$P = \left(x, 1/2, 1, \gamma, \lambda\right) = \begin{cases} \dfrac{\lambda}{2\sqrt{\pi}} x^{-3/2} \exp\left(-\dfrac{\lambda^2}{4\left(x-\gamma\right)}\right), & if \ x > \gamma \\ 0 & if \ x \leq \gamma \end{cases} \tag{6}$$

Figure 10. The family of stable distribution laws

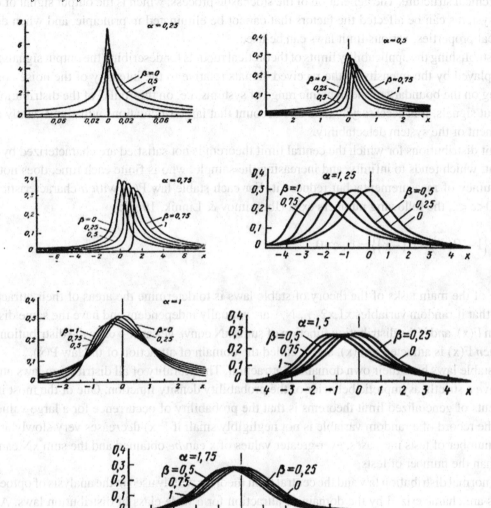

Non-Gaussian random processes are used to describe the charge transport on the surface of a semi-conductor, the interaction of optical radiation with glass. Non-Gaussian isotropic distributions are characterized by asymmetric properties. Flux variance and, consequently, all higher moments of stable distributions are infinite. This is due to the behavior of tails of probability density at large values of random variables. The distribution of time characteristics of occurring in the flux is not exponential, but inverse power function.

When using the central limit theorem, the elementary deviations that cause deviations of the sum of independent random variables (signal and noise fluctuations) are assumed to be comparable by the order of their influence on the dissipation of the sum. The probability of occurrence for high intensity values is very small. The use of generalized limit theorems allows us to describe the stochastic behavior of a random variable taking into account the fact that the probability of registration of large intensity values is not negligibly small. Elementary deviations that affect the sum of random variables can have a hierarchical structure. The generation of the stochastic process, which is the output signal of optoelectronic systems can be affected the factors that cannot be eliminated in principle, and when describing statistical properties, various limit laws can be used.

In establishing the applicability limits of theoretical models for describing the output signals, the main role is played by the intensity of the received signals relative to the intensity of the noise component. Working on the boundary of the dynamic range of systems, i.e. on the "tails" of the distribution density of output signals, it is necessary to take into account that large fluctuations will significantly affect the assessment of the system detectability.

Limit distributions for which the central limit theorem is not satisfied are characterized by a second moment, which tends to infinity, and increasing the sample, who is finite each time, does not increase the accuracy of measurements, but reduces it. For each stable law P(x) with *a* characteristic index α, where $0<\alpha<2$, the following numbers exist (Ibragimov & Linnik, 1965):

$$\lim_{x \to \infty} x^{\alpha} \left\{ 1 - P(x) + P(-x) \right\} = c, \; c>0. \tag{7}$$

One of the main tasks of the theory of stable laws is to determine the areas of their attraction. It is known that if random variables x1,x2,...,N, are mutually independent and have the same distribution function F(x), and if the distribution function of sums xN converges at $N \to \infty$ to the distribution function P(x), then F(x) is attracted to P(x), and is called the domain of attraction of the law P(x).

All stable laws have their own domain of attraction. The plurality of all distribution laws attracted to P(x) have a strictly asymptotic behavior of the probability density function. One of the most important statements of generalized limit theorems is that the probability of occurrence for a large value appearing in the record of a random variable is not negligibly small if P(x) decreases very slowly at larger x. As the number of tests increases, ever-greater values of x can be obtained and the sum xN can increase faster than the number of tests.

The normal distribution law and the central limit theorem widely used in the analysis of optoelectronic systems are characterized by the domain of attraction for a wide class of distribution laws. According to Theorem (), (Lifshits, 2007) the distribution function P(x) belongs to the domain of attraction of a normal law if and only if $N \to \infty$:

$$\frac{X^2 \int_{|x|>X} dP(x)}{\int_{|x|<X} x^2 dP(x)} \to 0. \tag{8}$$

In order for the distribution law P(x) to belong to the domain of attraction of a stable law with a characteristic index $0<\alpha<2$, it is necessary and sufficient that according to ():

$$\frac{P(-x)}{1-P(x)} \to \frac{c_1}{c_2} \quad when \quad x \to \infty, \tag{9}$$

at each constant $k>0$

$$\frac{1-P(x)+P(-x)}{1-P(kx)+P(-kx)} \to k^{\alpha} \quad when \quad x \to \infty. \tag{10}$$

Expressions (8) and (10) are the theoretical basis for determining the behavior of the distribution function, which allows us to analyze the asymptotic behavior of the "tails" of the distribution function in a statistically significant sample. The choice of the threshold value X (expression (8)) motivates the analysis of left and right tails of the distribution function, which characterizes the experimental results. The area under the curve of the distribution function at values of a random variable x that does not belong to the interval [-X; X] and a similar area for values of x which lies in the interval [-X; X] are compared. As this ratio tends to 0 when the threshold value $X \to \infty$, it indicates that P(x) belongs to the domain of attraction of the normal law.

The properties of stable laws (Levy's laws) are analyzed in (Goodman, 1988; Lifshits, 2007; Zolotarev, 1983; Zolotarev et al., 1999), and (Uchaikin & Sibatov, 2004).

1.   If the characteristic index $\alpha=2$, then P(x) is reduced to the Gaussian distribution.
2.   2. If the characteristic index $0<\alpha<2$ and $N \to \infty$, then P(x) decreases by the power law with the same exponent and is determined as:

$$P(x) \underset{x \to \infty}{\cong} \frac{\alpha}{x^{1+\alpha}} + O\left(\frac{1}{x^{1+2\alpha}}\right).$$

3.   If the characteristic index $1<\alpha<2$, the function P(x) describes the fluctuations X_N around the mean $\bar{X}$, and therefore extends to the range from $+\infty$ to $-\infty$.

Large Fluctuations (CLT – Central Limit Theorem).

In ordinary statistics, fluctuations obey the standard central limit theorem. The mean $\bar{X}$ and variance $D(X)$ are finite. Fluctuations of the sum X_N in the transition from one sample to another tend to zero when the sample size, i.e. the number of summands N, increases. Fluctuation $\sigma_r(X)$ relative to the mean value $\bar{X}$ is determined for a sample with the size $N \to \infty$ as:

$$\sigma_r(X) = \frac{\langle |(X_N/N)-\bar{X}| \rangle}{\bar{X}}$$

$$D(X)<\infty: \sigma_r(X) \cong \frac{\sigma}{\bar{X}\sqrt{N}}.$$

Therefore, fluctuations tend to zero when N tends to infinity.

Dependence of Levy sums on the number of summands. One of the most important statements of the generalized limit theorem is that for a characteristic index $\alpha<1$ the Levy sum XN is proportional to $X^{1/\alpha}$. For example, if $\alpha=0.5$, then the sum XN increases as $X2$; and if $\alpha=1/4$, it increases as $X4$. The smaller α, the greater the exponent of power-law dependence of XN o$_n$ N.

This behavior of a random variable differs from the behavior of a random variable X with a finite mean value $\overline{X}$, for which the sum XN obeys the law of large numbers and the central limit theorem.

This is because the probability of a large value X is not negligibly small if $P(X)$ decreases very slowly for larger N. As the number N of tests increases, ever greater values of X can be obtained, which explains why the sum X_N can increase faster than N.

THE THIRD SECTION. DETERMINATION OF LIMIT DISTRIBUTIONS OF OUTPUT SIGNALS OF OPTOELECTRONIC SYSTEMS. DOMAINS OF ATTRACTION

To justify the adoption of the model of output signals based on α-stable processes, it is necessary to analyze belonging the limit distributions that characterize the output signal of the optoelectronic system in accordance with expressions (8) and (10) to the domain of attraction of the normal law

$$P(x) = \frac{1}{\sqrt{2\pi}} \int_{-\infty}^{x} e^{-\frac{t^2}{2}} dx$$ at $x \to \infty$ which has *finite variance,* as well as their belonging to the domain of

attraction of stable distribution laws with *infinite variance* (), (Korzhenevsky & Kamzina, 1998), and (Glauber, 1966).

Accordingly, in the first stage of determining the domain of attraction of one of the limit laws, it is necessary to analyze the statistical behavior of the *variance* of a random variable (fluctuations of the output signal of optoelectronic systems with limited dynamic range) depending on the number of measurements.

According to the accepted statistical models, united by the central limit theorem, the process of convergence of the variance D of random value is given for the normal (Figure 11) and Poisson (Figure 12) distribution laws. Different curves on Figures 11 and 12 correspond to different realizations. As can be seen, according to the central limit theorem, the fluctuations of the noise component should tend to zero with increasing number of measurements ($N \to \infty$), i.e. stabilization of the studied value occurs.

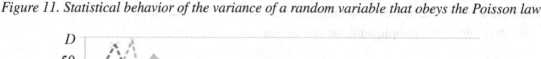

Figure 11. Statistical behavior of the variance of a random variable that obeys the Poisson law

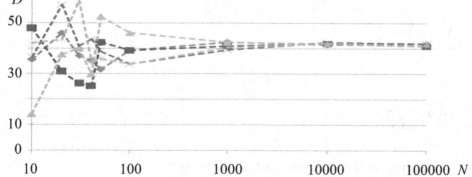

Figure 12. Statistical behavior of the variance of a random variable that obeys the normal law

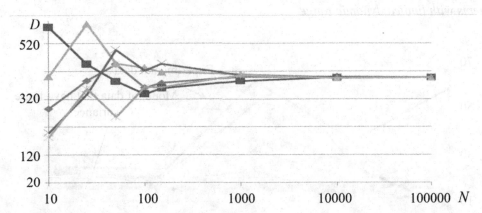

Figure 13 shows the dependence of the behavior of the variance D of random value on the number of analyzed values N specific for stable distribution laws with the characteristic coefficient $0<\alpha<2$.

Figure 13. Statistical behavior of the variance of a random variable that obeys the stable law $\alpha=1$

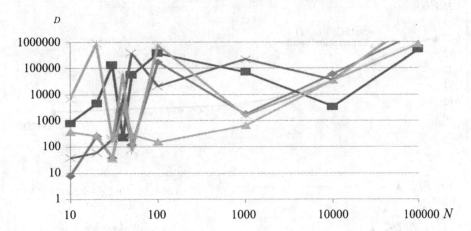

Experimental studies of the attenuation process in optoelectronic systems have shown that with increasing the number of measurements of the responses of the resolution element of the photodetector $N\rightarrow\infty$, stabilization of the variance of the fluctuation component was not observed (Figure 14).

This statistical behavior of the output signals is explained by the appearance in the samples such values of intensity which probability of occurrence is not determined by the central limit theorem. Since the variance of the fluctuation component is *not stabilized*, we can proceed to the next step.

In the second step of determining the domain of attraction for one of the limit laws, it is necessary to check the asymptotic behavior of the "tails" of the distribution of the studied random variable.

Figure 15 shows the distribution densities of experimental data and the limit distribution law with a characteristic index $\alpha=1$. The asymptotic behavior of the "tail" of the distribution density of experimental data in the range from 2 to 5 SD is well approximated by stable laws in terms of least squares criterion.

Figure 14. Statistical behavior of the variance D (in the grey scale units) of output signals of optoelectronic systems with limited dynamic range

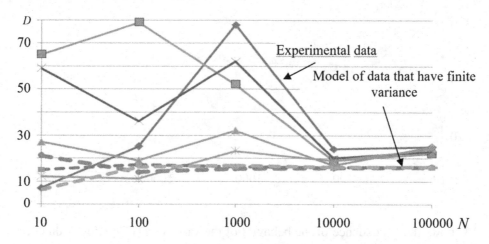

Figure 15. Approximation of experimental data by stable laws; a) attenuator k=5; b) attenuator k=10

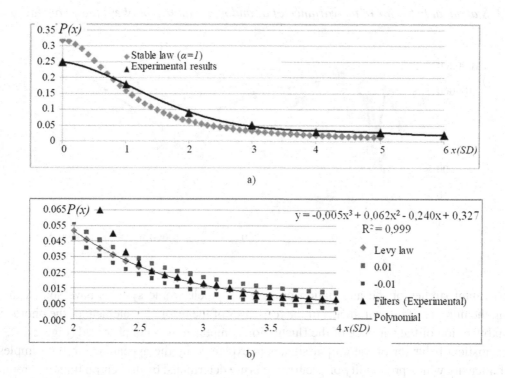

Thus, the analysis of the sum of independent random variables characterizing the output signal of optoelectronic systems that use an attenuator showed that the limit distribution laws of output signals belong to domains of attraction of stable laws.

The stochastic behavior of output signals can be determined on the basis of the model of the interaction of input radiation photons with the optical link of the optoelectronic system. Description and prediction of the properties of the optical link are mainly based on the determination of deterministic parameters, i.e. refractive, reflective, and absorption indices.

When describing the process of interaction of optical radiation with an optical link, optical glass is mainly considered in terms of macroscopic theory of order and symmetry. However, when describing the interaction of light quanta with optical glass material, the fact must be taken into account that in such materials spatial disorder can occur resulting by the presence of chaotically located defects as a consequence of atmospheric influences and aging processes, as well as stochastic layout of color centers in optically dense materials (for example, neutral filters). Disordered materials at the microscopic level are fundamentally heterogeneous and can be different in structure. In various parts of the macroscopically homogeneous optical glass, the processes of photon energy transformation can cause additional fluctuations.

In nonlinear statistically inhomogeneous media, the location and orientation of individual particles (defects) can be considered unknown. Characterizing the interaction of optical radiation with unregulated structures (for example, optical link), we can imagine that photons enter the amorphous material, scattered, reflected, absorbed, activate vibrational processes (photon-phonon interaction) inside and come out of the opposite side. It is rather difficult to describe the process of radiation passing inside the material (Figure 16)

Figure 16. Formation of the spatial distribution of optical radiation in the plane of the photodetector

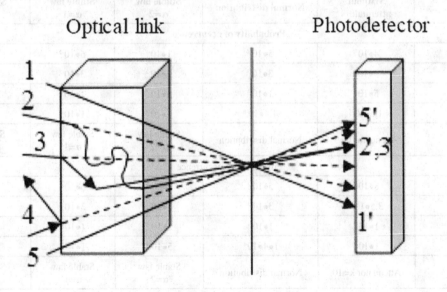

This process can be described as random walks, in which a photon, overcoming a distance in an amorphous substance, can lose some of its energy during absorption, change the direction of motion when interacting with glass defects. A photon passing through a lattice of amorphous substance can excite a phonon in it, discharging part of own energy; as a result, its frequency decreases, and quanta of other energy appear. If a phonon has already been excited in the substance, a flying photon can absorb it, thereby increasing own energy, as a result, photons with higher energy appear. So, due to the release of photons from the material, the optical radiation will have a lower intensity and a different spectral composition. Atmospheric influences and glass aging processes can change scattering and reflective indices and as a consequence lead to the fact that light enters the material, many times reflected inside, changes the trajectory and goes outside not in the expected place, breaking the principle of linear superposition.

Thus, the process of interaction of photons with the optical link of the system can be randomized not only by the intensity level and frequency composition of input and output signals, but also by the trajectory of its motion through the optical link and, consequently, space-time distribution of optical radiation in the photodetector plane.

Table 2 shows a comparative analysis of the probabilities of occurrence for normalized and centered values obtained during experimental studies and limit distribution laws.

Statistical models of output signals, based on Poisson and Gaussian laws and used to describe output signals, agree well with experimental data in the range of up to 4 SD for systems that do not use neutral filters as input attenuators. In the range from 5 SD and more the asymptote of the distribution law of output signal becomes closer to α-stable law with characteristic index $\alpha=2$. Obviously, the use of output signal models based on Poisson and Gaussian laws to detect objects can cause errors if the detection threshold is chosen in the range greater than 5 SD.

Table 2. Probabilities of occurrence for different normalized and centered distribution laws and experimental data

X, SD	Without attenuators	Normal distribution	Stable law $\alpha=2$	Stable law $\alpha=1$	Stable law $\alpha=0.5$
Probability of occurrence					
2σ	$4 \bullet 10^{-2}$	$5 \bullet 10^{-2}$	$1 \bullet 10^{-1}$	$6 \bullet 10^{-2}$	$5 \bullet 10^{-2}$
3σ	$3 \bullet 10^{-3}$	$3 \bullet 10^{-3}$	$3 \bullet 10^{-2}$	$3 \bullet 10^{-2}$	$1 \bullet 10^{-2}$
4σ	$5 \bullet 10^{-4}$	$1 \bullet 10^{-4}$	$5 \bullet 10^{-3}$	$1 \bullet 10^{-2}$	$6 \bullet 10^{-3}$
5σ	$1.6 \bullet 10^{-4}$	$1.4 \bullet 10^{-6}$	$5 \bullet 10^{-4}$	$1 \bullet 10^{-2}$	$3 \bullet 10^{-3}$
X, SD	Attenuator k=5	Normal distribution	Stable law $\alpha=2$	Stable law $\alpha=1$	Stable law $\alpha=0.5$
Probability of occurrence					
2σ	$8 \bullet 10^{-2}$	$5 \bullet 10^{-2}$	$1 \bullet 10^{-1}$	$6 \bullet 10^{-2}$	$5 \bullet 10^{-2}$
3σ	$3.5 \bullet 10^{-2}$	$3 \bullet 10^{-3}$	$3 \bullet 10^{-2}$	$3 \bullet 10^{-2}$	$1 \bullet 10^{-2}$
4σ	$1 \bullet 10^{-2}$	$1 \bullet 10^{-4}$	$5 \bullet 10^{-3}$	$1 \bullet 10^{-2}$	$6 \bullet 10^{-3}$
5σ	$1 \bullet 10^{-2}$	$1 \bullet 4 \bullet 10^{-6}$	$5 \bullet 10^{-4}$	$1 \bullet 10^{-2}$	$3 \bullet 10^{-3}$
X, SD	Attenuator k=10	Normal distribution	Stable law $\alpha=2$	Stable law $\alpha=1$	Stable law $\alpha=0.5$
Probability of occurrence					
2σ	$8 \bullet 10^{-2}$	$5 \bullet 10^{-2}$	$1 \bullet 10^{-1}$	$6 \bullet 10^{-2}$	$5 \bullet 10^{-2}$
3σ	$1 \bullet 10^{-2}$	$3 \bullet 10^{-3}$	$3 \bullet 10^{-2}$	$3 \bullet 10^{-2}$	$1 \bullet 10^{-2}$
4σ	$8 \bullet 10^{-3}$	$1 \bullet 10^{-4}$	$5 \bullet 10^{-3}$	$1 \bullet 10^{-2}$	$6 \bullet 10^{-3}$
5σ	$5 \bullet 10^{-3}$	$1.4 \bullet 10^{-6}$	$5 \bullet 10^{-4}$	$1 \bullet 10^{-2}$	$3 \bullet 10^{-3}$

The asymptotic behavior of the output signal distribution function in optoelectronic systems with a limited dynamic range varies depending on the neutral absorption coefficient of the filter. With increasing an absorption coefficient, the probability densities of output signals enter the domains of attraction of stable laws with a characteristic index of $0<\alpha<2$.

Thus, when registering optical radiation from optoelectronic systems, the model of output signals based on stable laws can be accepted. The use of such a model will avoid conflict between experimental data and accepted mathematical models of the output signals, and can be a theoretical basis for creating highly efficient signal processing algorithms.

As a result of the experiment, the distribution densities were obtained for output signals without the use of attenuators and with the use of neutral filters in the system (Figure 17).

Figure 17. Probability densities of output signals a) normal distribution law; b) stable law $\alpha=1$; c) experimental results of signal attenuation in optoelectronic systems that use neutral filters

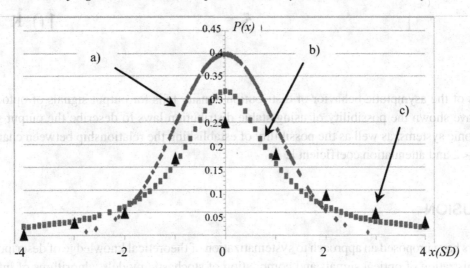

The results of the analysis of experimental studies of the attenuation of optical fluxes in optoelectronic systems, which included the study of the statistical characteristics of the output signals of optoelectronic systems using various neutral filters to attenuate the input optical fluxes, make it possible to characterize the relationship between the limit distribution law of fluctuations of the output signal and the attenuation coefficient of neutral filter k, which is a part of the optical link of the system (Figure 18).

Figure 18. Experimental relationship between characteristic index 0<α<2 and attenuation coefficient of optical link k

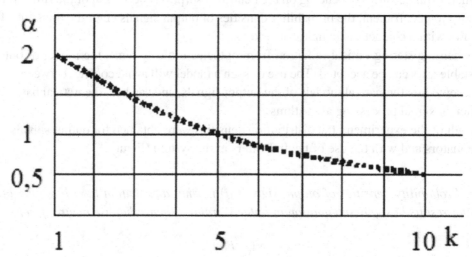

Studies of the asymptotic behavior of distribution density tails for output signals of optoelectronic systems have shown the possibility of using stable distribution laws to describe the output signals of optoelectronic systems, as well as the possibility of establishing the relationship between characteristic index $0<\alpha<2$ and attenuation coefficient k.

CONCLUSION

The authors have proposed an approach to systematization of theoretical knowledge of description methods for registration of optical signals and compilation of stochastic models, algorithms of information processing in optoelectronic systems using combined corpuscular and wave description of optical signals.

Consideration of the development strategy in optoelectronic technologies which is aimed at improving the physical mathematical and stochastic models of signal description, assumes that its main task is to optimize the energy calculations of optical signals. Therefore, there is a need to develop a new methodology for describing stochastic signal behavior in systems.

The application of this methodology will facilitate the transition from fundamental studies in the field of signal reception and processing to applied problems in the fields of applied optics, astronomy, biology, medicine, ecology, pharmacy, optical control and microscopy, creation and research of the latest nanomaterials, information technologies of data processing in optoelectronic systems.

Stochastic and deterministic theory of signal reception and processing in optoelectronic systems includes issues related to the study and consideration of the processes of stochastic signal transformation by the basic system elements, as well as optimization of the processes of signal detection against a background of noise. The stochastic process of output signal generation in optoelectronic systems can be considered taking into account the photonic nature of electromagnetic radiation and related to the Poisson law. However, stochastic absorption and transmission processes may affect the timing and spatial coordinates of the photo detector response and may not correspond to the input stream. In this case, the deterministic relationship between the stochastic input stream and the stochastic output stream is broken. Changing photon statistics can have a significant impact during application run. From a mathematical point of view, stochastic systems with controllable parameters create new problems. To overcome these problems, the authors propose to use structural approximation methods, when a complex system is replaced by a sequence of simpler ones, which analysis allows approximating the characteristics of the original system. This approach has proven its effectiveness in the implementation of several international projects dedicated to the study of different classes of stochastic systems.

REFERENCES

Baouche, F. Z., Hobar, F., & Hervé, Y. (2013). Level Modeling in Optoelectronic Systems; Transmission Line Application. *International Journal of Advances in Engineering and Technology*, 6(1), 392–404.

Bardoux, F., Bushko, J.-F., Aspe, A., & Cohen-Tannuji, K. (2006). *Levy Statistics and Laser Cooling. How rare events stop atoms*. FIZMATLIT.

Barthelemy, P., Bertolotti, J., & Weirsma, D. S. (2008). Levy flight for light. *Nature*, 453(7194), 495–498. doi:10.1038/nature06948 PMID:18497819

Bedard, G. (1967). Analysis of light fluctuations from photon counting statistics. *Journal of the Optical Society of America*, 57(10), 1201–1203. doi:10.1364/JOSA.57.001201

Chandrasekhar, S. (1943). Stochastic Problems in Physics and Astronomy. *Reviews of Modern Physics*, 15(1), 1–89. doi:10.1103/RevModPhys.15.1

Cox, D., & Lewis, P. (1969). *Stochastic analysis of sequences of events*. Moscow. WORLD (Oakland, Calif.), 310.

Davis, A. B., Marshak, A., & Pfeilsticker, K. P. (1999). Anomalous/Lévy Photon Diffusion Theory: Toward a New Parameterization of Shortwave Transport in Cloudy Columns. *Ninth ARM Science Team Meeting Proceedings*, 1-13.

Davis, A. B., Suszcynsky, D. M., & Marshak, A. (2000). Shortwave Transport in the Cloudy Atmosphere by Anomalous/Lévy Diffusion: New Diagnostics Using FORTÉ Lightning Data. *Tenth ARM Science Team Meeting Proceedings*, 1-18.

Denisov, V. I., & Timofeev, V. S. (2011). Sustainable Distributions and Estimation of Parameters of Regression Dependencies. Izvestia, Tomsk Polytechnic University, 318(2), 10-15.

Engel, K. J., Steadman, R., & Herrmann, C. (2012). Pulse Temporal Splitting in Photon Counting X-Ray Detectors. *IEEE Transactions on Nuclear Science*, 59(4), 1480–1490. doi:10.1109/TNS.2012.2203610

Fedoseev, V. I. (2011). Reception of spatio-temporal signals in optoelectronic systems (Poisson model). Moscow, Russia: University Book.

Fedoseev, V. I., & Kolosov, M. P. (2007). Optoelectronic devices for orientation and navigation of spacecraft. Academic Press.

Field, D. J. (1987). Relations between the statistics of natural images and the response properties of cortical cells. *Journal of the Optical Society of America. A, Optics and Image Science, 4*(12), 2379–2394. doi:10.1364/JOSAA.4.002379 PMID:3430225

Flores-Fuentes, W., Rivas-Lopez, M., & Sergiyenko, O. (2013). Digital Signal Processing on Optoelectronic for SHM. *Proceedings of the World Congress on Engineering and Computer Science.*

Gikhman, I. I., & Skorokhod, A. V. (1977). *Introduction to the theory of random processes.* Science.

Glauber, R. (1966). *Optical coherence and photon statistics.* World.

Gnedenko, B. V. (1950). Areas of attraction of the normal law. *Reports of the USSR Academy of Sciences, 71*(3), 425–428.

Gnedenko, B. V., & Kolmogorov, A. N. (1949). *Limit distributions for sums of independent random variables* (Vol. M). GITTL.

Goodman, J. (1988). *Statistical Optics.* World.

Gracheva, M. E., Gurvich, A. S., & Kallistratova, M. A. (1970). Measurement of the dispersion of" strong "fluctuations in the intensity of laser radiation in the atmosphere. *Izv. Universities. Radiophysics, 13,* 56.

Habasaki, J., Okada, I., & Hivatari, Y. (1997). Fraction excitation and Levy flight dynamics in alkali silicate glasses. *Physical Review, 55*(10), 6309–6315. doi:10.1103/PhysRevB.55.6309

Harvey, J. E., Choi, N., Krywonos, A., Peterson, G., & Bruner, M. (2010). *Image degradation due to scattering effects in two-mirror telescopes.* https://opticalengineering.spiedigitallibrary.org/article.aspx?articleid=1096366

Ibragimov, I. A., & Linnik, Yu. V. (1965). *Independent and stationary related quantities.* Science.

Imai. (1986). Statistical Properties of Optical Fiber Speckles. *Bulletin of the Faculty of Engineering, Hokkaido University, 130,* 89-10.

Johnson, D. H. (n.d.). *Statistical Signal Processing.* http://www.ece.rice.edu/~dhj/courses/elec531 / notes.pdf

Khoroshun, G., Riazantsev, A., Ryazantsev, O., Popiolek-Masajada, A., & Strelkova, T. (2020). Estimation of the image quality and noise forinformation system. *Scientific and practical conference "Now Technologies In Nauts Ta Osviti", 107.*

Klyatskin, V. I. (1975). Statistical description of dynamical systems with fluctuating parameters. *Science, 239.*

Klyshko, D. N. (1996). The nonclassical light. *Успехи физических наук, 166*(6), 613–638. doi:10.3367/UFNr.0166.199606b.0613

Kolyadin, V. L. (2002). Distributions with infinite dispersion and the limitations of classical statistics. *Wuxiandian Gongcheng,* (2), 4–11.

Korzhenevsky, A. L., & Kamzina, L. S. (1998). Anomalous diffusion of light in ferroelectrics with a diffuse phase transition. *Solid State Physics, 40*(8), 1537–1541. doi:10.1134/1.1130566

Kumar, N., Harbola, U., & Lindenberg, K. (2010). *Memory-induced anomalous dynamics: emergence of diffusion, subdiffusion, and superdiffusion from a single random walk model.* https://journals.aps.org/pre/abstract/10.1103/PhysRevE.82.021101

Kupchenko, L. F. (2009). *Acousto-optical effects with strong interaction. Theory and experiment (Continuous fraction method for solving acousto-optic problems).* EDENA.

Lappa, A. V., Bakhvalov, E. V., & Anikina, A. S. (2004). The method of characteristic functions in estimating the mathematical expectation of random variables with infinite dispersion. *Izvestia Chelyabinsk Scientific Center, 2*, 1–6.

Levy, P. (1972). *Stochastic processes and Brownian motion.* Science.

Lifshits, M. A. (2007). *Stable distributions, random variables and processes.* SPb.

Lytyuga, A. P. (2009). A mathematical model of signals in television systems when observing low-orbit space objects in the daytime. *Scientific Works of Kharkiv National Air Force University, 4*(22), 41–46.

Lytyuga A. P. (2010). Algorithms for Detecting Optical Signals from Low-Orbit Space Objects in the Daytime. *Information Processing Systems, 4*(22), 41-46.

Metzler, R., & Klafter, J. (n.d.). *The random walk's guide to anomalous diffusion: a fractional dynamics approach.* https://www.tau.ac.il/~klafter1/258.pdf

Milyutin, E. R., & Gumbinas, A. Y. (2002). *Statistical Theory of the Atmospheric Channel of Optical Information Systems.* Radio and Communications.

Mosyagin, G. M., Nemtinov, V. B., & Lebedev, E. N. (1990). *Theory of Optoelectronic Systems.* Engineering.

Nikitin, V. M., Fomin, V. N., Nikolaev, A. I., & Borisenkov, I. L. (2008). *Adaptive Interference Protection of Optoelectronic Information Systems.* BelSU Publishing House.

Noguez, C., & Ulloa, S. E. (1997). First-principles calculations of optical properties: Application to silicon clusters. *Physical Review. B, 56*(15), 9719–9725. Advance online publication. doi:10.1103/PhysRevB.56.9719

Nowak, R. D., & Kolaczyk, E. D. (2000). A Statistical Multiscale Framework for Poisson Inverse Problems. *IEEE Transactions on Information Theory, 46*(5), 1811–1825. doi:10.1109/18.857793

Parfenov, V. I., & Kirillov, V. S. (2012). Detection of optical signals when receiving a stream of photoelectrons with an unknown density form. *Computer Optics, 36*(4), 618–622.

Pedersen, K. S., Duits, R., & Nielsen, M. (2005). On α Kernels, Levy Processes, and Natural Image Statistics. In *Scale Space and PDE Methods in Computer Vision* (pp. 468–479). SpringerVerlag. doi:10.1007/11408031_40

Prokhorov, Yu. V., & Rozanov, Yu. A. (1973). *Probability Theory.* Science.

Romanovsky M. Yu. (2007). Analytical representations of non-Gaussian random walks. *Actual problems of statistical physics (Malakhovsky digest), 63*, 56-81.

Romanovsky, M. Yu. (2009). Levy Functional Walks. *Proceedings of the Institute of General Physics named after A.M. Prokhorova, 65,* 20-28. 10.3103/S1541308X09030078

Rozovsky, L. V. (n.d.). Probabilities of large deviations of sums of independent random variables with a common distribution function from the domain of attraction of the normal law. *Probability Theory and Its Application, 4,* 686-705.

Ruderman, D. L., & Bialek, W. (1994). Statistics of natural images: Scaling in the woods. *Physical Review Letters, 73*(6), 814–817. doi:10.1103/PhysRevLett.73.814 PMID:10057546

Rudoi, Yu. G. (2007). Random walks and anomalous diffusion of Levi-Khinchin in the physical chemistry of polymers. *Science - the foundation for solving technological problems of the development of Russia, 2,* 74-101.

Ruelle, D. (2001). *Chance and Chaos.* Izhevsk: Research Center. *Regular and Chaotic Dynamics.*

Sabathil, M. (2005). *Opto-electronic and quantum transport properties of semiconductor nanostructures. In Selected Topics of Semiconductor Physics and Technology (* Vol. 67). Verein zur Förderung des Walter Schottky Instituts der Technischen Universität München.

Šanda, F., & Mukamel, S. (2006). Anomalous continuous-time random-walk spectral diffusion in coherent third-order optical response. *Physical Review E: Statistical, Nonlinear, and Soft Matter Physics, 73*(1), 1–14. doi:10.1103/PhysRevE.73.011103 PMID:16486118

Seitz. (2011). *Single-Photon Imaging.* Springer – Verlag Berlin Heidelberg.

Sibatov, R. T., & Uchaikin, V. V. (2007). Fractional differential kinetics of charge transfer in disordered semiconductors. *Physics and Technology of Semiconductors, 41*(3), 346–357.

Strelkov, A. I., Lytuga, A. P., & Strelkova, T. A. (2013). Prospects for the development of special-purpose optoelectronic systems using the methods of stochastic-deterministic processing of optical signals. *Int. scientific-practical conf. Actual problems and prospects of development of radio engineering and info-communication systems,* 50-53.

Strelkov, A. I., Moskvitin, S. V., Lytyuga, A. P., & Strelkova, T. A. (2010). *Optical location. The theoretical basis of the reception and processing of optical signals.* Apostrophe.

Strelkov, A. I., Stadnik, A. M., Lytyuga, A. P., & Strelkova, T. A. (1998). Comparative Analysis of Probabilistic and Determinate Methods for Attenuating Light Flux. *Telecommunications and Radio Engineering, 52*(8), 54–57. doi:10.1615/TelecomRadEng.v52.i8.110

Strelkova, T., Kartashov, V., Lytyuga, A., & Strelkov, A. (2016). Theoretical methods of images processing in optoelectronic systems. In *Developing and Applying Optoelectronics and Machine Vision* (pp. 181-206). https://www.igi-global.com/book/developing-applying-optoelectronics-machine-vision/147652

Strelkova, T. A. (2014a). Studies on the Optical Fluxes Attenuation Process in Optical-electronic Systems. *Semiconductor Physics, Quantum Electronics & Optoelectronics (SPQEO), 4,* 421–424.

Strelkova, T. A. (2014b). Statistical properties of the output signals of optical-television systems with a limited dynamic range. *East European Journal of Advanced Technologies, No., 2/9*(68), 38–44.

Strelkova, T. A. (2015). The use of stable distribution laws in evaluating the efficiency of signal processing in optoelectronic systems. *East European Journal of Advanced Technologies, No.*, 2/9(74), 4–9.

Strelkova, T. A., & Sautkin, V. A. (2013). A stochastic approach to assessing the quality of optical glass. Scientific-practical. conf. Technologies for processing optical elements and applying vacuum coatings, 85-86.

Strelkova, T. A., Sozonov, Yu. I., & Yanovsky, Yu. A. (2012). Study of the statistics of spatio-temporal signals in optoelectronic systems. *All-Ukrainian Interdepartmental Scientific and Technical Journal Radiotehnika, 170*, 185–188.

Strikova, T.O., Lytyuga, O.P., Skorupski, K. & Bugubayeva, A. (2019). Stochastic deterministic methods for processing signals and images in optical electronic systems. *Proc. SPIE 11176, Photonics Applications in Astronomy, Communications, Industry, and High-Energy Physics Experiments.* doi:10.1117/12.2536605

Strilkova, T. O., & Lytyuga, O. P. (2018). *Stochastic-deterministic methods of signals and images processing in optpelectronic systems. Optoelectronic information technology PHOTONIKA-ODS-2018.* Vinnitsya.

Takao, A., Yasuda, M., & Sawada, K. (1995). Noise Suppression Effect in an Avalanche Multiplication Photodiode Operating in a Charge Accumulation Mode. *IEEE Transactions on Electron Devices, 42*(10), 1769–1774. doi:10.1109/16.464420

Tatarsky, V. I. (1967). *Wave propagation in a turbulent atmosphere.* Science.

Uchaikin V. V. (2003). Self-similar anomalous diffusion and stable laws. *Uspekhi Fizicheskikh Nauk., 173*(8), 847-876.

Uchaikin, V. V., & Korobko, D. A. (1999). On the theory of multiple scattering in a fractal medium. *Letters in ZhTF, 25*(1), 34–40. doi:10.1134/1.1262508

Uchaikin, V. V., & Sibatov, R. T. (2004). One-Dimensional Fractal Walking with a Finite Speed of Free Motion. *Letters in ZhTF, 30*(8), 27–33.

Vetnzel, E. S., & Ovcharov, L. A. (2000). *Theory of random processes and its engineering applications.* Higher. Shk.

Vidov, P. V., & Romanovsky, M. Yu. (2007). Analytical representations of non-Gaussian laws of random walks. Actual problems of statistical physics (Malakhovsky digest), 63, 3-19.

Wentzel, E. S., & Ovcharov, L. A. (1988). *Theory of random processes and its engineering applications.* Science.

Wetzel, B., Blow, K. J., Turitsyn, S. K., Millot, G., Larger, L., & Dudle, J. M. (2012). Random walks and random numbers from supercontinuum generation. *OSA Optics Express, 20*(10), 11143–11152. doi:10.1364/OE.20.011143 PMID:22565737

Yang, F., Lu, Y. M., Sbaiz, L., & Vetterli, M. (2012). Bits From Photons. Oversampled Image Acquistion Using Binary Poisson Statistics. *IEEE Transactions on Image Processing, 21*(4), 1421–1436. doi:10.1109/TIP.2011.2179306 PMID:22180507

Zolotarev, V. M. (1983). *One-dimensional stable distributions.* Science.

Zolotarev, V. M. (1984). *Stable laws and their application*. Knowledge.

Zolotarev, V. M., Uchaykin, V. V., & Saenko, V. V. (1999). Superdiffusion and Stable Laws. *Soviet Physics, JETP, 115*(4), 1411–1425.

Chapter 6
Informational Model of Optical Signals and Images in Machine Vision Systems

Oleksandr Ryazantsev
Volodymyr Dahl East Ukrainian National University, Ukraine

Ganna Khoroshun
Volodymyr Dahl East Ukrainian National University, Ukraine

Andrii Riazantsev
Volodymyr Dahl East Ukrainian National University, Ukraine

Tatyana Strelkova
Kharkiv National University of Radio Electronics, Ukraine

ABSTRACT

The rapid development and use of optical systems in measurement, navigation, and space technology to obtain accurate and detailed information about an object of observation is accompanied by the problems of transmitting high quality information through the optical system and processing of the obtained data. Integration of artificial intelligence systems in industry requires the creation and improvement of objective assessment and self-assessment systems. This is especially designed for automated recognition and classification systems. The problem of the object movement registration also contains some peculiarities such as background and main signal separation, noise influence and main objects selecting. Information about data quality is a set of properties that reflects the degree of suitability of specific information. It contains the data about objects and their relationship to achieve the goals of user requirements.

The chapter is focused on a wide class of problems associated with signal propagation through optical systems and material media, including an interaction of a laser radiation with the environment. The employment of Machine Vision Systems (MVSs) (Sinha, 2012; Sergiyenko et al., 2020; Flores-Fuentes et al., 2020) for acquisition, inspection, evaluation, and processing of optical images offers a great solu-

DOI: 10.4018/978-1-7998-6522-3.ch006

tion providing the highest accuracy of measurements and reliability of the data obtained. The MVS is a useful tool for independent laboratories and laboratories owned by research institutes and universities. It is considered such topic as the design of optical setup for direct measurements of laser light features and the parameters of micro objects in the reflected light. The software fabrication for micro particles is represented; hardware for optical elements managing is discussed. The implementation of the model of service rendering in using optical laboratory in the conditions of individual needs of the customer is analyzed through the chapter. Also, the ways of the future development of MVSs is considered with application in different areas of science and technology, such as encryption (Alfalou & Brosseau, 2009), holography (Nehmetallah & Banerjee, 2012; Testorf & Lohmann, 2008), identification and tracking (Alfalou & Brosseau, 2010; Alfalou & Brosseau, 2013; Rosen et al., 2009).

The Machine Vision System represents a complete toolbox for solving the automated vision problems and can be considered wider than the automated optical information systems described earlier (Riazantsev et al., 2019). The MVS (Figure 1) contains the optical source which is a laser or an incoherent light source, or represents a combination of several sources; optical elements, including lenses, filters and spatial light modulator for the formation of the required light-field pattern and intensity; a CCD or CMOS camera for acquisition and conversion of optical radiation into electrical signals; a PC for the data visualization; a software for data processing and analyzing; a protocol suite with algorithms what to do for the software and hardware with a final decision of conceptual framework of the system's operation. Actually, the MVS is an adaptive system with feedback due to which the iterative improvement of the signal processing can be realized. The quality of the image experiences the influence of fluctuations, instabilities, and the system aberrations which makes the accuracy of measurements not higher than 10%.

Figure 1. A general structure of the Machine Vision System for optical research realization

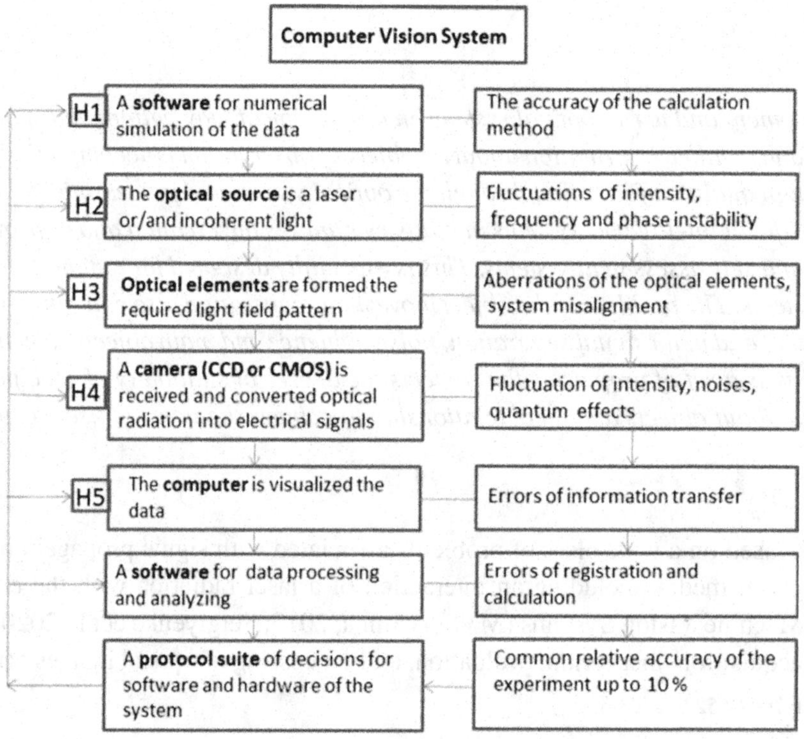

164

MODEL OF SERVICE RENDERING IN USING OPTICAL LABORATORY IN THE CONDITIONS OF INDIVIDUAL NEEDS OF THE CUSTOMER

Development of data object design with application to the intensity distributions for controlling micro- and nanoparticles was the goal of some previous studies (Khoroshun et al., 2020a; Bekshaev et al., 2017). A model of providing services relating to the use of an optical laboratory in view of the individual customer needs is presented in (Khoroshun, 2019).

Consider a possible model for the MVS arrangement in an optical laboratory with full-service package including the following components: theoretical calculations, experiment, comparative analysis of theoretical and experimental data, the decision of the price/quality relation according to the customer's requirements. The features of the service system can be formalized in terms of linguistic variables similar to the system described in works (Pronina, 2018a; Pronina, 2018b). Consider the common optical task of the image registering using laser radiation.

Formally, the order can be represented by a tuple:

$$Z = \left\langle \{C\}, \left\{p_l^C\right\}_{l=1}^5, \{T\}, \left\{p_b^T\right\}_{b=1}^7, \{E\}, \left\{p_d^E\right\}_{d=1}^8, \right.$$
$$\left. \{D\}, \left\{p_f^D\right\}_{f=1}^6, \{S\}, \left\{p_g\right\}_{g=1}^4 \right\rangle \tag{1}$$

$\{C\}$ the set of input data "customers"

$\left\{p_l^C\right\}_{l=1}^5$ the set of parameters describing the customer's requirements

$\{T\}$ the set of "theory"

$\left\{p_b^T\right\}_{b=1}^7$ the set of parameters characterizing the theoretical calculations

$\{E\}$ the set of "experiment"

$\left\{p_d^E\right\}_{d=1}^8$ the set of parameters describing characteristics of the experiment

$\{D\}$ the set of "data analysis"

$\left\{p_f^D\right\}_{f=1}^6$ the set of parameters describing characteristics of the data processing

$\{S\}$ the set of output data "services"

$\left\{p_g\right\}_{g=1}^4$ the set of parameters describing the services

The set of "customers" $\{C\}$. includes the following parameters $\left\{p_l^C\right\}_{l=1}^5$:

p_1^C the image required features for production

p_2^C the catalog of the system parameters with their physical characteristics

p_3^C the time necessary for the MVS completion

p_4^C the quality of the data visualization

p_5^C the cost range of MVS procedure

The set of performers consists of three types of performers who carry out theoretical studies, experimental measurements and comparative analysis of the processed data. In particular, the set of "theory", $\{T\}$. contains the parameters $\left\{ p_b^T \right\}_{b=1}^{7}$:

p_1^T the name of the person who makes calculations

p_2^T the method of calculation

p_3^T error caused by approximations

p_4^T theoretical images or results

p_5^T recommendation for the experimenter

p_6^T time of doing job

p_7^T cost of the theoretical investigation

The features $\left\{ p_d^E \right\}_{d=1}^{8}$ of the "experiment" set $\{E\}$.are as follows:

p_1^E the name of the person who is responsible for conducting experiment

p_2^E unites the element names

p_3^E costs of the elements

p_4^E experimental images or values of some quantity

p_5^E the total relative error of the experiment

p_6^E time necessary for conducting the experiment

p_7^E hardware managing system

p_8^E cost of the experimental investigation

The set $\{D\}$ of the "data processing" contains the following features $\left\{ p_f^D \right\}_{f=1}^{6}$:

p_1^D the algorithm of experimental data processing

p_2^D the special software for image analysis

p_3^D the information about the characteristics of the image quality

p_4^D the information of the image quality criteria

p_5^D the time expenses necessary for the service development

p_6^D the service price

The tuple completes a plurality of "services"$\{S\}$ with a set of service features $\left\{ p_g^S \right\}_{g=1}^{4}$:

p_1^S the cost of the MVS creation

p_2^S the quality of the data visualization process

p_3^S the efficiency of the particle registration by MVS

p_4^S the MVS creation time

We formalize the basic features of services $p_1^S, p_2^S, p_3^S, p_4^S$, using linguistic variables $\beta1, \beta2, \beta3, \beta4$, which are defined by tuples $\langle \beta1, T(\beta_1), X\rangle, \langle \beta2, T(\beta2), X\rangle, \langle \beta3, T(\beta3), X\rangle, \langle \beta4, T(\beta4), X\rangle$ accordingly. For $\beta1$, "cost of service", $T(\beta1) = \{$"low", "medium", "high"$\}$, X=[Xmin,Xmax] where values Xmin and Xmax are determined by the customer. For other variables: $\beta2$, "visualization quality", $T(\beta2)=\{$"low", "medium", "good", "excellent"$\}$; $\beta3$, "efficiency of the particle registration", $T(\beta3)=\{$"low", "medium", "high"$\}$; $\beta4$, "MVS creation time", $T(\beta4)= \{$"slow", "average", "fast"$\}$. For every variable X=[Xmin,Xmax] where values Xmin and Xmax are defined by the client.

It is important to emphasize, that automatization of every set performing and artificial intelligence implementations are the fruitful ways of modern machine vision system development.

CALCULATED AND EXPERIMENTAL DATA FOR THE CASE "A" AND CASE "B"

It is important to know that the acquisition of the optical image can be organized in two ways which are named "case A" and "case B". The first one, case A, is the direct measurement of a laser radiation profile by a camera matrix without an objective lens. It is used for registration of the intensity pattern including the interference and diffraction patterns of the perturbed field. The measurement is very accurate because the objective lens, which usually induces a sort of aberration, is absent. But the method requires that the measuring unit (matrix) be put immediately "inside" the beam cross-section in Figure 2.

The case B is based on observation of illuminated objects by detectors positioned "outside" the beam as in Figure 3. Such kind of registration needs an objective lens or some other optical system, which provides a high-quality image of the laser beam cross-section at a desirable position under the necessary viewing angle and with necessary resolution. In this chapter, optical images obtained by both techniques "case A" and "case B" are analyzed.

Figure 2. The scheme of the field registration for the case "A" directly by the CCD without objective lens.

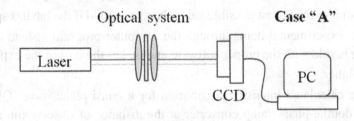

Figure 3. The scheme of the micro object registration for the case "B" by the CCD with objective lens.

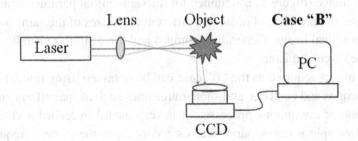

The required optical image is characterized by different types of physical parameters p_1^C, which are beam intensity and phase distributions and by the system parameters p_2^C identifying the laser radiation wavelength, the beam waist radius, initial distribution of the beam phase and intensity and the image observation range of distances z.

Let us consider the problem of a laser beam propagation through an optical system. The method of calculation is described by the parameter p_2^T. The diffraction is the standard phenomenon which occurs when the beam passes via a series of optical elements; the corresponding beam transformation is expressed by the transverse complex amplitude distribution A(x1,y1,z1) and can be analyzed using the Kirchhoff integral (Matveev, 1988):

$$A(x_1, y_1, z) \sim \iint A_0(x, y) \exp(i\Phi_0(x, y) + if(x - x_1, y - y_1)) dxdy \tag{2}$$

Where A0(x,y) is an initial amplitude distribution in the transverse plane XY where the diffraction element is placed, Φ0(x,y) is the initial phase in the plane of the diffraction element, f(x–x1, y–y₁) is the function describing the diffraction phenomenon, and the diffraction pattern is observed at a distance z behind the screen – at the plane X1Y1. The error of the method p_3^T is depend on the way of numerical integration and the number of points Variation of the initial conditions and the distance of registration allow to obtain set of p_4^T - theoretical images and results. Due to analysis of the results it is easy to find the optimal parameters for setup and to fix the recommendation p_5^T to the experimenter. In a physical experiment, an immediately measurable quantity is the intensity which equals to the squared amplitude:

$$I(x_1, y_1, z) = \left[A(x_1, y_1, z) \right]^2 \tag{3}$$

Distributions of the intensity or values of some quantity are the main result of the research that is corresponded to p_4^E parameter. The total relative error of the experiment p_5^E should be less than 10% and is strived in the limit to the lowest possible value. It is very good if the lab has special hardware p_7^E to shift and rotate the experimental items through the computer program instead of manual job. Usually special hardware is sold with the proper software and makes the cost of the experimental investigation p_8^E essentially higher.

Now consider the calculated intensity distribution for a quasi-plane-wave (QPW) laser radiation passing through the double-phase-ramp converter at the distance of observation z=40cm (Figure 2). The figure contains maxima – bright peaks, minima – dark-grey areas, and isolated zeroes of intensity looking as black elliptical spots on the central horizontal line (Figure 4).

The experimental image (Figure 5) is obtained for the same initial parameters as the theoretical one (Figure 4). One can see the presence of noise and differences in sizes of the same topological objects as are present in the theoretical image. Generally, figures 4 and 5 are similar visually, and more details of their structure will be discussed later.

Any image of an object acquired as the "B" case can be obtained using optical devices: long-range devices such as telescopes and cameras, and short-range ones such as magnifiers and microscopes. For the optical observation of continuous processes, it is very useful to record a video-file during some period of time and then split it into separate frames according to the camera frequency. For example,

Figure 4. Calculated diffracted field at a distance z=40 cm behind the diffractive element.

Figure 5. Experimental image at a distance z=40 cm behind the diffractive element.

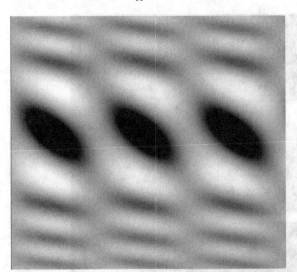 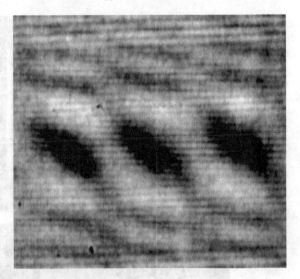

equipment for registration of the microparticle motion contains the laser, optical system, a cell with the particles, micro-objective, filters, and camera, as well as holders for fixing each element on the single optical stage. The video files of the microparticles movement are recorded by the camera. The snapshot

Figure 6. The frame of the video file demonstrating a part of the experimental setup with the laser and the cell containing suspended microparticles inside.

Figure 7. The frame of the video file of the experimental setup with the laser and the cell containing suspended microparticles inside is recorded using a filter

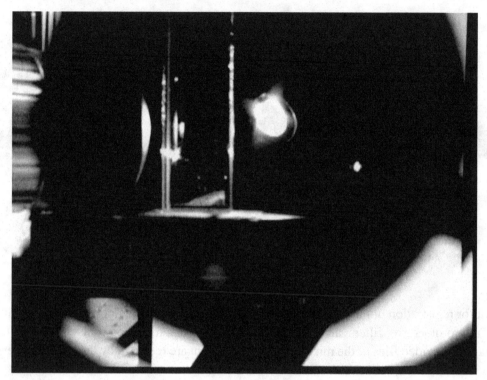

presented in Figure 6 is obtained in the setup without an additional source of light and filter. In contrast, the snapshot of Figure 7 is acquired with using additional external illumination and a filter.

DATABASES FOR THE STRUCTURE DESCRIPTION AND THE PROCESS PRESENTATION

The tuple describing the common optical task of the image registering using laser radiation is presented in formula (1) contains big volume of information. Received data during the experiment implementation and the data relating the initial parameters of the setup and its environment have to be saved in a conveniently and simply way for users. This is realized using the databases with the main purpose to store information in the sets of input $\{C\}$ data and output $\{S\}$ data, process of making research using theoretical $\{T\}$, experimental $\{E\}$ and data processing $\{D\}$ tools. The additional software is allowed to retrieve, transmit, and manipulate information according to observed results.

It is considered as an example of the experimental data for the optical field and its features in the Table I. The main topological objects of the field – the intensity maxima, minima, and zeroes – are presented. The total number of topological objects is expressed by the set N={N1, N2, N3}. It can be found via the preliminary express analysis based on the image dividing into segments S of the same size. The size of the segment is determined by the value of a significant area. To complete the analysis process, time T, exact coordinates B and E, and the objects' shape characteristics H and A at the initial and final moments of the event are necessary. The object's shape can be described by the degree of the radial symmetry of the intensity distribution (Khoroshun, 2010), the ellipticity, and the orientation angle of equal-intensity ellipses (Khoroshun et al., 2016) in the object area, and by the topological charge m. The processing of the image, recording its characteristics, and searching of the diffraction field topological objects with a

Table 1. Database for the main topological objects of the light field structure description

Object Feature	Max	Min	Ov
1. Total number, N	N1	N2	N3
2. Segment, S	$\left\{(I1)_k^{max}\right\}_{k=1}^{N1}$	$\left\{(I2)_l^{min}\right\}_{l=1}^{N2}$	$\left\{(I3)_p^{ov}\right\}_{p=1}^{N3}$
3. Process duration time dt, T	$\left\{(tb1,te1)_k^{max}\right\}_{k=1}^{N1}$	$\left\{(tb2,te2)_l^{min}\right\}_{l=1}^{N2}$	$\left\{(tb3,te3)_p^{ov}\right\}_{p=1}^{N3}$
4. Coordinates of the event begun at t_0, B	$\left\{(xb1,yb1,zb1)_k^{max}\right\}_{k=1}^{N1}$	$\left\{(xb2,yb2,zb2)_l^{min}\right\}_{l=1}^{N2}$	$\left\{(xb3,yb3,zb3)_p^{ov}\right\}_{p=1}^{N3}$
5. Coordinates of the event finished at t_1, E	$\left\{(xe1,ye1,ze1)_k^{max}\right\}_{k=1}^{N1}$	$\left\{(xe2,ye2,ze2)_l^{min}\right\}_{l=1}^{N2}$	$\left\{(xe3,ye3,ze3)_p^{ov}\right\}_{p=1}^{N3}$
6. Shape parameters of the object at t_0, H	$\left\{(b1,b1,b1)_k^{max}\right\}_{k=1}^{N1}$	$\left\{(b2,b2,b2)_l^{min}\right\}_{l=1}^{N1}$	$\left\{(b3,b3,b3)_p^{ov}\right\}_{p=1}^{N3}$
7. Shape parameters of the object at t_1, A	$\left\{(e1,e1,e1)_k^{max}\right\}_{k=1}^{N1}$	$\left\{(e2,e2,e2)_l^{min}\right\}_{l=1}^{N1}$	$\left\{(e3,e3,e3,m)_p^{ov}\right\}_{p=1}^{N3}$

given criterion should be automated.

A similar table can be constructed for any characteristic objects of the optical image taking into account the specific description task and the particular aim of the experiment.

THE DATA PROCESSING

The set {D} contains the data about the steps of image or video processing in p_1^D and special software for image analysis in p_2^D. As the result of the data processing should be retrieved the information p_3^D about the characteristics of the image quality.

The video file obtained during an experiment is usually divided into separate shots which are chosen in agreement with the several steps according to the scheme of Figure 4 before the final information is extracted from them. The preliminary analysis of the image, which can be called as "zero steps", is based on its comparison with the expected calibrated image. The calibrated image is acquired from the proper experiment or can be calculated theoretically with high similarity to the required one. To combine the data extracted from images obtained in different ways, the signal is quantized according to the characteristics of the used experimental equipment. The analogous image should be digitalized. The received theoretical or digitalized image is quantized by intensity into the range from 0 to 255 with step 1, which corresponds to the usual features of the CCD camera. More details about providing this operation for the optical image can be seen in (Ryazantsev et al., 2019). It is worth mentioning that due to the quantization, the results are represented with an absolute error of the intensity level about 0.5.

The **first step** is a definition of the criteria for image quality estimation. These criteria should be suitable and available for the physical task and are based on the usual requirements of relevance, objectivity, measurability, and completeness according to the standards (The AICPA Assurance Services Executive Committee, 2018). There are three criteria chosen based on the experimental skills and standards requirements. Two first criteria follow from the most common parameters: signal-to-noise ratio (SNR) and contrast ratio (CR). The third one, not so common, is the structural similarity index (SSIM). So, for the image quality estimation, there are used the Rose criterion (Rose, 1973) for the signal/noise ratio, the resolution Rayleigh criterion after the definition of the contrast between adjacent pixels (Born & Wolf, 1999), and the manually selected criterion for structural similarity index.

The statistical analysis is the **second step** of an optical image processing. It is very convenient to have statistical parameters for experimental and theoretical data and to define the image quality parameters SNR, CR, and SSIM. The **last step of the preliminary analysis** is determining the quality of the image according to the choosing criteria and making a decision to stop or continue the image analysis.

Figure 8. The information model of the image processing. First part is the preliminary analysis.

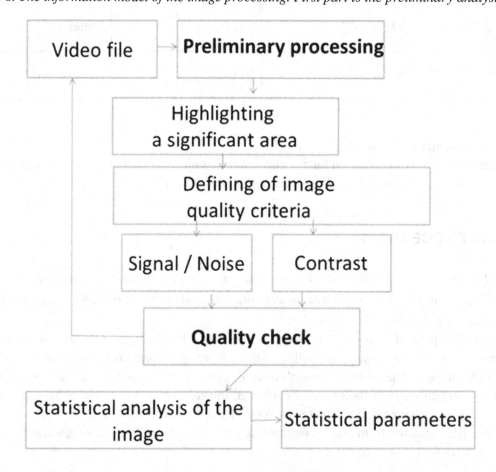

Figure 9. The information model of the image processing. Second part describes the analysis of the background contributions and revelation of the moving microparticles.

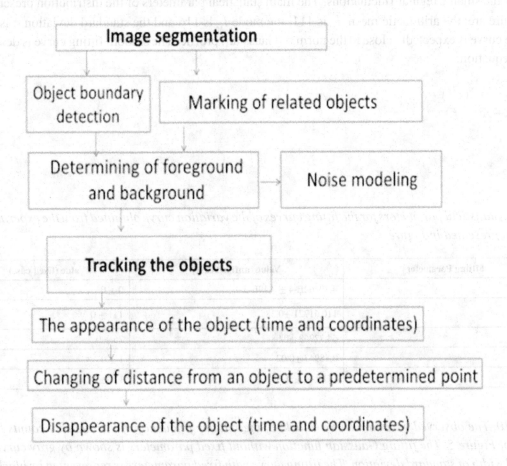

The main stage includes the segmentation of the picture and observation of the particles' motion. Here, the questions of the noise modeling and of the background-associated distortions are paid attention to. Background modeling (Bouwman et al., 2008) is often used in various systems to model backgrounds and further to detect moving objects in a scene, such as a video surveillance, optical motion capture, and multimedia.

The approach of the background modeling, developed for the case when the camera is stationary, can be also used for the case of weak camera shaking. All points of the image that deviate significantly from the background model are treated as foreground objects. Thus, the problem of detection and maintenance of the object can be solved.

STATISTICAL ANALYSIS

Statistical methods are used for a preliminary analysis of the image (Ryazantsev et al., 2019). Their application can be illustrated by statistical analysis of the image intensity for the pattern presented in Figure 2. The experimental distribution of the intensity levels taken over the whole area of Figure 3 (variation curve) is shown by the black solid line in Figure 8, and it contains distinct "noise" contribution seen by the small irregular fluctuations. The main statistical parameters of the distribution presented in the Figure are: the arithmetic mean $\bar{I}$ .is 111, the mode I_M is 116 and the standard deviation σ is about 38. The curve is expectedly close to the normal (Gaussian) profile. The smooth fitting curve is described by the equation:

$$y = y_0 + Ae^{-\frac{(x-x_c)^2}{2^2}} .$$
(4)

Table 2. Statistical parameters for the fitting curves of the variation curve obtained from the experimental image represented in Figure 3.

Fitting Parameter	Value (unfixed case)	Value (fixed case)
y_0	4,29362E-4 ± 7,49E-5	0 ± 0
$\bar{I}$	110,61721 ±0,38	111 ± 0
	64,77968 ±0,98	76 ± 0
A	0,88947 ±0,02	1 ± 0
x_c	111,144 ± 0,58	111 ± 0

Figure10. The observed variation curve (black) of the relative frequency of the intensity counts for the image of Figure 5. The fitting Gaussian function without fixed parameters is shown by gray curve with marked width of standard deviation. The fitting curve with fixed parameters is represented by black dots.

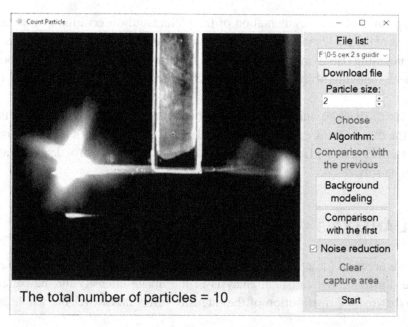

Figure 11. Variation curves for the experimental images presented in Figure 4 (dotted line) and Figure 7 (solid black line)

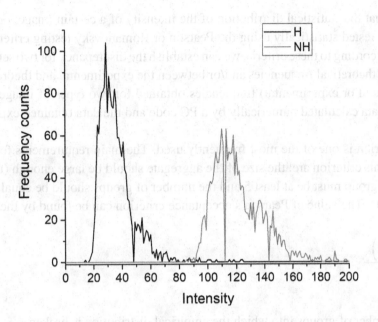

Table 3. Statistical parameters for experimental images represented in Figure 4 and Figure 5

Statistical Parameters	Shot 0 for Fig. 4	Shot 0 for Fig. 5
Mean, $\bar{I}$	121,38	36,14
Standard Deviation, 6	17,81	12,00
Median, I_{Me}	118	34

where the parameter y_0 is the shifting of the graph along Y-axis, A is the height of the curve's peak, x_c is the position of the peak center and σ is the standard deviation.

The fitting curves which are built by two methods are demonstrated in Figure 5. The fitting Gaussian distribution determined under the condition of the maximum number of coincident points of the experimental and fitting data is indicated by the gray curve. The second curve (dotted) has been built based on the two calculated statistical parameters: mean value of the intensity (111) and the intensity standard deviation (38) that is represented by the dotted curve. The fitting parameters for the two cases are presented in Table 2.

The initial conditions for recording the motion of microparticles can be studied by the statistical analysis too. The picture of Figure 4 differs from that of Figure 3b by higher brightness of the cell and has more pixels with high intensity. Accordingly, the median value is much higher for the more illuminated case, which can be seen clearly from the data of Table 3. The variation curves for images of Figure 4 and Figure 5 are essentially different from each other (Figure 9). They allow one to easily recognize the case of recording video due to the difference in the curves' locations along the intensity axis. Accordingly, the statistical analysis of both cases can be performed simultaneously.

ESTIMATION OF THE CONSENT CRITERIA

The assumption that the statistical distribution of the intensity of a certain image, or the laser field, is normal, should be tested statistically using the Pearson or Romanovsky testing criteria (Greenwood & Nikulin, 1996). According to these criteria, we can establish the discrepancy for two sets of data: between the calculated and theoretical frequencies and/or between the experimental and theoretical frequencies. Empirical (calculated or experimental) frequencies obtained for two types of images are investigated in this work: the data calculated numerically by a PC code and the data obtained experimentally using a laser setup.

Pearson's criterion is one of the most frequently used. The main requirements for the data properties according to this criterion are: the size of the aggregate should be large enough (more than 50), the frequency of each group must be at least 5 and the number of groups should be equal for empirical and theoretical data sets. The value of Pearson's acceptance criterion can be found by the formula:

$$\chi^2 = \sum_{j=1}^{k} \frac{\left(F_j - F_j'\right)^2}{F_j'} \qquad (5)$$

Where k is the number of groups into which the empirical distribution is broken,

F_j - the observed frequency of the trait in the j-th group,
F_j' – the theoretical frequency.

For the distribution, tables were compiled showing the critical value of the criterion (5) for the chosen level of significance and given number of the degrees of freedom df.

The significance level is the probability of an erroneous declination of the hypothesis, i.e. the probability that the correct hypothesis will be rejected.

The number of degrees of freedom df is defined as the number of groups in the distribution series minus the number of bonds (regular relations between the groups): $df = k - z$. The number of relations is understood as the number of indicators of the empirical series used in the calculation of theoretical frequencies, i.e. indicators linking empirical and theoretical frequencies. For example, when aligning with a normal distribution curve, there are three links demonstrated the equality of arithmetic mean, standard deviation, and number of groups in empirical and theoretical data:

$$\bar{I} = \bar{I}' \quad \sigma = \sigma' \quad \sum F_j = \sum F'_j, \ldots \qquad (6)$$

Therefore, when aligning along the normal distribution curve, the number of degrees of freedom is defined as $df = k-3$. For estimation, the differences of the theoretical and empirical data the table for Pearson's consent criterion is used.

In our case the value of k is about 55 that are much bigger than k value in ordinary table for Pearson's consent criterion determination, so we can use at the next step the Romanovsky criterion. The Romanovsky criterion based on the use of the Pearson criterion $\chi 2$ and the number of degrees of freedom df is defined by the formula:

$$c = \frac{\left| \chi^2 - df \right|}{\sqrt{2df}}. \tag{7}$$

If $c < 3$, then the differences between the distributions are random, if $c > 3$, then they are not random and the theoretical distribution cannot serve a model for studying the empirical distribution.

The fitting curves in Fig. 5 for unfixed main statistical parameters supply a higher value of the Romanovsky criterion than for the fixing of mean value and standard deviation. The values of c are about 2.12 and 2.8 for unfixed and fixed parameters correspondingly.

CHARACTERISTICS AND CRITERIA OF THE IMAGE QUALITY

There are several basic parameters used for the characterization of the image quality. In connection with them, the several criteria of the image quality are introduced. The image quality criteria p_4^D including value of noise are follow from the image required features for production in p_1^C for the optical task.

1. **Signal-to-noise ratio** (abbreviated *SNR*) is a standard criterion used in science and technology that shows the relationship between signal and noise average intensities ("Signal-to-noise ratio", 2020). The signal and noise both have to be measured in a representative number of points or equivalent points of the image for the same frame. The signal-to-noise ratio is defined as:

$$SNR = \frac{P_{signal}}{P_{noise}} = \left(\frac{A_{signal}}{A_{noise}} \right)^2. \tag{8}$$

For optical images it is modified into the expression

$$SNR = \left(\frac{\overline{I}_{signal}}{\overline{I}_{noise}} \right). \tag{9}$$

To estimate the image quality using the *SNR* parameter, the Rose criterion (Rose, 1973) can be used. It is assumed that for reliable resolution of the image details, the Rose criterion must be not less than 5. For the experimental optical images in Figure 2b, the *SNR* values for the vertical and horizontal cross sections are 13.65889 and 28.52748, which means the non-significant noise according to the Rose criterion.

2. **The contrast ratio** is a property of the image, defined as the ratio of the brightest color intensity (white) to that of the darkest color (black). A high contrast ratio is a desired aspect for any image; its numerical characterization can be defined by the Michelson formula:

$$\tilde{N}_{ij} = \frac{I_{max} - I_{min}}{I_{max} + I_{min}} \qquad (10)$$

Where $\tilde{N}_{ij}$ is the contrast value at the point of image matrix with indices [i,j]; I_{min} and I_{max} are the minimum and the maximum values of the pixel brightness correspondingly of the whole image or its chosen part. For the experimental optical image in Figure 3, the contrast value is 0.91011, which is close to one, therefore means high enough quality of the image.

3. **The structural similarity index metric** (*SSIM*) is one of the methods for measuring the similarity between two images. The *SSIM* index employs a full-mapping method; in other words, it measures the quality based on the original image (not compressed or distorted). The *SSIM* index represents a development of traditional methods such as *PSNR* (peak signal-to-noise ratio) and the *MSE* error rate method, which are incompatible with the physiology of human perception.

A distinctive feature of this method, in addition to the previously mentioned (*MSE* and *PSNR*), is that the method takes into account the "perception of error" due to the structural change of information. The idea is that different pixels of the image are not independent and have a strong relationship, especially when they are spatially close. These relations carry important information about the structure of the objects and the scene as a whole.

The *SSIM* metric is designed for different window sizes. The difference between the two windows is expressed by the formula:

$$SSIM(x, y) = \frac{\left(2\mu_x\mu_y + c_1\right)\left(2\sigma_{xy} + c_2\right)}{\left(\mu_x^2 + \mu_y^2 + c_1\right)\left(\sigma_x^2 + \sigma_y^2 + c_2\right)}. \qquad (11)$$

Where μx is the average value for the first image;

μy is the average value for the second;
σx is the root mean square deviation for the first image;
σy is the root mean square deviation for the second picture;
σx_y is already a covariance

$$\sigma_{xy} = \mu_{xy} - \mu_x\mu_y,$$

c_1 and c_2 are the correction coefficients we need due to the small denominator, and they are can be defined in "Structural similarity" (2020).

$c_1 = (k_1 L)^2$, $c_2 = (k_2 L)^2$ are variables,

L   - dynamic pixel range, $k_1 = 0.01$ and $k_2 = 0.03$ are constants

The formula (6) is applicable only to the brightness of the image for which the quality assessment takes place. The SSIM index can range from −1 to +1. The value of +1 is achieved only when the samples are fully authenticated.

The structural similarity index values were calculated using the experimental images in Figure 5 and the theoretical one obtained at the same initial conditions in Figure 4. The value for the whole image is 0.514, which means a satisfactory similarity. Taking into account that the image has mirror symmetry with respect to the X-axis, one can calculate SSIM just for a half of the image. As a result, for the upper half, this value is 0.264, which is low, but the lower part shows a much higher SSIM (its value is 0.739), so the images are very similar to each other. The reason for the distinction is the misaligned position of an optical element in the system.

Figure 12. Simulation of the Gaussian-beam intensity distribution modulated by the noise given by a random function with the coefficient 0.5.

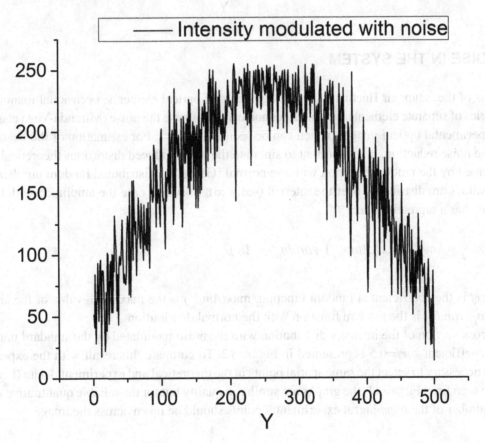

Figure 13. Experimental intensity distribution along the Y-axis is shown by gray circles and the theoretical calculations for the same points are marked by black squares.

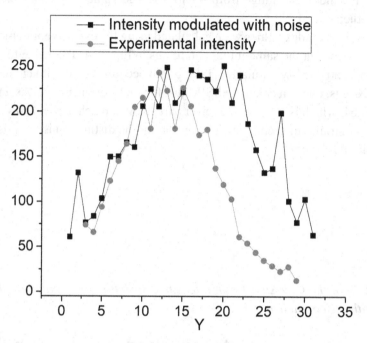

THE NOISE IN THE SYSTEM

Vibrations of the setup, air fluctuations, misalignments of optical elements, occasional manufacturing inaccuracies of separate elements, light sources and cameras cause the noise (Miranda-Vega et al., 2019) in the experimental optical images, which can be seen in Figure 5. For estimation of the beam quality and for the noise reduction, it is important to simulate the noise-induced distortions theoretically. It can be performed by the random function with the normal (Gaussian) distributed random numbers whose element values are distributed over the interval $(-\infty, +\infty)$. In such a case the amplitude with the noise $Ampnoise_{i,j}$ has a representation:

$$Ampnoise_{i,j} = koef \cdot \max\left(Amp_{i,j}\right) \cdot randn_{i,j} + Amp_{i,j} \tag{12}$$

Where *koef* is the coefficient at random function, $\max(Amp_{i,j})$ is the maximum value of the amplitude array $Amp_{i,j}$. $randn_{i,j}$ is the random function with the normal distribution.

The cross-section of the intensity distribution with the noise modulated by the standard normal law with the coefficient *koef*=0.5 is presented in Figure 12. To compare this result with the experimental noise it is necessary to select the same spatial points in the theoretical and experimental data (Figure 13). As can be seen from Figure 11, the graphs are similar in quality but for the reliable quantitative analysis, a larger number of the meaningful experimental points should be taken across the image.

SOFTWARE APPLICATION FOR COUNTING OF MICROPARTICLES

Different strategies for the video analysis, including artificial intelligence and motion science, are applied (Lin & Lin, 2006). Recognition of the shifting objects and distinguishing them from resting ones in the video file is an actual and widespread practical task (Gustafsson & Lanshammar, 1977; Jarret et al., 1974; Lanshammar, 1985; Lindholm, 1974; Mitchelson, 1975), varying from the automated control of production to the construction of robotic mechanisms. In the section it is presented the developed software p_2^D for counting microparticles.

There are three basic steps of the video analytics:

1. Identifying of the required objects;
2. Tracking the change in position and status of these objects between frames;
3. Object behavior analysis.

It should be noted that the object detection and tracking are two very closely related processes, as tracking often begins with the detection of the required objects, and the detection of the object again in the next sequence of frames is necessary to verify the accuracy of tracking.

Figure 14. The view of the interface of the developed application software after the video file is uploaded

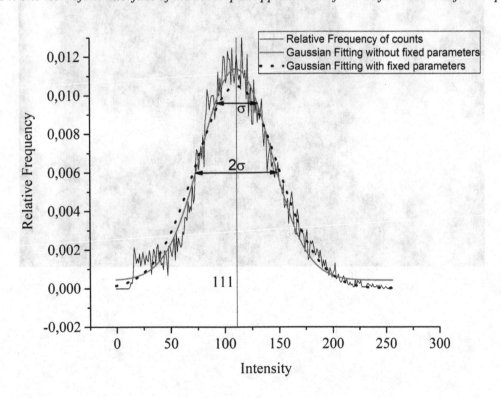

The example of microscopic particles guided or highlighted by laser light is considered. Some distributions of light enable to control of each atom individually or a group of atoms. Here, the application of the video file processing with the goal to count the number of moving micro objects is described.

The whole procedure includes processes of obtaining, storing, transforming, presenting, and transmitting information on the registration of a number of particles of a certain size. The registration of particles occurs due to three observable effects: a scattering of the laser light by the particles, guiding the particles along the laser beam, and fixing the particles on the wall of the cell. A light pulse is detected by a photodetector - a camera that records the passage of a sample through the beam. Therefore, there is a need to create software for processing the data obtained from the photodetector. So, it is important to have the software applicable for the automatic registration of a microparticle motion.

The developed software application (Khoroshun et al., 2020b) represents the main stage of the frame processing and counting of moving microparticles using the AForge.NET library, which has ready-made implementations of the motion detection algorithms. The convenient interface was developed for particle counting (Figure 14. It includes several icons, as "File list"; "Download file"; "Particle size"; "Algorithm"; "Noise reduction"; "Clear capture area"; "The total number of particles".

Program operation starts from the selection of the video file using the "Download File" button. After selecting the video file, the first video frame will be displayed in the program interface (Figure 14). Then, the areas of interest are selected by the rectangles. There is the option to choose the particle

Figure 15. A video frame with areas containing microparticles.

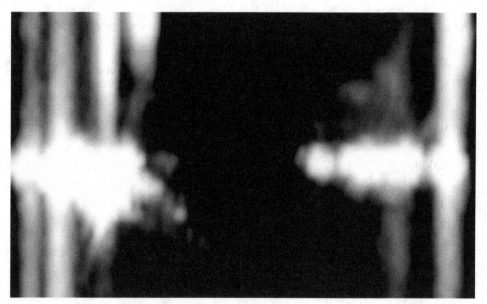

Figure 16. Histogram built with a bin 20 over the range of intensity values for the case of images obtained at average contrast (grey bars) and at high contrast (white bars)

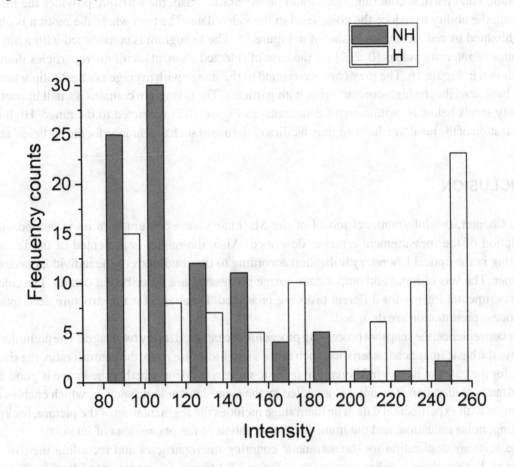

size in the range of 1 to 10 pixels; in our research, the choice 2-3 pixels has been made. In order to filter out unnecessary sections of the frame, one can select separate areas with the mouse where the motion is sought; they will be displayed in the frame only when the video processing procedure does not work. So we can clear the selection list and set new ones. The program calculates the number of particles, which is equal to 10 at the moment presented in Figure 14. After finishing the video processing or pressing the "Stop" button, it is an opportunity to resume searching or download another video. The particle counter then will be reset.

There are three search algorithms, which can be chosen for video processing: comparison with the previous one, comparison with the first frame and background modeling, which are built in AForge. NET library. It is used in the example the last approach background modeling, which can be applied if the camera is stationary, and a background changes a little. All points of the image that deviate significantly from the background model are considered as foreground objects. Thus it is possible to solve the problem of object detection and maintenance. The background simulation is often used in various systems to simulate the background and subsequent detection of moving objects in the scene, optical motion capture and multimedia. The easiest way to simulate a background is to get a background image that does not include a moving object.

The algorithm for comparison with the previous frame is selected by default. To start the motion registration and particle counting, one should press "Start". Also, the software provides the means for choosing the ability to reduce the noise level in the video data. The area where the motion is observed is highlighted by red rectangles as shown in Figure 15. The histogram is constructed with a bin 20 and the range of intensity values [0, 260] for the case of selected elements with microparticles illuminated as is shown in Figure 16. The grey bars correspond to the image with average contrast ratio whereas the white bars describe the high-contrast areas with particles. The histogram contains a small impact for the intensity levels below 80 within significant areas, so Figure 10 is restricted to the range. High-quality visualization of the results in the histogram facilitates the use of such a technique for the particles' analysis.

CONCLUSION

In the Chapter, the Informational model of the Machine Vision System with basic components and description of the measurement errors is described. Also, the model is presented of the full service rendering in the optical laboratory utilization according to the conditions of the individual needs of the customer. The sets of input and output data, image processing are described in details. The calculated and experimental images for different tasks are presented. Databases for the structure description and the process presentation are designed.

For convenience, the snapshot processing procedure is characterized by two stages. The preliminary one employs the basic image characteristics, such as the signal-to-noise ratio, the contrast ratio, the similarity index, for the cases of an example image in order to make a decision whether the image is good enough for further consideration. The noise is modeled by the Gaussian random function, which enables a good agreement with experimental data. The main stage includes the segmentation of the picture, background modeling, noise reduction, and the immediate observation of the phenomena of interest.

The software application for the automatic counting microparticles and recording their motion is presented. The developed application uses the AForge.NET library for processing video files that enables the allocation and counting of the moving object. It possesses the possibility of choosing an algorithm of counting, adapting different particle sizes, and the ability to filter noise in the video.

Further developments of the presented approach will be useful for elaborating the IT for the laser equipment with applications in optical diagnostics, information transfer, micromanipulation and, metrology purposes.

In the future, it is planned to employ artificial intelligence based on the deep learning system for counting the particles separately, depending on their features and trajectories.

REFERENCES

Alfalou, A., & Brosseau, C. (2009). Optical image compression and encryption methods. Advances in Optics and Photonics, 1, 589. doi:10.1364/AOP.1.000589

Alfalou, A., & Brosseau, C. (2010). Exploiting root-mean-square time-frequency structure for multiple-image optical compression and encryption. Optics Letters, 35, 1914–1916.

Alfalou, A., Brosseau, C., & Alam, M. S. (2013). Smart pattern recognition. Proceedings of SPIE 8748, optical pattern recognition XXIV, 874809. doi:10.1117/12.2018249

Bekshaev A., Chernykh A., Khoroshun A., & Mikhaylovskaya L. (2017). Singular skeleton evolution and topological reactions in edge-diffracted circular optical-vortex beams. *Optics Communications, 397*, 72–83. doi:10.1016/j.optcom.2017.03.062

Born, M., & Wolf, E. (1999). *Principles of Optics*. Cambridge University Press. doi:10.1017/CBO9781139644181

Bouwmans, T., El Baf, F., & Vachon, B. (2008). Background Modeling using Mixture of Gaussians for Foreground Detection. A Survey. Recent Patents on Computer Science, 1(3), 219-237.

Flores-Fuentes, W., Rivas-Lopez, M., Hernandez-Balbuena, D., Sergiyenko, O., Rodríguez-Quiñonez, J. C., Rivera-Castillo, J., . . . Basaca-Preciado, L. C. (2020). Applying Optoelectronic Devices Fusion in Machine Vision: Spatial Coordinate Measurement. In Natural Language Processing: Concepts, Methodologies, Tools, and Applications (pp. 184-213). IGI Global.

Greenwood, P. E., & Nikulin, M. S. (1996). *A guide to chi-squared testing*. Wiley.

Gustafsson, L., & Lanshammar, H. (1977). *Enoch - An integrated system for measurement and analysis of human gait* (Ph.D. thesis). UPTEC 7723 R.

Jarret, M. O., Andrews, B. J., & Paul, J. P. (1974). Quantitative analysis of locomotion using television. *Proceedings of lSPO World Congress*.

Khoroshun, A., Ryazantsev, A., Ryazantsev, O., Sato, S., Kozawa, Y., Masajada, J., Popiołek-Masajada, A., Szatkowski, M., Chernykh, A., & Bekshaev, A. (2020). Formation of an optical field with regular singular-skeleton structure by the double-phase-ramp converter. *Journal of Optics*, *22*(2), 025603. doi:10.1088/2040-8986/ab61c9

Khoroshun, A. N. (2010). Optimal linear phase mask for the singular beam synthesis from a Gaussian beam and the scheme of its experimental realization. Journal of Modern Optics, 57(16), 1542–1549.

Khoroshun, A. N., Chernykh, A. V., Tsimbaluk, A. N., Kirichenko, J. A., Yezhov, P. V., Kuzmenko, A. V., & Kim, J. T. (2016). Properties of an Axial Optical Vortex Generated with the use of a Gaussian Beam and Two Ramps. *Journal of Nanoscience and Nanotechnology*, *16*(2), 1–3. doi:10.1166/jnn.2016.12029 PMID:27433739

Khoroshun, G. (2019). Model of service providing on the use of optical laboratory in conditions of individual customer needs. Visnik of the Volodymyr Dahl East Ukrainian National University, 8(256), 118-122.

Khoroshun, G., Luniakin, R., Riazantsev, A., Ryazantsev, O., Skurydina, T., & Tatarchenko, H. (2020). The Development of an Application for Microparticle Counting Using a Neural Network. *Proceedings of the 4th International Conference on Computational Linguistics and Intelligent Systems (COLINS 2020)*, 1186-1195.

Lanshammar, H. (1985). Measurement and analysis of displacement. *Gait Analysis in Theory and Practice Proceedings of the 1985 Uppsala Gait Analysis Meeting*, 29 – 45.

Lin, H. Y., & Lin, J. H. (2006). A visual positioning system for vehicle or mobile robot navigation. IEICE Transactions on Information and Systems, 89(7), 2109-2116. doi:10.1093/ietisy/e89-d.7.2109

Lindholm, L. E. (1974). An optoelectronic instrument for remote on-line movement monitoring. In Biomechanics IV. University Park Press.

Matveev, A. N. (1988). *Optics*. Mir Publishers.

Miranda-Vega, J. E., Rivas-Lopez, M., Flores-Fuentes, W., Sergiyenko, O., Rodríguez-Quiñonez, J. C., & Lindner, L. (2019). Methods to reduce the optical noise in a real-world environment of an optical scanning system for structural health monitoring. In *Optoelectronics in machine vision-based theories and applications* (pp. 301–336). IGI Global. doi:10.4018/978-1-5225-5751-7.ch011

Mitchelson, D. (1975). Recording of movement without photography. In Techniques for the Analysis of Human Movement. London Lepus Books.

Nehmetallah, G., & Banerjee, P. P. (2012). Applications of digital and analog holography in three-dimensional imaging. Advances in Optics and Photonics, 4, 472–553.

Pronina, O. I. (2018). Model of service presentation in the context of individual customer needs. Computer Integrated Technologies: Education, Science, Production, 33, 128– 133.

Pronina, O. I. (2018). Formalization of the order organization in the conditions of individual needs of the client. *Proceedings of All-Ukrainian scientific-practical conference of higher education seekers and young scientists "Computer Engineering and Cybersecurity: Achievements and Innovations"*, 93 – 95.

Riazantsev, A.O., Khoroshun, G.M., & Ryazantsev, O.I. (2019). Statistical image analysis for information system. *Visnik of the Volodymyr Dahl East Ukrainian National University, 5*(253), 84-86. DOI: doi:10.33216/1998-7927-2019-253-5-84-86

Rose, A. (1973). *Vision: Human and Electronic*. Plenum Press.

Rosen, J., Katz, B., & Brooker, G. (2009). Fresnel incoherent correlation hologram—a review. Chinese Optics Letters, 7, 1134–1141.

Ryazantsev, O., Khoroshun, G., Riazantsev, A., Ivanov, V., & Baturin, A. (2019). Statistical Optical Image Analysis for Information System. In *Proceedings of 2019 7th International Conference on Future Internet of Things and Cloud Workshops (FiCloudW), Istanbul, Turkey*, (pp. 130-134). IEEE. 10.1109/FiCloudW.2019.00036

Sergiyenko, O., Flores-Fuentes, W., & Mercorelli, P. (Eds.). (2020). *Machine Vision and Navigation*. Springer Nature. doi:10.1007/978-3-030-22587-2

Signal-to-noise ratio. (2020). In *Wikipedia*. https://en.wikipedia.org/wiki/Signal-to-noise_ratio

Sinha, P. K. (2012). *Image Acquisition and Preprocessing for Machine Vision Systems*. SPIE Press. doi:10.1117/3.858360

Structural similarity. (2020). In *Wikipedia*. https://en.wikipedia.org/wiki/Structural_similarity

Testorf, M., & Lohmann, A. W. (2008). Holography in phase space. Applied Optics, 47, A70–A77.

The AICPA Assurance Services Executive Committee's SOC 2® Working Group Description. (2018). *Criteria for a Description of a Service Organization's System in a SOC 2*. https://www.aicpa.org/content/dam/aicpa/interestareas/frc/assuranceadvisoryservices/downloadabledocuments/dc-200.pdf

Chapter 7
Indoor Navigation Aid Systems for the Blind and Visually Impaired Based on Depth Sensors

Fernando Merchan

(iD) https://orcid.org/0000-0001-8555-5424

Universidad Tecnologica de Panamá, Panama

Martin Poveda

Universidad Tecnologica de Panamá, Panama

Danilo E. Cáceres-Hernández

Universidad Tecnologica de Panamá, Panama

Javier E. Sanchez-Galan

(iD) https://orcid.org/0000-0001-8806-7901

Universidad Tecnologica de Panamá, Panama

ABSTRACT

This chapter focuses on the contributions made in the development of assistive technologies for the navigation of blind and visually impaired (BVI) individuals. A special interest is placed on vision-based systems that make use of image (RGB) and depth (D) information to assist their indoor navigation. Many commercial RGB-D cameras exist on the market, but for many years the Microsoft Kinect has been used as a tool for research in this field. Therefore, first-hand experience and advances on the use of Kinect for the development of an indoor navigation aid system for BVI individuals is presented. Limitations that can be encountered in building such a system are addressed at length. Finally, an overview of novel avenues of research in indoor navigation for BVI individuals such as integration of computer vision algorithms, deep learning for the classification of objects, and recent developments with stereo depth vision are discussed.

DOI: 10.4018/978-1-7998-6522-3.ch007

INTRODUCTION

The World Health Organization (WHO) reports that globally the number of blind or visually impaired (BVI) individuals amounts to roughly 2.2 billion, of which at least 1 billion are in conditions that could have been prevented or are still pending to be addressed (a.e. uncorrected refractive errors or cataracts). Moreover, with a vast majority of vision impaired individuals being of 50 years or over (World Health Organization 2013). Bourne et al. (2017) provides estimates on how these numbers conditions were for each condition, for instance: 36 million people were blind, 188 million had mild visual impairment, 1,095 million people aged 35 years and older had functional presbyopia, and 667 million of those were 50 years of age or older. Blindness and visual impairment are growing problems that go side-by-side with an increasing aging world population (He, Goodkind, and Kowal 2016). Estimates suggest that by 2050 the total population will amount to 9.7 billion with almost 22% having a form of visual impairment (Ackland, Resnikoff, and Bourne 2017).

In a review from Köberlein et al. (2013), the authors set to calculate the economic burden (per person) of treating a BVI individual, they achieve this by doing a systematic analysis of publications on the subject. Results suggested that the direct cost were attributed to the hospitalization and medical services used for diagnosis and treatment; and most indirect cost were attributed to the use of caregivers (often pay by the hour). Noteworthy, to mention that both costs were increasing proportionally with the degree of visual impairment.

Traditionally the most popular methods to assist the visually impaired people are the use canes and guide dogs. However today there are several technologies that can be used to provide information about the environment and assist their navigation. These technologies are grouped under the umbrella term *assistive technologies* (AT), which can represent any software system or hardware device that is used to maintain or improve the functionality and promote the general well-being of an individual by augmenting or enhancing their possibilities due to a disability (World Health Organization 2018; Cook and Polgar 2014).

Research in assistive technologies for BVI is a very open and dynamic. Contributions and new developments come often from the engineering, computational (software) and electronic (hardware) fields. The main objective of this field is to provide technologies to bring comfort to the individual in their daily lives, assessing their own limitation and enhancing their re-entrance to normal life (Cook and Polgar 2014; Assistive Technology Industry Association 2020).

For instance, assistive technologies for BVI encompasses different aspects of the daily life of an individual. Tasks include the continuous monitoring of the progress of the condition, also attention is given to mobility and navigation issues, legislation on the adequation of the environments (a.e. appropriate buildings access), but also information retrieval and presentation (including digital and traditional braille readers, and digital enhancements to normal monitors), adaptation of devices to daily life activities (a.e. adaptation to the use of kitchen utensils) and enhancement of human-human interaction (Hersh and Johnson 2010).

The field of assistive technologies for BVI is very broad, however, our main objective in this document is to review general ideas and implementations on the subject of navigation aid systems for the BVI, first focusing on generals research ideas found in literature and then presenting results from an indoor navigation aid system based on Microsoft Kinect depth sensors.

The structure of this document is as follows: in Section 2, background information about assistive technologies for BVI, describing the most used technologies and examples of aids developed are given.

Also, navigation systems and specially vision based systems are discussed in detail. In Section 3 the details of the development of a Microsoft Kinect depth sensor based system for indoor navigation, are provided. In Section 4 resulting limitations of the Kinect system and recommendations on how to address them are discussed. In Section 5 future research directions, including the latest deep learning based systems are discussed. Finally, in Section 6 conclusions of the use of depth sensor for indoor navigation and a few ideas on what can be next in this field are discussed.

ASSISTIVE TECHNOLOGIES FOR BLIND AND VISUALLY IMPAIRED INDIVIDUALS

In a recent review by Manjari, Verma, and Singal (2020), the authors describe broad classification of assistive technologies for BVI, based on its purpose these can technologies are aimed to provide: visual substitution, visual enhancement and visual replacement. The first referring to the substitution of the visual cortex system for a camera or sensor which captures the environment, processed it and returns commands via sounds or vibrations (or both) to the user. The second and third referring to a similar capture and processing system, but in the second the feedback is given to the user via images (a.e. as in augmented reality system) and in the third the feedback is given as direct visual cortex stimulation in the brain of the subject.

Moreover, in accordance to Hersh and Johnson (2010) (and many other authors), Manjari, Verma, and Singal (2020) states that the first of the three, namely visual substitution, can be further divided into three types of devices that help achieve the purpose of the technologies: *Electronic Travel Aids* (ETA), *Electronic Orientation Aids* (EOA) and *Position Locator Devices* (PLD). A general description of these systems is given in Figure 1.

Figure 1. General view of a Navigation System. Source: The Authors

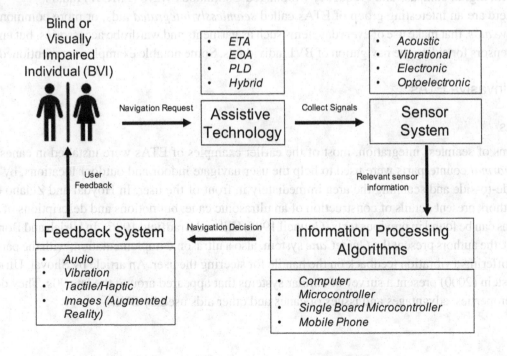

Some authors, for convenience, further categorize these devices attending on what context they are used, for instance used as an indoor or as an outdoor navigation device. For example, the first group of devices being mostly for indoor navigation and the last two types of devices, mostly used for outdoor navigation. However the reader will find that the devices often can be used for both contexts. Characteristics of each of these visual substitution aids and examples found in literature will be further explained in the next subsections.

Electronic Travel Aids (ETA)

ETA groups the systems that captures information from the immediate environment using sensors and provide a feedback to the user to alert of possible obstacles. These can be accounted for the most popular visual substitution systems. The most common example of this systems include the use of canes and garments enhanced with sensors to make the user aware of the distance to nearby objects and providing instruction to the user in order to avoid obstacles in the immediate surroundings.

For ETA other types of subdivision can be made, in fact, in a review by Elmannai and Elleithy (2017) focused on both wearable and portable assistive devices, the authors end up dividing them by: analysis type (online or offline), coverage (indoor, outdoor or both), time of usage (day, night or both), range (short ≤1m, 1m <medium ≤5m and long >5m), and object type avoidance (static, dynamic or both). To these categories Islam et al. (2019) adds descriptions as: capturing device (camera, sensor or both), feedback to the user (audio, tactile or both), weight (high or low) and cost (high or low).

In Hu et al. (2019), the authors present an overview of different types of assistive devices but limiting to canes, glasses, hats and gloves. They are described in terms of function, context of use, behaviour of user, principle that is based on, state and structure. Also, in Hu et al. (2019) a categorization, is made based on sensor type, among them used for ETAs: Ultrasonic sensor, Global Positioning System (GPS) receiver, Red-Green-Blue (RGB) camera (including RGB-Depth cameras and wide angle (270° lens cameras), inertia measurement unit, lasers, infrared sensors, Radio-Frequency Identification (RFID) network, digital compass and Frequency-Modulated Continuous Wave (FMCW) radars.

There are an interesting group of ETAs called *seamlessly integrated* aids, or more commonly *non-invasive aids*, that make use of everyday items such as garments and wardrobe accessories, but enhanced with sensors for aiding the navigation of BVI individuals. Some notable examples are mentioned below.

Non-Invasive ETAs

Canes

In terms of seamless integration, most of the earlier examples of ETAs were installed in canes and as their *normal* counterparts were used to help the user navigate indoor and outdoor locations, by sweeping side-to-side and scanning the area immediately in front of the user. In Hoydal and Zelano (1991), the authors present details of construction of an ultrasonic cane, but notions and descriptions of similar systems can be found in previous texts (Council 1986; Hill and Bradfield 1986). In Ulrich and Borenstein (2001), the authors present the *GuideCane* system, also a ultrasonic implementation, with the particularity of offering a vibration feedback on the handle for steering the user. An article by Shoval, Ulrich, and Borenstein (2000) present a survey of similar systems that appeared around the late 90s. They describe their properties, advantages and issues of canes and other aids used by BVI.

Recently this cane systems have been upgraded by incorporating multiple sensors to detect obstacles and provide feedback by different mediums. In Lodh, Subramaniam, and Paswan (2016), the authors discuss such a system but based around an ultrasound and also a sonar. It was coupled with a vibration motor capable of generating different vibration patterns to alert the user, which function in relation the distance on which the obstacle is found. In Wahab et al. (2011), the *Smart cane* is presented, this cane communicates with the user via voice alert and also via vibration. In Bousbia-Salah, Bettayeb, and Larbi (2011), a more refined version of the cane is presented, with multiple sensor system. The systems used an ultrasound sensor, an accelerometer, a foot-switch and keypad for route decision, finally a micro-controller for digital speech output is used as feedback mechanism.

Belts

Others systems are installed in purpose-made belts that can be carried naturally by the user. A notable earlier example of functioning of a belt system is the *Navbelt* by Shoval, Ulrich, and Borenstein (2003). The *Navbelt* consisted on a belt, made of an array of eight ultrasonic sensors each one covering a section of the forward space on a 15° (with a total range of 120°. The feedback was provided to the user via an acoustic signal to stereo headphones displaying a a virtual acoustic image and a sense of possible navigation direction. A more recent iteration of a similar belt system is presented in Bhatlawande, Mukhopadhyay, and Mahadevappa (2012), instead of a heavy computer and ultrasound sensor array, the system is based on portable sensors and a microprocessor suitable to process the ultrasound signals. The feedback in this system consist of a pre-recorded audio message that is played to the user via a headphone.

Eyeglasses

Many authors have contributed to the design of glasses for BVI for task of navigation in both indoor and outdoor environments. In Velazquez et al. (2006), the authors present *The Intelligent Glasses* a visual-tactile system, with two cameras mounted on the user glasses and a device to build a tactile map representation of the scenery which can be used for navigate amongst obstacles. Dakopoulos, Boddhu, and Bourbakis (2007), presents the *Tyflos* system, consisting of cameras mounted cameras on dark glasses, with a microphone and a ear piece. As a novelty this systems had a two dimensional (2D) feedback system, that was made of a 4 *x*. vibration array that were place on the chest of the user. The systems builds a three dimensional (3D) representation of the space which is translated to the user into the vibrators in correspondence to the distance to obstacles.

More recently, in Bai et al. (2017), the authors present a system based on the fusion of multiple sensors (ultrasonic and depth sensor) to detect small and transparent obstacles (such as doors). The feedback system was adapted to the level of visual challenge, for blind users auditory signals were used to communicate the direction clear of objects. For user with a milder visual impairment, Augmented Reality (AR) is used to convey the clear course direction through the glasses. The system proved to be useful in a complicated indoor environment.

In (Pardasani et al. 2019), the authors present two aid systems based on simple and cheap sensors. The first is a pair of multi-functional smart glasses that have an integrated camera and micro-controller. This glasses can be used to detect and classify objects, also are capable of detecting object labels in a fixed setting and provide the user feedback via audio. Also, the authors present a *smart shoe*, basically a working prototype of a shoe with an integrated sensor which can be used for navigation.

Finally, although Simultaneous Localization and Mapping (SLAM) algorithms have been used in navigation and mapping to help unmanned aerial or ground mobile robots, there is still remains work to be done in supporting BVI persons. Real *et al.* 2020 conclude that BVI needs are not well met by the current approaches, mostly due the fact that approaches are tackling separately. Table 1 shows some BVI based SLAM approaches. It can be noticed that researches solved the problem without taken into account for example the robustness of the application, processing time, speed.

Table 1. BVI Based SLAM Approaches. Source: The Authors

Authors	Method	Prototype	F-Measure	Speed
Nguyen et al. (Nguyen et al. 2014)	SLAM	—	80.39%	1.25 ft/s 0.381 m/s
Zhang et al. (Zhang et al. 2015)	SLAM	Yes	—	—
Zhang et al. (Zhang et al. 2016)	SLAM	Yes	—	—
Endo et al. (Endo et al. 2017)	SLAM	Yes	—	—

Gloves

As noted in Kunhoth et al. (2020), aid systems for indoor navigation can be used by any person unfamiliar with a new environment. In contrast with outdoor environments which can be easily traversed with the help of Global Positioning System (GPS) signals, indoor environments are confined with no line-of-sight for traditional positioning systems. Moreover, indoor environments are difficult to navigate, having a collection of individuals and objects. This is issues can be solved implementing Radio frequency (RF) signals, or a combination of computer vision and sensors useful for tracking the navigation of users in indoor environments.

In the case of BVI, Ganz et al. (2012) presents the *PERCEPT* glove system provides a indoor navigation system for health facilities as hospitals and clinics. The system is based on the use of passive radio frequency identification (RFID) sensors, which are deployed in the environment and a custom-designed handheld glove unit and a smartphone carried by the user. Basically, the system stores building information via RFID tags and when the user is near one of such tags, it position is known and can be directed to the appropriate destination following the next RFID tags.

Hybrid Systems

In the context of navigation with a definite purpose authors of Advani et al. (2016) present *Third-Eye prototype* an hybrid system geared towards enhancing the experience of a BVI user in the context of assistance in grocery shopping. The system consist of many aid subsystems, it has a pair of android-based smart glasses with a camera and headset capabilities. It also paired to a glove with a camera and a set of vibration motor. Finally, the whole system can be paired with a smart shopping cart equipped with a laptop computer and sensor for indoor location. Every subsystem is connected to the internet, and using the resources of a high-performance computing system for providing real-time response to the user. This system enables the user to shop and receive feedback to its actions via a augmented reality in the glasses or audible and/or vibration as feedback cues, which adapt to the visual challenge of the user.

Electronic Orientation Aids (EOA) and Position Locator Devices (PLD)

EOA and PLD are sometimes place together being that they make use of complementary technologies. In general, EOA groups the systems that help the user navigate through in predetermined path. PLD groups the systems specifically make use of a Global Positioning System (GPS) to find the users location and from there help the navigation of the environment. In both groups the navigation could be achieved via a combination of wearable sensors, a satellite-based Global Positioning System (GPS), or a Geographic Information System (GIS). In Manjari, Verma, and Singal (2020), furthers enhances this definition by categorizing EOAs and PLDs into wearable and handheld devices.

For both groups of devices, the process of orientation is similar. Basically, adjusting and taking into account the visual challenge, guiding the user to follow an expected trajectory by providing appropriate instructions and alerts about possible object in the path.

The design and implementation of an EOA system, can be summarized into 3 steps: 1) calculation the optimal travel route, 2) tracking the user to estimate its location and 3) provide a guide or instructions on how the user can navigate Kammoun et al. (2011; Manjari, Verma, and Singal 2020).

In Kammoun et al. (2011), the authors furthers enhances the description to develop an EOA system:

1. The guidance system can achieve the calculation of the optimal travel route by: determining the availability appropriate sidewalks and crossings network, qualify the inclinations and the surface or soil texture of the possible path and finally determining points-of-interest (POI) for the user.
2. The tracking of the user and the estimation of the location can be made via GPS, with the common problems of losing the line connection with the satellite. In terms of in-place localisation, it can be made using Bluetooth beacons and internet-relying technologies such as RFID tags. Also, in some cases when the system are vision based, it can be programmed to recognize environmental landmarks (natural, artificial or man-made) and achieve localisation.
3. One notion is that visual guide or instructions should be bi-directional or should have the possibility for the user to approve or not a calculated route according to this decision. The feedback as in many systems can be turn-by-turn instructions via binaural audio.

In (Katz et al. 2012), the authors describe the Navigation Assisted by artificial Vision and Global Navigation Satellite System (NAVIG). The system tries to bring autonomy to the user by building a virtual augmented reality system, which integrates GIS and landmark objects recognition on a fixed map, and provides the needed feedback for navigation as audio clips. This can be very helpful to BVI users for outdoor positioning.

In Sáez et al. (2019) and Vejarano, Henriuez, and Montes (2018) authors go in a complementary direction to the use of GPS and GIS systems for public transport navigation for BVI individuals. Arguing that in the context of BVI individuals and their interaction with public transport, GPS systems lacks accuracy. Based on this the authors provides an implementation based on the use of Radio Frequency (RF) communication systems. The systems works having three modules, one located on the bus, one located on the bus stop and other that remains with the user. The systems is activated by the proximity of the three modules, basically when the bus approaches and stops the user module transmit an RF signal to the bus module and a transaction regarding the routes will be conveyed to the user. The authors propose that feedback system to be via a vibration buzzer to avoid the sound interference than can occur in crowded urban environments.

In Dhod et al. (2017), the *smart white cane* is presented. This system function as a normal white cane, as described earlier in this section, but enhanced with GPS and a global system for mobile communication (GSM) to ensure the accuracy of navigation in both indoor and outdoor environments. The system has integrated sensors to avoid obstacles, and more interestingly also sensors for depth and water detection. All feedback is provided via a playback recorder circuit.

Indoor Positioning Systems

The newest iteration of EOAs and PLDs are geared to serve as indoor positioning systems (IPS), basically enabling the tracing of human beings and pets in a confined interior space (Silva and Wimalaratne 2017). The difference with traditional positioning is the need to rely on existing infrastructure in the space, and need to adapted for the use in a BVI navigation system.

Wi-Fi, Bluetooth and other solutions that rely on signals, are based on the idea that the closer you are to the emitting device the stronger the signal the receiver will perceive (Cheng et al. 2014). Some commonly used technologies for this task are: Wi-Fi, Bluetooth beacons, RFID, Ultra-wideband (UWB). The characteristics and accuracy of these technologies are presented in Table 2.

Table 2. Technologies used for Indoor Navigation Systems. Source: The Authors

Technology	Energy Consumption	Measurement Accuracy	Known Caveats
Bluetooth	Low	1-5 m	Security Issues
Wi-Fi	Mid	5-10 m	Low SNR in high-density
UWB	High	0.1 m - 0.5 m	Only work in open spaces

Other indoor positioning systems make use of a magnetic field, also called magnetic "fingerprint", basically making a map of the magnetic potential of different areas in a space (Smirnov, Panyov, and Kosyanchuk 2010; Pérez-Navarro et al. 2019). One such system is described in Kasmi, Norrdine, and Blankenbach (2015). with positioning via a magnetic sensor, which measures the strength of the magnetic field generated by a number of coils (beacons) situated in specific places in the test environment.

Main Disadvantages of Indoor Navigation Methods

In Perez-Navarro et al. (2019), the authors summarize clearly the main challenges faced by existing methods for location based systems (LBS), challenges that are extended to all indoor navigation systems. Probably the most obvious one is that GPS or any other satellite based system needs a clear line of sight. Therefore positioning via cellular networks became a solution, with the limitation of only working in areas well-covered by the cellular network, and with an error of over 100 meters. As an alternative to signal based technologies for indoor navigation, many authors point to the use of vision systems as will be discussed in the next section of the chapter.

Machine Vision Based Assistive Technologies for Navigation

An important segment of the research made in assistive technologies for navigation has focused on the creation of vision systems to enhance BVI capabilities. The increasing access to cheap micro-computers has made the possibility to make this process in real-time. In fact, Li et al. (2018) states that the advances in computer vision software for visual odometry (or the use of data provided from motion sensors to estimate change in position over time of an object or the user). Also, the use of graphic processing units (GPUs) has provided the suitable scenario for the development of vision based systems for BVI, and more importantly capable of Simultaneous Localization and Mapping, also shortened as SLAM. SLAM is very important for BVI, since it centers on basically solving the problem of constructing a map on an unknown environment, and also keeping track of the approximate location user that is sensing the environment. All this is achieved by sensing or capturing images while traversing the unknown environment. However, given that SLAM approaches are computational intense, they have not been used widely in BVI real time applications.

Normally, vision based navigation developments are usually done with mounted cameras, for instance in smart glasses as discussed earlier, but there exist a great body of work done in intraocular (Liu et al. 1997) and retinal prosthesis (Weiland, Liu, and Humayun 2005; Mills, Jalil, and Stanga 2017).

Nonetheless, most common implementations are done mounted with stand-alone cameras, that the user will carry in a head mount, glasses, chest or shoulder. Among the cameras and lenses combination there are systems made of monocular cameras (Apostolopoulos et al. 2014), stereo vision cameras (Pradeep, Medioni, and Weiland 2010), omni-directional or panoramic cameras (Murali and Coughlan 2013) and wide-angle multiplexing cameras (Peli et al. 2009).

Navigation Systems Based on Smartphone Cameras

Recently the development of ETA have shifted the focus into using the capabilities of smartphones and specially in the development of mobile apps to use the functionalities of the phones. For instance in Caldini, Fanfani, and Colombo (2015), the authors present the development of a system to extract three dimensional (3D) representation of a scene for the detection of nearby obstacles using a smartphone camera, processing the images with a Structure from Motion (SfM) algorithm and the information provided by the gyroscope, built-into the smartphone. For the obstacle detection and avoidance the algorithm assess the proximity to the user by estimating the ground-plane and identifying points in the 3D space. From the determination of the ground-plane a classification can be made. That is marking all points that are not on the ground-plane as obstacles, and using the level of depth to the determine the warning level (smaller warning for far objects and bigger warnings for closer objects).

Staircase Detection

An important issue to solve in vision based navigation systems for BVI is obstacle avoidance. This is necessary to provide the user autonomous movement around the surroundings. In that regards staircases present an interesting challenge that has been addressed with various approaches for both robots (Westfechtel et al. 2018) and BVI individuals (Shimakawa et al. 2017; Carbonara and Guaragnella 2014).

Stairs detection method can be divided into three major approaches or "techniques": distance, vision, and vision-depth based analysis. In case of the distance based methods, ultrasonic and laser schemes have been proposed. Those approaches are mostly focused on measuring or estimating distance to the ground or objects located in front of the user. Pallejà et al. (2010) proposed an electronic white cane using a laser device while Bouhamed, Kallel, and Masmoudi (2013) proposed a ultrasonic based assistive device. For the case of vision application there are mono and stereo vision research. In both methods, researchers are focused on detecting and recognizing the stair candidate region mostly based on the primary features. One example of these features are the set of line segments formed into the image due to the transformation from the real world coordinates to the 2D image coordinate.

The detection of stairs using single camera comes down to a segmentation problem for both indoor and outdoor approaches, relying in the detection of the vertical and horizontal vanishing point (Hernandez and Jo 2010). Also, stairways has been treated with a combination of edge detection, directional filters, considering planar motion and tracking, to conclude with a stair reconstruction method (Hernández, Kim, and Jo 2011; Hernández and Jo 2011; Hernandez and Jo 2011).

Harms et al. (2015) propose a stair detection method based on stereo image and an inertial measurement unit (IMU) sensor. This method make use of the dense disparity data to fit the convex and concave line segments to finally fit a stair model using a tracking strategies. Vision depth based methods or RGB-D methods relies on combining RGB and depth data. Guerrero et al. (2015) present a stairs detection framework based on range data. To do this, the authors uses a region growing strategy to classify them as a planar or non-planar regions, using a sample consensus method. Then a segmentation method, that included a normal estimation, planar analysis, Euclidean extraction and concludes with a plane classification, was implemented. As a result a set of connected candidates regions of the stairs are extracted. Then, the algorithm uses the step width, tread length, riser height and number of steps as a parameter within the model to validate the candidate as a stair or not.

Recently, researchers have been merging or combining the use of these strategies with deep learning, combining the segmentation with an abstraction to determine meta-features of the stairs, using pre-trained Convolutional Neural Networks (CNN) Sanchez-Galan, Jo, and Cáceres-Hernández (2020). Also recently, in Habib et al. (2019), the authors present an hybrid system for the detection of staircase and the ground, based on a pre-trained Faster Region Convolution Neural Network (R-CNN) object detection model and an ultrasonic sensor. Alluding to a new era of staircase detection with deep learning and convolutional neural networks as features extractors.

In Shimakawa et al. (2019) the authors present a smartphone and mobile application based system for the detection of stairs from RGB images by using Convolutional Neural Network (CNN). The mobile application is useful for processing images and detecting obstacles and stair, and the smartphone provides feedback to the user for safety walking.

As mentioned earlier in this section, for vision based navigation systems an important body of work is based on RGB-Depth cameras, that we want to further discuss in the next subsection.

Working with RGB-Depth Cameras

While 2D image acquisition system has been available to researchers for decades, a few caveats make them cumbersome in the task of extracting depth information. One of them is that 2D images can be sensible to occlusions, illumination problems and shadows, thus unusable to extract depth information. As Li et al. (2018) suggests, the decrease in the cost on GPUs have made possible great advances in the field of computer vision. Systems based of parallel processing power are now accessible, and working with 3D features and depth became standard.

Earlier work by Mian, Bennamoun, and Owens (2006) focused on 3D modelling for the segmentation of objects on cluttered scenarios, based on multiple unordered images forming overlapping views (represented as tensors). Also, in Breitenstein et al. (2008) present a 3D construction of faces for pose estimation. In both cases the images were captured in a static manner, with a normal camera.

Wearable RGB-D Navigation Systems

In Lee and Medioni (2014) the authors describe a glass-mounted wearable RGB-D based BVI navigation system, comprised of a laptop running the software, a smartphone app which provides both audio and controls the vibration of the haptic interface system that is located in a vest, worn by the user. The orientation information and navigation algorithm is Six Degrees of Freedom (6-DOF) for odometry based on feature extraction extracted from images captured by the RGB-D camera. The algorithm is also used to create a 3D map of the environment that is used to analyze and determine the correct traverse path. In Lee and Medioni (2016), the authors presents an enhanced version of the previously described system, with substantial changes made in the computer vision algorithm, now taking into account both sparse features and dense cloud points. Moreover, the capabilities of making a probabilistic 2D map and planning capacities for determination of the shortest navigation path, providing feedback in real-time while the user is moving. More importantly, this systems have an interesting characteristic. the map created is stored into an XML file, that can be reloaded when the user needs it, saving time in recalculations when revisiting and environment.

In Li et al. (2018) authors presents a system for mobile indoor navigation aid for BVI, called Intelligent Situation Awareness and Navigation Aid or ISANA. The system is based on a Google Tango mobile device with an embedded RGB-D camera with a wide-angle camera for visual motion tracking and 9-DOF for odometry, thus making it a ready-made vision based mobile solution. Also it has an electronic cane for providing haptick feedback in the navigation. One particularity of the system is that it not only focuses on indoor mapping, but also on spatial context-awareness, the combination of the two is often called the creation of a semantic map. This semantic map is constructed using geometrical information from architectural (CAD) models and construct a 2D map for path planning. Among the many features of the system, it features obstacle detection and avoidance and adjusting of path in real-time.

Microsoft Kinect RGB-D Sensor

One of the main developments on the field of RGB-D sensor and of 3D mapping and reconstruction, came with the launch of Microsoft Kinect systems for the XBOX 360 gaming platform in November 2010. Kinect, like other RGB-D sensors available in the market provides RGB (color information) and estimation of depth for each pixel, however it was initially cheaper than its competitors. The Kinect, like other RGB-D sensors, provides color information as well as the estimated depth for each pixel (Litomisky 2012). In Firman (2016), the author makes an important review data sets of RGB-D images, providing a view into the earlier work of depth sensors, and focusing in the non-trivial task of collecting and working with RGB-D. It is clear that using the Microsoft Kinect brings advantage for research since is includes an array of sensors, such as: a 3D depth sensor; a color camera and a four-microphone array used to capture the full-body motion, and capabilities for facial and voice recognition (Zhang 2012).

In Henry et al. (2012), the authors present a review on the use of inexpensive (Kinect and Kinect-like) RGB-D cameras for indoor 3D mapping. They achieve this by detecting visual features and shape alignments in the visual information and combining it with depth information. Key aspects they found by using these cameras is that they can actually be used for building a 3D map of an indoor environment. However they are keen in highlighting that using RGB information or depth information by themselves will not yield a sufficiently robust model, both sensors have work in unison and be well calibrated, to provide information about the environment.

Noteworthy, is that the 3D depth cameras on the Kinect were built mostly to enhance the game-playing experience for Xbox games, however the capabilities they provide and applications can go beyond that main purpose. As an example, one of the main areas of interest of development using the Kinect is rehabilitation and physical therapy involving clinical patients (Mousavi Hondori and Khademi 2014).

Microsoft Kinect for BVI Navigation

In terms of BVI and RGB-D sensors, a number of authors have addressed the navigation problem by making use of the array of sensors that the Microsoft Kinect provides, more specifically using the depth sensor (Filipe et al. 2012). Here, the authors build the systems make use of neural networks to extract important features from the image and detect obstacles. It was based in classifying line or gradient profiles patterns using Neural Networks. It was developed using Microsoft SDK and Fast Artificial Neural Network (FANN) library for neural networks.

In Kanwal et al. (2015), the authors present a RGB-D fusion indoor/outdoor navigation system using Kinect. Both the camera and distance sensors are used to determine the proper navigation path. From the RGB camera, the systems obtains corner information and the depth sensor provides distances. With this information it detects the position of nearby obstacles and estimating their distance creating a depth map. As this implementation was tested on both blindfolded and visually impaired individuals, they share their conclusions about the system performance and more importantly details of situations in which the systems works less well. For instance, using the Kinect on outdoor locations decreased the performance of the mapping, due to sensor saturation caused by sunlight and shadow effects.

INDOOR NAVIGATION SYSTEM BASED ON KINECT DEPTH SENSOR: A WORKING EXAMPLE

In this section we expose the inner workings of a proposed navigation system based on the Kinect V.1 depth sensor to assist blind and visually impaired individuals (BVI). The Kinect depth sensor is used to detect obstacles and other elements in the scene such as walls and stairs and provide a safe direction to walk free of obstacles. Similar to the work in Filipe et al. (2012) and unlike the systems proposed by Kanwal et al. (2015; Lee and Medioni 2014), the proposed systems uses only information provided from the depth sensor. The proposed systems combines two approaches. In one hand, 1-D line or gradient profile analysis based in the patterns of changes of its derivatives, leading to detection of a walkable path at a very low computational cost. In the other hand, a second 2-D approach provides additional information of the surfaces in the scene such as wall, provided their orientation. Thus, the system provides path and surfaces (i.e., 3D mapping) detection similar to RGB-D based systems using only in the depth sensor (i.e., without needing ilumination) at low computational cost.

Image Acquisition

The principal component of our system is the Microsoft Kinect. This device includes an infrared emitter, an RGB camera and a depth camera or sensor. The proposed system uses mainly the infrared emitter to project an infrared light pattern in the room or environment. This pattern is modified in contact with the surfaces encountered. The Kinect V.1 depth sensor provides depth images that are coded in 24 bits by pixel, with an image resolution of 640×80 pixels. This sensor operates in the range between 0.80m to 4m of distance. Objects located in farther distances are represented as whiter gray tones. In the other hand, object located closer to the sensor are represented in darker gray tones. The black tone can represent object located out of the range of the sensor (closer than 0.80m or farther than 4m). Also, surfaces that deflect or absorb the infrared lights appear in the black tone.

For this study, a set of RGB images (and videos) with its corresponding depth images were acquired indoors, particularly in a normal household setting. These images include scenes of short corridors outside the apartment, areas near stairs and of different living areas inside the apartment. In order to capture the information of the walking surface (floor) the Kinect was positioned in the chest of the user with an inclination of 30° (towards the floor). Since our main focus is to provide a tool for BVI speed is not an issue, and moreover, the bottom half of the images are analyzed.

In Figure 2 a RGB image and its correspond depth image is shown. Figure 3 shows a close-up of the details at floor level of the same image.

Figure 2. Example RGB-D Input indoor scene images from Kinect cameras. Image color on the left shows an example region where the data was collected. Grey image color on the right shows the depth image from the left image using a Kinect, black pixels where no depth could be estimated.
Source: The Authors

For this study two kind of depth image analysis were considered:

- The columnwise approach: consists in analyzing the columns of the matrix representation of the depth images considering two aspects: the depth of each pixel and the evolution of the depth values along the column (i.e. slope and patters of the depth values). This approach provides information about walkable paths in the environment with a low computational cost.

- The two-dimensional approach: provide a more robust complementary information of the elements in the scenes by analyzing the planes formed by walls, floor and other surfaces.

Hardware and Software

The Kinect V.1 was connected to a 2.5 Ghz Core i5 64-bits processor computer with 8 GB of RAM memory. For the adquisition, Matlab and its image acquisition toolbox (MathWorks Inc, Natick MA) were used. Some additional supports package of the Matlab adquisition toolbox and Windows were also required to be installed. This set up was able to capture depth images for the indoor environemnt. These captured images were processed offline on the Matlab environment to test the two implemented approaches that will be described in the next sections.

In the following sections, more details of each approach is presented.

Columnwise Approach

Several aspects have been considered to choose the areas where the user can walk without obstacles. One aspect to take into account is the range to be considered in the analysis. As mentioned earlier, the depth sensor provides information in the range of 0.80m to 4m. For instance, Figure 3 presents the same image presented in Figure 2 but only considering the range between 0.8m and 2.5m. Objects located at farther distances are eliminated (seen as plain black pixels).

Figure 3. Columnwise approach region of analysis. Image on the left shows the depth image. Image on the right shows the binary image results after using the columnwise approach. Object located within the range between 0.8m to 2.5m are extracted (white pixels) while objects out of the range are removed (black pixels). Source: The Authors

In order to define a possible path for users, the system can consider the presence or absence of elements in each column of the image matrix which are below a given level of depth. For instance, Figure 3 presents a binary image indicating pixels with a depth inferior to 1m.

For this study a slow walking speed is considered. Consequently, it was opted to reduce the area of analysis of the image to the bottom section of the depth image as shown in Figure 2. The selected lower section of the image is presented in Figure 3.

To select a suitable path for the user it has to detect and classify columnwise the elements and objects in the image. Indeed, besides defining that there are no elements closer than 1m to the user in a given segment of the image. It is also important to determine if a given section of the images corresponds to a floor or more importantly if the user is approaching a structure such a wall or a set of stairs. Analyzing the slope patterns of the depth values of the elements of the column allows to infer if it corresponds to a floor and consider that section of the image as walkable path candidate.

Figure 4 shows an example of how slope pattern is represented in a depth image. As expected the floor like slope presents a small positive slope (shown in Figure 4a). For upward stairs, the slope presents a triangular wave like pattern (shown in Figure 4b), and for walls (i.e. floor then a wall), slopes presents a small positive then negative (shown in Figure 4c).

Figure 4. Typical slope patters in different indoor scenes,(a) floor-like surface, (b) stair (upward), and (c) floor and wall. Source: The Authors

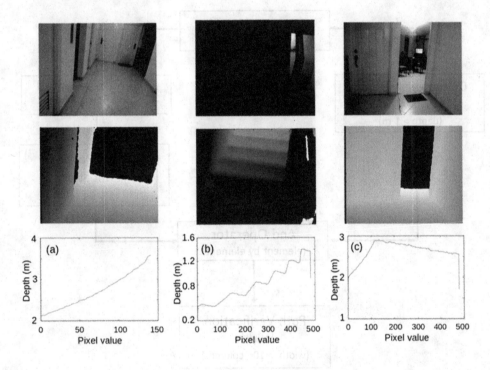

Implementation Details

To determine if a path is walkable or to detect the presence of structures of a wall or upstairs the following algorithms were implemented.

- **Walkable path detection**: To steps to detect walkable path are shown in Figure 5:
    1. Analyzing the depth of the elements of each column of the image If the depth is below the selected threshold (1m), the column is keep as walkable path candidate

2. Verifying the floor-like slope of the path: To determine the evolution of the depth of the elements in the column, the slope of segments of the columns are evaluated. This analysis only considers elements in the bottom part of the image. For this implementation, the 80 lowest elements of the column are considered. To minimize the effect of acquisition noise, a 3-by-3 2D median filter is applied to the depth image. Also finite impulse response (FIR) mean filter of length 3 is applied to each column. The slope is evaluated in intervals of 10 elements. To consider that a column corresponds to a floor the slopes in each segment must be in given range of value. For this implementation, the range is between 0 and 7 degrees.

3. Verifying the width of the path: To be considered a walkable path it is necessary that a given number of contiguous columns verify the depth and slope condition. For our implementation, we considered a minimum of 200 columns that corresponds to approximately to 1m of width.

Figure 5. Flowchart of the Walkable Path Detection Method. Source: The Authors

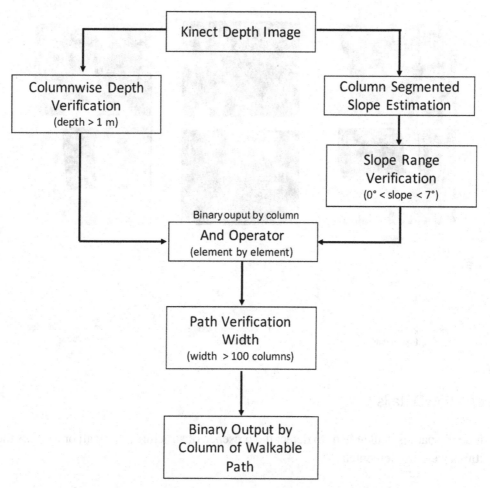

- **Stair and wall detection**: To detect if a wall or stair is in a given path of the user, the following steps are considered. An overall depiction the steps of the algorithm can be seen Figure 6:
    1. Obtaining values of the slope of different segments of a given column. For this analysis all the elements of the column are considered. For this implementation, segments of 10 elements were considered for the stair detection, while segments of 30 elements were considered for the wall detector.
    2. Binarizing the segments using 0 degrees as threshold to determine the sequence of segments with positive and negative slopes.
    3. Analyzing the binary pattern to verify the corresponding pattern. For the case of upstairs, we expect several changes of sign of the slope, while for the wall we expected only one change of sign from positive to negative.

Figure 6. Flowchart of the Column-wise Stair and Wall Detection Algorithm. Source: The Authors

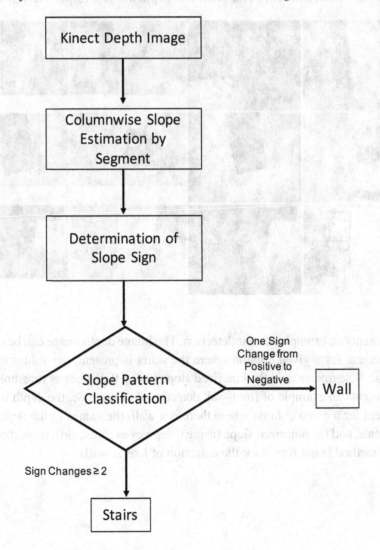

Results of the Columnwise Approach

A video sequence with the working implementation of the walkable path detection can be found in http://tinyurl.com/NavigationAidUTP. Some screenshots of this video are presented in Figure 7. The video presents a divided screen with the real time the RGB image and the depth image (in green tones) in the upper section and the same image with vertical red bands in the down section. The red bands indicate the path of the image where the user is suggested to avoid. The only object in this example that was not able to properly detect was the metallic elevator door. Indeed, for objects that are prone to absorb infrared light (e.g. metallic objects) the sensor is not able to provide depth information.

Figure 7. Image captions of the working implementation of the system tested under different indoor scenes from a Kinect. The first column shows the input RGB image. The second column shows the depth images information represented by monochrome green color. The third column and fourth columns present walkable path regions (not colored in red) in the RGB and depth images, respectively. Source: The Authors

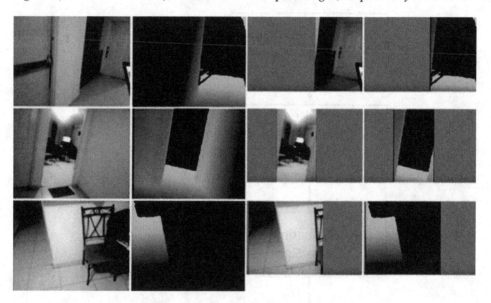

Figure 8a-c presents an example of stair detection. The image depth image can be observed in Figure 8b. Figures a-b present for a given column where the stairs is present, the values of the depth of the elements, the slope of segments, and the binarized slope (using 0 degrees as threshold), respectively.

Figure 8d-f presents an example of front-wall detection. In Figure 8e, the depth image is presented. Figures 9d-f present for a given column where there is a wall, the values of the depth of the elements, the slope of segments, and the binarized slope (using 0 degrees as threshold), respectively. We observed, that the proposed method is not robust for the detection of lateral walls.

Figure 8. Detection of stairs and wall by column wise approach. (a) Stairs RGB image (b) Stairs depth image (b) Output of the stairs detector by column (d) Wall RGB image (e) Wall depth image (f) Output of the wall detector by column. Source: The Authors

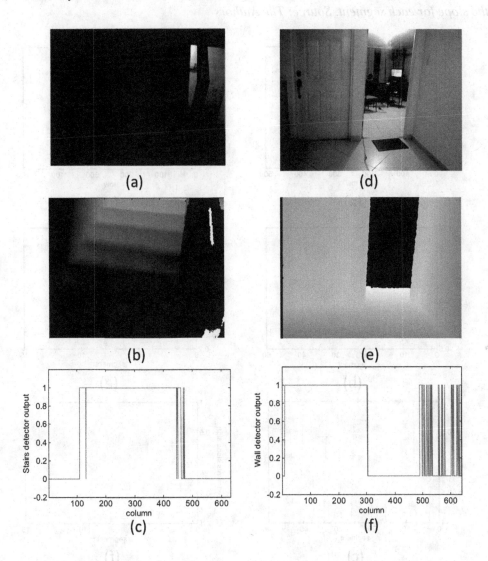

Two-Dimensional Approach

The columnwise approach is not exploiting the whole spatial information of the depth image, considering the depth image as 2D array of data allows to provide additional and more robust information of the elements or structures in the environment. An indoor environment presents elements with plane surfaces such as floor, wall in several positions, doors, chairs, refrigerators, etc. The two dimensional approach aims to provide information concerning the plane surfaces of the environment in order to provide complementary details to the user. The approach consists in three steps: surface detection, surface orientation estimation and structure classification. The overall depiction of the steps taken in this approach can be seen Figure 10.

Figure 9. Analysis of pixel evolution for stair (column 316, seen in Figure 8b) and wall detection (colums 213, seen in Figure 8d). For the stairs: (a) Depth of the pixel in each row (b) Slope of segments (c) Sign of the slope for each segment. For the wall: (d) Depth of the pixel in each row (e) Slope of segments (f) Sign of the slope for each segment. Source: The Authors

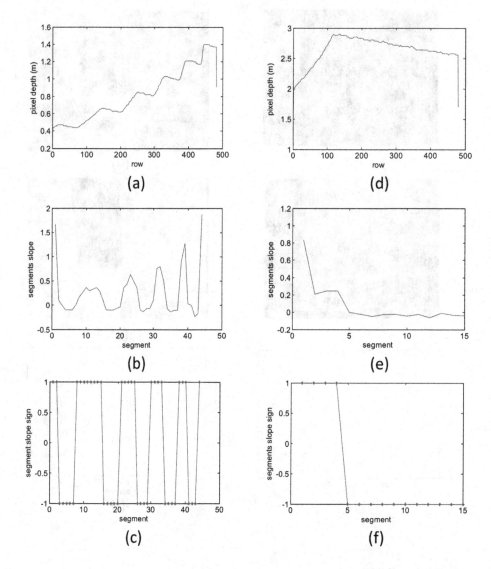

Implementation Details of the Two-Dimensional Approach

In the following subsection implementation details of the steps of the approach are presented:

- **Surface detection**: The depth image must be divided in several segments that correspond to the different surfaces in the environment. This is not a trivial task since the sensor does not provide enough resolution and precision to provide clear edges between the different surfaces of the image. However, elements in a given surface present some common features such gradient orienta-

Figure 10. Flowchart of the two dimensional approach. Source: The Authors

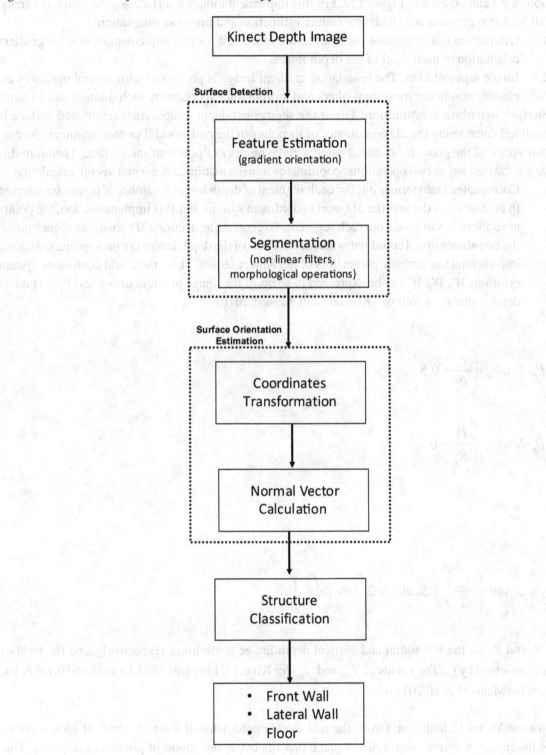

tion as we can observe in Figure 11c. For this implementation the surface segmentation is carried out with two operations: orientation feature estimation and image segmentation.

1. Orientation feature estimation: The feature of interest for this implementation is the gradient orientation of each pixel of the depth image.

2. Image segmentation: The orientation gradient image is processed with several operators including non-linear maximum filters and morphological operators such dilation and erosion.

*   **Surface orientation estimation:** Given the segmented depth image, each segmented surface is analysed considering the 3D coordinates of its points in the real world. For each segment, the normal vector of the plane is estimated using a given number of points of the surface. Therefore this step is carried out in two operations: coordinates transformation and normal vector calculation

1. Coordinates Transformation: For each segment of depth image a number of points are selected to be converted the into the 3D world coordinate system. For this implementation, the points are uniformly sampled from each segment. To generate an accurate 3D positions of each point, the two-dimensional coordinates and depth value of the depth image are back-projected taking into account the intrinsic parameters of the Kinect sensor. Thus, the world coordinate system positions W_x. W_y. W_z can be expressed in terms of the image pixels coordinates (P_y, P_x) and its depth value P_z as follows (Kowsari and Alassaf 2016):

$$W_x = P_z \cdot Scale_x \cdot \left(\frac{P_x}{R_x} - 0.5 \right)$$

$$W_y = P_z \cdot Scale_y \cdot \left(\frac{P_y}{R_y} - 0.5 \right)$$

$$W_z = P_z$$

with,

$$Scale_x = 2 \cdot \tan \cdot \left(\frac{f_{ov}^x}{2} \right) \quad Scale_y = 2 \cdot \tan \cdot \left(\frac{f_{ov}^y}{2} \right) ..$$

where R_x and R_y are the horizontal and vertical depth image resolutions, respectively, and the field of view is represented by f_{ov}. The values of f_{ov}^x and f_{ov}^y for Kinect V1 are equal to 1.014468 and 0.7898094, respectively (Mankoff et al. 2011).

2. Normal Vector Calculation: Given the real world coordinates of a set of points of each segment of the image, a normal vector of the plane that fits better this cloud of points is calculated. This normal vector must be adjusted taking into account the inclination the sensor.

○ **Structure Classification:** The obtained normal vectors provides information of the orientation of the surfaces. Thus, surfaces corresponding to floors provides normal vectors oriented upwards; front walls provide normal vectors oriented frontward or backward and lateral walls provide normal oriented to the sides.

Results of the Two-Dimensional Approach

Figure 11 presents an example of orientation feature estimation and image segmentation for the image presented. In Figure 12, the results of the surface detection orientation stage, is shown. This image illustrates how the estimated orientation of each surface allow to infer the kind of element in the image, such as wall, front wall and lateral walls.

Figure 11. Depth image surface detection: (a) RGB image, (b) Depth image, (c) Gradient orientation of the depth image and (d) Segments obtained from the gradient orientation image. Source: The Authors

Figure 12. Segmented surfaces orientation estimation. Source: The Authors

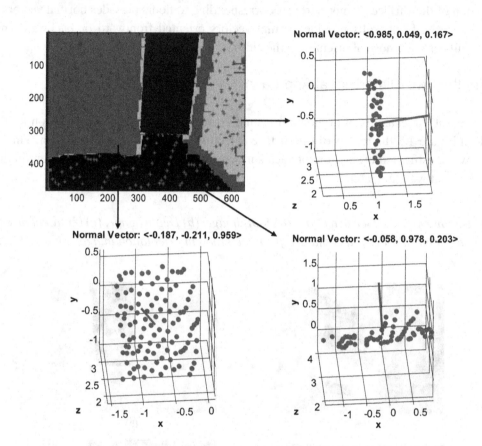

Discussion of the Results Obtained

Two kinds of approaches are propose proposed to exploit the data provided by the depth sensor in a Kinect Device for BVI to be able to navigate in an unknown environment. The columnwise approach provides a preliminary detection of walkable path with a low computational cost. A video showing the working of this implementation is provided. The columnwise information can also provide some insights with respects other elements in the scene such stairs and wall by analysing the evolution of the depth values of the columns of the depth image. This approach can be extended to detect ramps or other elements with distinctive slope patterns. Also, for the case of the stairs, it could adapted to estimate the dimensions of the steps. However, this element detection approaches are limited by the angle of approximation to the structure. They can detect properly structures in front of the user, but they are not robust with structures appearing in parallel to the trajectory of the user.

The two-dimensional approach provides complementary and robust information about elements in the scene with plane surfaces. For this implementation the gradient orientation of the image was exploited to provide features to distinguish and segment the surfaces in the scene as we can see in Figure 13 and Figure 14. However, the segmentation task from this image is not trivial and could lead to the fusion of surfaces of the image as shown in Figure 14. Other features and segmentation approaches could be explore to improve the robustness of this approach.

Figure 13. Corridor image surface detection: (a) RGB image, (b) Depth image, (c) Gradient orientation of the depth image and (d) Segments obtained from the gradient orientation image. Source: The Authors

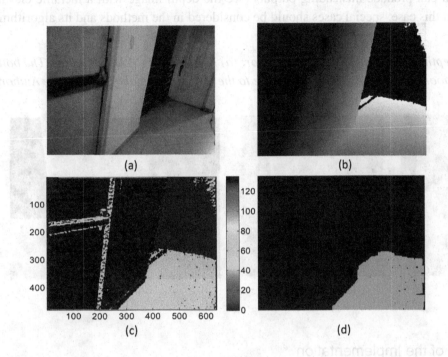

Figure 14. Living room image surface detection: (a) RGB image, (b) Depth image, (c) Gradient orientation of the depth image and (d) Segments obtained from the gradient orientation image. Source: The Authors

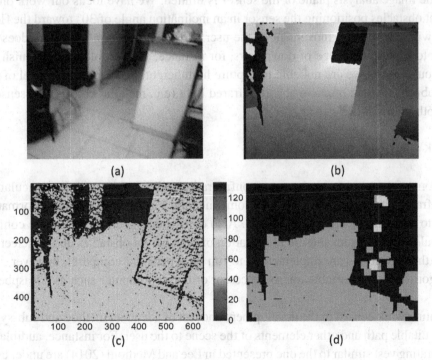

Both approaches are constrained by the limitations of the sensing device. Thus, elements that absorb infra-red light can produce misleading outputs (see the depth image with a metallic elevator door in Figure 15). In this case, special cases should be considered in the methods and its algorithms.

Figure 15. Depth image of metallic elevator door: (a) RGB image, (b) Depth image. The bottom part of the elevator door provides misleading response to the infra-red signals. Source: The Authors

(a)   (b)

Limitations of the Implementation

The proposed system has some limitations given the properties of the depth sensor and its configuration. In one hand the image analysis plane of the sensor is limited. We have focus our work on near ground and low height obstacles positioning the sensor in an inclination angle of 30° toward the floor, that may vary slightly with the walking movement of the user. However, this configuration, does not allow us to distinguish to capture the slope of downstairs, for instance, and be able to distinguish them from a hole in the ground. Also, we are not able to capture high height obstacle (e.g. at level of the head). In addition, for objects that are prone to absorb infrared light (e.g. metallic objects) the sensor is not able to provide depth information.

Future Work

Some limitations of the current depth sensor configuration will be addressed. In particular, to avoid the issues with infrared absorbing materials such some metallic surfaces, we can use information from the RGB camera to correctly identify such objects. Also other kind of sensor or rotative configuration for the sensor would let the system capture information of high height obstacles and to detect downstairs.

Currently, the algorithms are implemented to run on a personal computer, however, it is aimed to embed the algorithms on a portable micro-processor or micro-computer such as a Raspberry Pi to run in real-time.

Another future work consist in the development and test of adequate method and sub-system to communicate the suitable path and other elements of the scene to the user. For instance, audible sub-systems and haptic vibrating vest similar to the one presented in Lee and Medioni (2014) are under consideration.

The selection of such systems requires adequate test of the user experience, to confirm the approaches and methods that are best suited to provide the information to the user.

SOLUTIONS AND RECOMMENDATIONS

Limitations of Working with a RGB-D Camera

As described in Mallick, Das, and Majumdar (2014), there are implicit limitations of working with depth sensors. Issues arise in distance estimation, also noise problems that can surge because of objects that does not reflect well infrared wavelengths. Also, occlusions between objects become shadows from which no distance can be measured. Moreover, there is a major saturation effect when measuring in presence of strong light sources (such as sunlight). Finally, the depth values obtained from sensors might not be stable even when the camera is fixed and the object is not in movement. In the case of Kinect the authors go to say, that *"one has to be careful while using Kinect to develop navigation systems as the sensor"*, alluding to the fact that Kinect systems does not escape these reflection and errors in sensing of other RGB-D cameras.

Andersen et al. (2012) made an early report on the usability and limitations of the Kinect for Xbox 360 platform in terms depth sensor in general terms of: availability of software frameworks and spatial and depth resolution. More importantly, it provides a notion of structural noise of the system, for instance, is noteworthy that the depth estimation in the image is calculated based on the distance from the point to the sensor plane, instead of the distance from the point to the sensor. If this is notion applied correctly then if the sensor is pointing at a planar surface just in front of the Kinect, then all points in that plane will results in having similar depth estimates for the entire image, but variations are notably present on the depth image. The authors attribute it to The automatic transformation on the sensor over the depth data, also they attribute it to the sensors spatial resolution being not sufficient enough to record the infrared pattern correctly. Similar to (Mallick, Das, and Majumdar 2014), the authors also describe issue with getting depth information from low reflectivity materials, which is something common to all RGB-D systems. They also give an advice about a saturation of the Infrared sensor, and provide a tip, in few words using computer vision to complement the image depth image with the RGB image and vice-versa. However it is important to note that this claims only apply for the Kinect for Xbox 360, and not for the Kinect for Windows or Azure Kinect, subsequent versions that have reach the market in the last years.

Computer Vision Algorithms to Complement Depth Sensor Analysis

An important token of information of the limits of RGB-D based systems presented in Mallick, Das, and Majumdar (2014) is that there no only problems with low reflectivity objects but in terms complete surface, for instance when dealing with a floor or surface with a high refractive index. In these cases the authors allude to the use computer vision algorithms to understand the complete image content. Moreover, it is known that computer vision analysis can provides rich information of the user environments (Peters 2017), that can put together in a device for real-time navigation for BVI individuals (Li et al. 2018).

FUTURE RESEARCH DIRECTIONS

Looking of what coming ahead in the field of indoor navigation systems based on depth sensors for BVI three big areas can be distinguished: machine learning and deep learning techniques, combination of multiple systems (based on multiple sensor arrays) and portable systems.

Machine Learning and Deep Learning Techniques

Outside of navigation systems machine learning has been proposed ans used for other issues suffered by BVI individuals. On a review article by Barros et al. (2020) the authors present a list results in the area of retinal image processing for glaucoma diagnosis and detection.

Deep Learning

Deep learning techniques has now become the tool of choice for many researchers and product engineers, one of the reasons is that its performance is far superior to conventional artificial neural networks (Nagata et al. 2020; Nielsen 2015; LeCun, Bengio, and Hinton 2015). In the case of working with images and images classification convolutional neural networks (DCNN) extend the concept of convolutional neural networks by applying a set of predefined filters and masks to extract relevant features of the image (Merchan et al. 2020).

For indoor navigation many authors now complement the use of RGB and RGB-D images not only with computer vision algorithms, but with deep learning based subsystems. Moreover these methods are now easy to implement on smartphones directly (Ignatov et al. 2019). For instance using Convolutional Neural Networks (CNNs) for object recognition or Recurrent Neural Networks (RNNs), for scene classification (Lin et al. 2019).

In (Breve and Fischer 2020) authors center on the issue of BVI navigation and propose a low computational cost framework geared towards smartphones, which make use of convolutional neural networks (CNN), transfer learning (TL), and semi-supervised learning (SSL). In terms of which neural network architectures, they tested 17 different CNN architectures pre-trained on the ImageNet dataset, with VGG16 and VGG19 architectures resulting as best classification accuracy. More interesting is the semi-supervised learning (SSL) approach presented, in which the systems is capable of using the best resulting pre-trained CNN architectures (VGG16 and VGG19) as feature extractors. Then using Particle Competition and Cooperation (PCC) as semi-supervised learning algorithm. Hinting into the idea of systems that learn on the spot from the user, that is correcting the outputs given as feedback. For instance, when a predicted "clear path" was really am "unclear path" (and vice-versa), correcting not only the output of the algorithm, but adding generating a example from the user.

Combination of Multiple Sensors

In recent years, a new RGB-D camera, the Intel RealSense, has entered the market revolutionizing it by improving problems suffered by earlier versions of Microsoft Kinect. It offers a higher depth sensing capabilities. For instance the device can be used well in both indoor and outdoor environment, the error is in the order of millimeters and it has a longer range of 6 meters. More importantly it has dual (stereo) infrared (IR) receivers. A few examples of its use in computer vision research applied to BVI can be found in literature.

For instance in, Yang et al. (2016) introduce the use of RealSense RGB-D cameras for improving and expanding the free navigation path, helped by the large-scale matching of the IR sensor combined with RGB filtering. The navigation path is selected via RANdom SAmple Consensus (RANSAC) segmentation and the normal vector of the surface in the scene is also obtained. This navigation path is later expanded via a seeded growing region algorithm that takes advantage of the information from the depth and RGB images. To provide a larger navigation path, it takes into account immediate objects and thus allowing for a longer path planning and navigation. The authors present a prototype that provides feedback via an audio interface, used with BVI invididuals, with interesting results.

The same group also presented a similar approach in Yang et al. (2017), with their results suggesting that their algorithm and implementation can provide better path discovery in close range (165 mm) and long range (5000 mm). In Yang et al. (2018) the same authors present an enhanced version of the system that relies in a pair of smart glasses equipped with a RealSense R200 (RGB-D sensor), and a RealSense RS410 sensor worn in the waist specifically for location of low and close to the ground objects. The feedback is given by both sensors to the user via a bone-conduction headphone.

Similar efforts are presented Chen, Wang, and Yang (2018), authors present an algorithm that is based on the fusion of the images coming from two RealSense R200 devices, thus making stereo vision plus depth. It works by basically building a color and infrared stereo vision, that is be used among other things, to improve the depth estimation on transparent, reflecting and absorbing objects. Something that is quite necessary when dealing with indoor navigation problem for BVI individuals.

Portable Systems

Of one of the most obvious area of development is the use of portable systems. In that sense several ETA, describe earlier in the document, are already implemented to have small, efficient at portable systems as the basis. As examples implementations that make use of mobile applications and smartphones. Scanning the research literature it becomes evident that more recent systems are using the stand-alone microprocessors, smartphones, tablets and embedded computing boards as main computing agent. Putting aside laptops that area usually carried as a backpack in many ETA systems.

Example of this ideas, grouped by the umbrella name Embedded 3D for Vision is given in Poggi, Nanni, and Mattoccia (2015) and Poggi and Mattoccia (2016), where authors design a wearable system based on Convolutional Neural Networks. They use it for the detection and categorization of crosswalks and general real-world obstacles that can be an issue for BVI individuals. In both cases the system consisted of, among other things, a pair glasses (with RGB-D camera), a glove to provide vibration feedback, a pocket battery, a smartphone and a Field-programmable gate array (FPGA) integrated circuit for processing the image set.

CONCLUSION

Global estimation figures tell us that near 2.2 billion people are BVI, with a great majority being 50 years or older and more likely to be living in a low or a middle-income country. Navigation is still an issue to be solved for BVI individuals.

Many assistive technology solutions in has been proposed. These devices are usually categorized as: Electronic Travel Aids (ETA), Electronic Orientation Aids (EOA) and Position Locator Devices (PLD). However, the lines separating these categories are blurred and many devices provide a combination of services which can make them fit in various categories. Without a doubt, these devices have changed radically as time passes and new technologies become more mainstream. What began as simple sensors built into a cane, to bulky array of sensors and equipment that needs to be carried by the user, to wearable or non-invasive devices that seamlessly integrate into the users daily life.

This document focuses on vision based assistive technologies for BVI navigation. A great number of systems have been build around images from RGB cameras be it still cameras, smartphones cameras or even real-time video captures. These system have addressed well the problems of indoor and outdoor navigation, with systems even addressing the problem posed by staircases, a definitive challenge for a BVI individual.

The proposed indoor navigation system is able to provide walkable paths relying only in the depth sensor. It uses the combination of two approaches. In one hand, a 1-D analysis of the line or gradient profile provides a low computational cost detection of walkable path. Another 2-D approach provides information about the surfaces (3-D mapping of the scene). The system present some inherent limitation due to the depth sensor characteristics.

Special interest has been place in RGB-D cameras, or cameras able of acquiring not only color, but also depth of the image. There are many cameras of this type in the market, but for many years the Microsoft Kinect has been used as a tool for research, as well as for gaming, which was its original purpose. In this regards, through this document a first-hand experience on the use of Kinect is shared. Moreover, advances on the development of an indoor navigation aid system for BVI individuals, is presented. The proposed system takes RGB-D images provided by the Kinect and are enhanced using computer vision algorithms. Results suggest that the system as a whole can be used by BVI individuals to be able to navigate in an unknown environment. Results also suggest a few limitations that can be encountered in building such a system. These limitations are listed and manners to overcome them, found in literature are discussed.

The field of indoor navigation aids for the BVI is still providing new avenues of research. In the last years new systems integrating computer vision and machine learning algorithms are flourishing. This phenomenon is probably tied to the fact that newer embedded computing boards are geared toward rapid computing of images. Even some are ready build with deep learning processing capabilities, which is the case for Tensor Processing Units (Jouppi et al. 2018). In some case these capabilities are not build in, but can extended for this use via software updates. Moreover, since the arrival of RGB-D cameras dedicated to computer vision, researchers in the field are now focusing in solving the navigation of BVI individuals with stereo depth sensor. This will provide systems focused on creating larger and safer navigation paths, systems that will be well received by BVI individuals and their families.

FUNDING

This research received no specific grant from any funding agency in the public, commercial, or not-for-profit sectors.

ACKNOWLEDGMENT

The Sistema Nacional de Investigación (SNI) of the Secretaría Nacional de Ciencia, Tecnología e Innovación de Panamá (SENACYT) supports research activities by F.M. (Contract No. 17-2020), D.C. (Contract No. 168-2017) and J.E.S.-G. (Contract No. 129-2018). The authors acknowledge administrative support provided by CEMCIT-AIP and Universidad Tecnológica de Panamá.

REFERENCES

Ackland, Resnikoff, & Bourne. (2017). World Blindness and Visual Impairment: Despite Many Successes, the Problem Is Growing. *Community Eye Health, 30*(100).

Advani, S., Zientara, P., Shukla, N., Okafor, I., Irick, K., Sampson, J., Datta, S., & Narayanan, V. (2016). A Multitask Grocery Assist System for the Visually Impaired: Smart Glasses, Gloves, and Shopping Carts Provide Auditory and Tactile Feedback. IEEE Consumer Electronics Magazine, 6(1), 73–81.

Andersen, M. R., Jensen, T., Lisouski, P., Mortensen, A. K., Hansen, M. K., Gregersen, T., & Ahrendt, P. J. A. U. (2012). Kinect Depth Sensor Evaluation for Computer Vision Applications. Aarhus University.

Apostolopoulos, Fallah, Folmer, & Bekris. (2014). Integrated Online Localization and Navigation for People with Visual Impairments Using Smart Phones. *ACM Transactions on Interactive Intelligent Systems, 3*(4), 1–28.

Assistive Technology Industry Association (ATIA). (2020). *What Is at?* Assistive Technology Industry Association (ATIA). https://www.atia.org/at-resources/what-is-at/

Bai, Lian, Liu, Wang, & Liu. (2017). Smart Guiding Glasses for Visually Impaired People in Indoor Environment. *IEEE Transactions on Consumer Electronics, 63*(3), 258–66.

Barros, Moura, Freire, Taleb, Valentim, & Morais. (2020). Machine Learning Applied to Retinal Image Processing for Glaucoma Detection: Review and Perspective. *BioMedical Engineering OnLine, 19*, 1–21.

Bhatlawande, S. S., Mukhopadhyay, J., & Mahadevappa, M. (2012). Ultrasonic Spectacles and Waist-Belt for Visually Impaired and Blind Person. In *2012 National Conference on Communications (Ncc)*, (pp. 1–4). IEEE. 10.1109/NCC.2012.6176765

Bouhamed, S. A., Kallel, I. K., & Masmoudi, D. S. (2013). Stair Case Detection and Recognition Using Ultrasonic Signal. In *2013 36th International Conference on Telecommunications and Signal Processing (Tsp)*, (pp. 672–76). IEEE. 10.1109/TSP.2013.6614021

Bourne, Flaxman, Braithwaite, Cicinelli, Das, Jonas, Keeffe, Kempen, Leasher, & Limburg. (2017). Magnitude, Temporal Trends, and Projections of the Global Prevalence of Blindness and Distance and Near Vision Impairment: A Systematic Review and Meta-Analysis. *The Lancet Global Health, 5*(9), e888–e897.

Bousbia-Salah, Bettayeb, & Larbi. (2011). A Navigation Aid for Blind People. *Journal of Intelligent & Robotic Systems, 64*(3-4), 387–400.

Breitenstein, M. D., Kuettel, D., Weise, T., Van Gool, L., & Pfister, H. (2008). Real-Time Face Pose Estimation from Single Range Images. In *2008 IEEE Conference on Computer Vision and Pattern Recognition*, (pp. 1–8). IEEE. 10.1109/CVPR.2008.4587807

Breve, F., & Fischer, C. N. (2020). *Visually Impaired Aid Using Convolutional Neural Networks, Transfer Learning, and Particle Competition and Cooperation.* arXiv Preprint arXiv:2005.04473

Caldini, A., Fanfani, M., & Colombo, C. (2015). Smartphone-Based Obstacle Detection for the Visually Impaired. In *International Conference on Image Analysis and Processing*, (pp. 480–88). Springer. 10.1007/978-3-319-23231-7_43

Carbonara, S., & Guaragnella, C. (2014). Efficient Stairs Detection Algorithm Assisted Navigation for Vision Impaired People. In *2014 IEEE International Symposium on Innovations in Intelligent Systems and Applications (Inista) Proceedings*, (pp. 313–18). IEEE. 10.1109/INISTA.2014.6873637

Chen, H., Wang, K., & Yang, K. (2018). Improving Realsense by Fusing Color Stereo Vision and Infrared Stereo Vision for the Visually Impaired. *Proceedings of the 2018 International Conference on Information Science and System*, 142–46. 10.1145/3209914.3209944

Cook, A. M., & Polgar, J. M. (2014). *Assistive Technologies-E-Book: Principles and Practice.* Elsevier Health Sciences.

Council, National Research. (1986). *Electronic Travel Aids: New Directions for Research.* National Research Council.

Dakopoulos, Boddhu, & Bourbakis. (2007). A 2D Vibration Array as an Assistive Device for Visually Impaired. In *2007 IEEE 7th International Symposium on Bioinformatics and Bioengineering*, (pp. 930–37). IEEE.

Elmannai & Elleithy. (2017). Sensor-Based Assistive Devices for Visually-Impaired People: Current Status, Challenges, and Future Directions. *Sensors, 17*(3), 565.

Filipe, Fernandes, Fernandes, Sousa, Paredes, & Barroso. (2012). Blind Navigation Support System Based on Microsoft Kinect. *Procedia Computer Science, 14*, 94–101.

Firman, M. (2016). RGBD Datasets: Past, Present and Future. *Proceedings of the Ieee Conference on Computer Vision and Pattern Recognition Workshops*, 19–31.

Ganz, A., Schafer, J., Gandhi, S., Puleo, E., Wilson, C., & Robertson, M. (2012). PERCEPT Indoor Navigation System for the Blind and Visually Impaired: Architecture and Experimentation. *International Journal of Telemedicine and Applications, 2012*, 2012. doi:10.1155/2012/894869 PMID:23316225

Guerrero, J. J., Pérez-Yus, A., Gutiérrez-Gómez, D., Rituerto, A., & López-Nicolás, G. (2015). Human Navigation Assistance with a Rgb-d Sensor. *ACTAS V Congreso Internacional de Turismo Para Todos: VI Congreso Internacional de Diseno, Redes de Investigacion Y Tecnologia Para Todos Drt4all*, 285–312.

Habib, A., Islam, M. M., Kabir, M. N., Mredul, M. B., & Hasan, M. (2019). Staircase Detection to Guide Visually Impaired People: A Hybrid Approach. *Revue d'Intelligence Artificielle, 33*(5), 327–334. doi:10.18280/ria.330501

Harms, H., Rehder, E., Schwarze, T., & Lauer, M. (2015). Detection of Ascending Stairs Using Stereo Vision. In *2015 Ieee/Rsj International Conference on Intelligent Robots and Systems (Iros)*, (pp. 2496–2502). IEEE. 10.1109/IROS.2015.7353716

He, Goodkind, & Kowal. (2016). *An Aging World: 2015*. U.S. Census Bureau, International Population Reports, P95/16-1. U.S. Government Publishing Office.

Henry, Krainin, Herbst, Ren, & Fox. (2012). RGB-d Mapping: Using Kinect-Style Depth Cameras for Dense 3D Modeling of Indoor Environments. *The International Journal of Robotics Research, 31*(5), 647–63.

Hernandez, D. C., & Jo, K.-H. (2010). Outdoor Stairway Segmentation Using Vertical Vanishing Point and Directional Filter. In *International Forum on Strategic Technology 2010*, (pp. 82–86). IEEE. 10.1109/IFOST.2010.5667914

Hernandez, D. C., & Jo, K.-H. (2011). Stairway Segmentation Using Gabor Filter and Vanishing Point. In *2011 IEEE International Conference on Mechatronics and Automation*, (pp. 1027–32). IEEE. 10.1109/ICMA.2011.5985801

Hernández, D. C., & Jo, K.-H. (2011). Stairway Tracking Based on Automatic Target Selection Using Directional Filters. In *2011 17th Korea-Japan Joint Workshop on Frontiers of Computer Vision (Fcv)*, (pp. 1–6). IEEE. 10.1109/FCV.2011.5739752

Hernández, D. C., Kim, T., & Jo, K.-H. (2011). Stairway Detection Based on Single Camera by Motion Stereo. *International Conference on Industrial, Engineering and Other Applications of Applied Intelligent Systems*, 338–47. 10.1007/978-3-642-21822-4_34

Hersh & Johnson. (2010). *Assistive Technology for Visually Impaired and Blind People*. Springer Science & Business Media.

Hill & Bradfield. (1986). Electronic Travel Aids for Blind Persons. *Journal of Special Education Technology, 8*(3), 31–42.

Hoydal, T. O., & Zelano, J. A. (1991). An Alternative Mobility Aid for the Blind: The 'ultrasonic Cane'. In *Proceedings of the 1991 Ieee Seventeenth Annual Northeast Bioengineering Conference*, (pp. 158–59). IEEE. 10.1109/NEBC.1991.154627

Hu, M., Chen, Y., Zhai, G., Gao, Z., & Fan, L. (2019). An Overview of Assistive Devices for Blind and Visually Impaired People. *International Journal of Robotics and Automation, 34*(5). Advance online publication. doi:10.2316/J.2019.206-0302

Ignatov, A., Timofte, R., Kulik, A., Yang, S., Wang, K., Baum, F., Wu, M., Xu, L., & Van Gool, L. (2019). *AI Benchmark: All About Deep Learning on Smartphones in 2019*. arXiv Preprint arXiv:1910.06663

Islam, Sadi, Zamli, & Ahmed. (2019). Developing Walking Assistants for Visually Impaired People: A Review. *IEEE Sensors Journal, 19*(8), 2814–28.

Jouppi, Young, Patil, & Patterson. (2018). Motivation for and Evaluation of the First Tensor Processing Unit. *IEEE Micro, 38*(3), 10–19.

Kammoun, Macé, Oriola, & Jouffrais. (2011). Toward a Better Guidance in Wearable Electronic Orientation Aids. In *IFIP Conference on Human-Computer Interaction*, (pp. 624–27). Springer. 10.1007/978-3-642-23768-3_98

Kanwal, N., Bostanci, E., Currie, K., & Clark, A. F. (2015). A Navigation System for the Visually Impaired: A Fusion of Vision and Depth Sensor. *Applied Bionics and Biomechanics, 2015*, 2015. doi:10.1155/2015/479857 PMID:27057135

Katz, Kammoun, Parseihian, Gutierrez, Brilhault, Auvray, Truillet, Denis, Thorpe, & Jouffrais. (2012). NAVIG: Augmented Reality Guidance System for the Visually Impaired. *Virtual Reality, 16*(4), 253–69.

Köberlein, Beifus, Schaffert, & Finger. (2013). The Economic Burden of Visual Impairment and Blindness: A Systematic Review. *BMJ Open, 3*(11).

Kowsari, K., & Alassaf, M. H. (2016). Weighted Unsupervised Learning for 3D Object Detection. *Information and Organization*. Advance online publication. doi:10.14569/IJACSA.2016.070180

Kunhoth, J., AbdelGhani, K., Al-Maadeed, S., & Al-Ali, A. (2020). Indoor Positioning and Wayfinding Systems: A Survey. Human-Centric Computing and Information Sciences, 10, 1–41.

LeCun, Bengio, & Hinton. (2015). Deep Learning. *Nature, 521*(7553), 436–44.

Lee & Medioni. (2016). RGB-d Camera Based Wearable Navigation System for the Visually Impaired. *Computer Vision and Image Understanding, 149*, 3–20.

Lee, Y. H., & Medioni, G. (2014). Wearable Rgbd Indoor Navigation System for the Blind. In *European Conference on Computer Vision*, (pp. 493–508). Springer.

Li, Munoz, Rong, Chen, Xiao, Tian, Arditi, & Yousuf. (2018). Vision-Based Mobile Indoor Assistive Navigation Aid for Blind People. *IEEE Transactions on Mobile Computing, 18*(3), 702–14.

Lin, Y., Liu, Z., Huang, J., Wang, C., Du, G., Bai, J., & Lian, S. (2019). Deep Global-Relative Networks for End-to-End 6-Dof Visual Localization and Odometry. In *Pacific Rim International Conference on Artificial Intelligence*, (pp. 454–67). Springer. 10.1007/978-3-030-29911-8_35

Litomisky, K. (2012). Consumer Rgb-d Cameras and Their Applications. Rapport Technique, University of California 20.

Liu, W., McGucken, E., Vitchiechom, K., Clements, M., De Juan, E., & Humayun, M. (1997). Dual Unit Visual Intraocular Prosthesis. *Proceedings of the 19th Annual International Conference of the Ieee Engineering in Medicine and Biology Society. 'Magnificent Milestones and Emerging Opportunities in Medical Engineering'(Cat. No. 97CH36136), 5*, 2303–6. 10.1109/IEMBS.1997.758824

Lodh, Subramaniam, & Paswan. (2016). Ultrasound Based Assistive Mobility Devices for the Visually-Impaired. In *2016 Ieee 7th Power India International Conference (Piicon)*, (pp. 1–6). IEEE.

Mallick, Das, & Majumdar. (2014). Characterizations of Noise in Kinect Depth Images: A Review. *IEEE Sensors Journal, 14*(6), 1731–40.

Manjari, K., Verma, M., & Singal, G. (2020). A Survey on Assistive Technology for Visually Impaired. In Internet of Things. Elsevier. doi:10.1016/j.iot.2020.100188

Mankoff, K. D., Russo, T. A., Norris, B. K., Hossainzadeh, S., Beem, L., Walter, J. I., & Tulaczyk, S. M. (2011). Kinects as sensors in earth science: glaciological, geomorphological, and hydrological applications. *AGU Fall Meeting Abstracts*, C41D–0442.

Merchan, Guerra, Poveda, Guzmán, & Sanchez-Galan. (2020). Bioacoustic Classification of Antillean Manatee Vocalization Spectrograms Using Deep Convolutional Neural Networks. *Applied Sciences, 10*(9), 3286.

Mian, Bennamoun, & Owens. (2006). Three-Dimensional Model-Based Object Recognition and Segmentation in Cluttered Scenes. *IEEE Transactions on Pattern Analysis and Machine Intelligence, 28*(10), 1584–1601.

Mills, Jalil, & Stanga. (2017). Electronic Retinal Implants and Artificial Vision: Journey and Present. *Eye, 31*(10), 1383–98.

Mousavi Hondori, H., & Khademi, M. (2014). A Review on Technical and Clinical Impact of Microsoft Kinect on Physical Therapy and Rehabilitation. *Journal of Medical Engineering*. doi:10.1155/2014/846514 PMID:27006935

Murali & Coughlan. (2013). Smartphone-Based Crosswalk Detection and Localization for Visually Impaired Pedestrians. In *2013 IEEE International Conference on Multimedia and Expo Workshops (Icmew)*, (pp. 1–7). IEEE. 10.1109/ICMEW.2013.6618432

Nagata, Tokuno, Otsuka, Ochi, Ikeda, Watanabe, & Habib. (2020). Development of Design and Training Application for Deep Convolutional Neural Networks and Support Vector Machines. In *Machine Vision and Navigation*, (pp. 769–86). Springer.

Nielsen, M. A. (2015). *Neural Networks and Deep Learning*. Determination Press.

Pallejà, Tresanchez, Teixidó, & Palacin. (2010). Bioinspired Electronic White Cane Implementation Based on a Lidar, a Tri-Axial Accelerometer and a Tactile Belt. *Sensors (Basel), 10*(12), 11322–11339.

Pardasani, Indi, Banerjee, Kamal, & Garg. (2019). Smart Assistive Navigation Devices for Visually Impaired People. In *2019 IEEE 4th International Conference on Computer and Communication Systems (Icccs)*, (pp. 725–29). IEEE.

Peli, Luo, Bowers, & Rensing. (2009). Development and Evaluation of Vision Multiplexing Devices for Vision Impairments. *International Journal on Artificial Intelligence Tools, 18*(3), 365–78.

Peters, J. F. (2017). *Foundations of Computer Vision: Computational Geometry, Visual Image Structures and Object Shape Detection* (Vol. 124). Springer. doi:10.1007/978-3-319-52483-2

Poggi, M., & Mattoccia, S. (2016). A Wearable Mobility Aid for the Visually Impaired Based on Embedded 3d Vision and Deep Learning. In *2016 IEEE Symposium on Computers and Communication (Iscc)*, (pp. 208–13). IEEE. 10.1109/ISCC.2016.7543741

Poggi, M., Nanni, L., & Mattoccia, S. (2015). Crosswalk Recognition Through Point-Cloud Processing and Deep-Learning Suited to a Wearable Mobility Aid for the Visually Impaired. In *International Conference on Image Analysis and Processing*, (pp. 282–89). Springer. 10.1007/978-3-319-23222-5_35

Pradeep, V., Medioni, G., & Weiland, J. (2010). Robot Vision for the Visually Impaired. In *2010 Ieee Computer Society Conference on Computer Vision and Pattern Recognition-Workshops*, (pp. 15–22). IEEE.

Sáez, Muñoz, Canto, García, & Montes. (2019). Assisting Visually Impaired People in the Public Transport System Through Rf-Communication and Embedded Systems. *Sensors, 19*(6), 1282.

Sanchez-Galan, J. E., Jo, K.-H., & Cáceres-Hernández, D. (2020). Stairway Detection Based on Single Camera by Motion Stereo for the Blind and Visually Impaired. In *Machine Vision and Navigation* (pp. 657–673). Springer. doi:10.1007/978-3-030-22587-2_20

Shimakawa, M., Akutagawa, R., Kiyota, K., & Nakano, M. (2017). A Study on Staircase Detection for Visually Impaired Person by Machine Learning Using RGB-d Images. In *Proceedings of the 5th IIAE International Conference on Industrial Application Engineering 2017*. The Institute of Industrial Application Engineers. 10.12792/icisip2017.080

Shimakawa, M., Matsushita, K., Taguchi, I., Okuma, C., & Kiyota, K. (2019). Smartphone Apps of Obstacle Detection for Visually Impaired and Its Evaluation. *Proceedings of the 7th Acis International Conference on Applied Computing and Information Technology*, 1–6. 10.1145/3325291.3325381

Shoval, Ulrich, & Borenstein. (2003). NavBelt and the Guide-Cane [Obstacle-Avoidance Systems for the Blind and Visually Impaired]. *IEEE Robotics & Automation Magazine, 10*(1), 9–20.

Shoval, S., Ulrich, I., & Borenstein, J. (2000). Computerized Obstacle Avoidance Systems for the Blind and Visually Impaired. Intelligent Systems and Technologies in Rehabilitation Engineering, 414–48.

Ulrich & Borenstein. (2001). The Guidecane-Applying Mobile Robot Technologies to Assist the Visually Impaired. *IEEE Transactions on Systems, Man, and Cybernetics-Part A: Systems and Humans, 31*(2), 131–36.

Vejarano, Henriuez, & Montes. (2018). Sistema Para La Interacción Activa Con Autobuses de Rutas Urbanas de Panamá Para Personas Con Discapacidad Visual. *I+ D Tecnológico, 14*(2), 17–23.

Velazquez, Pissaloux, Guinot, & Maingreaud. (2006). Walking Using Touch: Design and Preliminary Prototype of a Non-Invasive Eta for the Visually Impaired. In *2005 IEEE Engineering in Medicine and Biology 27th Annual Conference*, (pp. 6821–4). IEEE.

Wahab, Talib, Kadir, Johari, Noraziah, Sidek, & Mutalib. (2011). *Smart Cane: Assistive Cane for Visually-Impaired People.* arXiv Preprint arXiv:1110.5156

Weiland, Liu, & Humayun. (2005). Retinal Prosthesis. *Annu. Rev. Biomed. Eng., 7*, 361–401.

Westfechtel, Ohno, Mertsching, Hamada, Nickchen, Kojima, & Tadokoro. (2018). Robust Stairway-Detection and Localization Method for Mobile Robots Using a Graph-Based Model and Competing Initializations. *The International Journal of Robotics Research, 37*(12), 1463–83.

World Health Organization. (2013). *Vision Impairment and Blindness.* https://www.who.int/en/news-room/fact-sheets/detail/blindness-and-visual-impairment

World Health Organization. (2018). *Assistive Technology.* https://www.who.int/news-room/fact-sheets/detail/assistive-technology

Yang, Wang, Hu, & Bai. (2016). Expanding the Detection of Traversable Area with Realsense for the Visually Impaired. *Sensors, 16*(11), 1954.

Yang, Wang, Zhao, Cheng, Bai, Yang, & Liu. (2017). IR Stereo Realsense: Decreasing Minimum Range of Navigational Assistance for Visually Impaired Individuals. *Journal of Ambient Intelligence and Smart Environments*, 9(6), 743–55.

Yang, K., Wang, K., Lin, S., & Bai, J. (2018). Long-Range Traversability Awareness and Low-Lying Obstacle Negotiation with Realsense for the Visually Impaired. *Proceedings of the 2018 International Conference on Information Science and System*, 137–41.

Zhang, Z. (2012). Microsoft Kinect Sensor and Its Effect. *IEEE Multimedia, 19*(2), 4–10.

Chapter 8
Detection of Landmarks by Mobile Autonomous Robots Based on Estimating the Color Parameters of the Surrounding Area

Oleksandr Poliarus

Kharkiv National Automobile and Highway University, Ukraine

Yevhen Poliakov
iD https://orcid.org/0000-0002-3248-7461
Kharkiv National Automobile and Highway University, Ukraine

ABSTRACT

Remote detection of landmarks for navigation of mobile autonomous robots in the absence of GPS is carried out by low-power radars, ultrasonic and laser rangefinders, night vision devices, and also by video cameras. The aim of the chapter is to develop the method for landmarks detection using the color parameters of images. For this purpose, the optimal system of stochastic differential equations was synthesized according to the criterion of the generalized variance minimum, which allows to estimate the color intensity (red, green, blue) using a priori information and current measurements. The analysis of classical and nonparametric methods of landmark detection, as well as the method of optimal estimation of color parameters jumps is carried out. It is shown that high efficiency of landmark detection is achieved by nonparametric estimating the first Hilbert-Huang modes of decomposition of the color parameters distribution.

DOI: 10.4018/978-1-7998-6522-3.ch008

INTRODUCTION

The mobile autonomous robots (MARs) navigation in unknown terrain in the absence of GPS requires, in particular, the remote detection of landmarks that are concentrated in surrounding space (concentrated landmarks). For this purpose, low-power radars, laser and ultrasonic rangefinders, night vision devices and technologies are used, and in the daytime, also video cameras, images of which are processed by methods of objects recognition. The video cameras installed at the robots are effective tools for the landmark detection, if its shape is known in advance. With such a restriction, no landmarks can be found at all, since only single objects have a strict shape and their number is usually not large. Even the landmarks that are perfectly recognizable under ideal conditions can be partially covered by tree branches, vegetation, etc. Robots must make their own decisions about an object detection and assigning it to the landmark of some type. Such systems are, in particular, radars, the physical basis of which is the radiation and reception of waves reflected from the surrounding area. However, together with the echo signal from the landmark the unwanted signals, reflected from the surrounding area, enter the receiver input.

In the navigation process, the mobile autonomous robot must determine its coordinates with respect to fixed or moving objects, the coordinates of which are always known. In the absence of GPS, the reference objects may be pillars or other objects placed in the natural environment (single tree, small rock, etc.). The first task of the MAR is to detect a landmark with probability of correct detection exceeding a certain level. The second task is to measure the range to the landmark and its azimuth with a given accuracy. This chapter of the monograph deals with the first task. In the broad sense, this is interpreted as the robot automatic detection of individual objects that are concentrated in terrain and are significantly different from the background in some parameters or features. The type of parameters depends entirely on the technical means used to solve the landmark detection task. A single landmark near which no other similar objects and background differs radically from the landmark (the first terrain type) is best identified.

Let's imagine a metal pillar in front of the forest or other inhomogeneous surrounding area. It is easy for a person to visually distinguish this object from the background. The use of radars, laser and ultrasonic rangefinders is not always advisable, since electromagnetic and ultrasonic waves are reflected from both the background and the landmark and are very slightly different in their parameters (amplitude, spectral structure, etc.). A radar with a frequency at which resonant scattering of waves from a pillar occurs has a better chance of detecting a landmark using the energy characteristics of electromagnetic waves. The systems that can detect landmarks by amplitude jumps in the process of scanning the space are complex and may not always provide the necessary probability of correct landmark detection (Poliarus et al., 2018). If the latter is a non-metallic pillar, then its probability detection on the background of the rugged terrain and forest is low. Measurement of thermal radiation of terrain elements does not significantly increase the probability of detecting landmarks.

In a real environment, it is very difficult to detect a fixed landmark against the background of stationary objects. However, from physical considerations it is clear that the criterion of detecting such a landmark against the background of the surrounding terrain in the daytime may be its color. Hence, there is the need of investigating the possibility of detecting such landmarks based on the color criterion (Abdellatif, 2013). The color parameters of real objects are random and heterogeneous in space. In the process of scanning the space with a camera, they change randomly, and if there is a concentrated landmark in the camera review sector, one or more parameters can be changed abruptly. This may be a sign of a decision for the landmark detection (Poliarus et al., 2018). The methods of detecting and estimating the parameters jumps of the signals reflected from the landmarks are considered in (Poliarus et al., 2018; Abdellatif, 2013; Poliarus et al., 2020; Poliarus & Poliakov, 2020a; Poliarus & Poliakov, 2020b).

For the reliable detection and recognition of landmarks, the systems that have different physical principles may be used at the same time. The benefits of each system differ in the various situations inherent in navigating MAR. The probability of a landmark detection by a few systems is significantly increased compared to the probability of detection by a single system. The methods of calculating the detection characteristics of landmarks for different systems will be similar. They are best described for radars. The theory of moving targets detection against the background of noise was created for them. The echo signal from a moving target has some Doppler shift, and the Doppler frequency shift of the signals reflected from the terrain is zero. The well-known theory requires some adjustments regarding the detection of landmarks, since there are almost no differences between the parameters of the signals reflected from the landmark and the background for the radar. In classic radar, signals reflected from the surface of the earth near the radar are suppressed. This also applies to ultrasonic rangefinders.

However, there are parameters that are significantly different for landmark and background. First of all, these are the color coordinates of a landmark image that can be described by three intensities (*R, G, B*) of red, green and blue colors. It should be expected that at least one of the three colors of the landmark will not match the color of the terrain near the landmark. So, it is necessary to transform the known theory of signals detection for cases of fixed landmarks detection on a motionless background. There is also an essential feature of detecting differences in neighboring colors from radar tasks. A radar uses a reference sounding signal with stable characteristics to build the optimal target detection system. The detection of landmarks by color is a special case of passive location, where no sounding signals are generated intentionally.

The authors believe that the detection of a landmark is carried out simultaneously by several different systems, for example, radar, laser and ultrasonic rangefinder, video camera. The chapter analyzes landmarks detection capabilities based only on the optimal processing of video camera signals with an estimation of color parameters. In classical radar, a strict sequence of two main stages is generally recognized: detecting targets and determining their coordinates. These stages are separated a short time or practically not separated. Using color information, the "binding" MAR to the landmark performs in the opposite order. First, the color coordinates of the entire image, which may or may not describe a landmark, are measured, and then, according to the developed algorithms, the results of its detection, recognition and determination of the azimuth are evaluated.

BACKGROUND

The problem of detection and recognition of landmarks that provides robots navigation, has been investigated in many scientific papers. All the landmarks are divided into two classes: artificial and natural. The artificial landmarks can be installed both indoors and outdoors. For example, they may be signs in the shape of concentric rings (Zhong et al., 2017) or with text to be decrypted. They can be mounted on the ceiling (Shih & Ku, 2016) or other places. The robot can recognize a qualitative description of scenes along the trajectory of its movement, which is created in advance (Zhengt et al., 1991). Cloud computing is increasingly used to store route maps and image processing (Ningombam et al., 2018) and elements of artificial intelligence (Elmogy & Zhang, 2009). Numerous studies have found that technical vision systems are best suited for detecting and recognizing landmarks that placed in unknown terrain

(Ramírez-Hernández et al., 2020; Sharma & Ingole, 2015; Sergiyenko, 2010). The natural landmarks are often detected using RGB sensors (Souto et al., 2017).

The theory of color and its measurement is outlined in many books and articles, for example, in (Krantz, 1975). Each color is described in practice by a combination of three colors (red R, green G and blue B) of varying intensity, which varies from 0 to 1. Color is not a physical property of an object but is determined by a sensation that depends on illumination, spectral reflectance of an object and the observer characteristics. During robot navigation the illumination of the landmark and the reflection coefficients of light can vary, and the values of the R, G, B components often also depend on the characteristics of the cameras installed at MAR. In (Mori et al., 1997) a model of normalized red and blue components is proposed to solve the problem of the dependence on external conditions and the effect of the shadow on the color measurement result. The authors (Bajscy et al., 1990) suggested a way to segment colors using hue and saturation. Determination of object color implies that the color parameters are approximately the same in the vision sector during the measurement process (Brainard & Wandell, 1986). There are many sources for determining chromaticity, and it should be remembered that color also determines hue, saturation, but does not determine illumination (Bouman & Shapiro, 1994). During the day, the color coordinates and hues change, and in the morning and evening they are similar (Romero et al., 2003). Multiple cameras may be used to improve the accuracy of color coordinate measurements (Varghese et al., 2014). Color information is important for segmenting individual parts of an object (Onta et al., 1980). Much attention in publications is devoted to the processing of information about the coordinates of chromaticity in dynamic measurements. In (Johnson et al., 1980), the analysis of the results of linear filtering of the image difference using the orthogonality of the components of the color space has been done. RGB or CMYK color models have been developed to build filtering algorithms that need to be associated with images functions in color space (Nishad & Chezian, 2013). Image processing provides maximum consideration of human vision (Stockman & Sharpe, 2006). Metrological aspects of measuring chromaticity are given in (Ellison & Williams, 2012).

Wide-angle video cameras are used for landmark detection, and narrow-angle ones for its recognition. The angle of view of the camera β depends on the focal length f of the lens and the physical size of the camera matrix (Gasvik, 2002). The greater the focal length f, the smaller β and vice versa. With a large angle of view, more objects get into the frame of the camera. The focal length of the cameras lens (the distance between the lens and the cameras matrix) is chosen based on the distance r to the landmark and its largest horizontal or vertical dimensions. The typical dimensions of the CCTV camera matrix are in mm: horizontally — 3.2, 4.8, 6.4, and vertically respectively — 2.4, 3.6, 4.8. Typical landmark distances for different focal lengths of the cameras vary within: f=2.8mm, β=86°, r=0...5m, and if f=16mm, β=17°, r=35...50m. These camera characteristics are useful in estimating the capabilities of technical vision systems of MAR for detecting and recognizing the landmarks. These opportunities depend on many factors, including type of landmark, range, surrounding area, light, weather, and more. Typical cameras detection range will not exceed some tens of meters.

THE PHENOMENOLOGICAL DESCRIPTION OF LANDMARK MODELS AND BACKGROUND AREAS

Landmark detection and recognition capabilities should be estimated for typical concentrated landmarks and typical terrain. After finding the regularities, it is necessary to generalize and extend them to other practical situations. The most accurate determination of the landmark coordinates is achieved if this landmark is distributed vertically. The most convenient model of such a landmark is a vertical cylinder or pillar (metal, plastic or wooden). A metal pillar can be detected by a radar with resonant scattering of electromagnetic waves, and by a camera, if the color of the pillar is different from the color of the background. The pillars of other material are best detected by a camera with a special image processing algorithm.

As a type of the first typical terrain, we choose a monotonous terrain with approximately the same color structure (Figure 1a), on which the landmark is resolved visually well. Such terrain includes a steppe, a desert, a plain coast of the sea, a flat terrain of other planets etc. In the second type (Figure 1b), the background is a dense picture of natural objects such as forest. In it, the terrain creates a monochrome colored background, on which the landmark is not always well distinguished. The third type includes variety rough terrains, as shown in Figures 1, c, d. There is still a fourth type — a cityscape with many surrounding objects, but in the chapter this model is considered only partially.

Figure 1. Illustration of three types of environment with landmark: a – the example of the first type of image; b – the example of the second type; c, d – the examples of the third type of images

The color of any image consists of colors of three types R,G,B (red, green, blue) and is determined by their relative coordinates, the values of which (the color intensities of the camera matrix pixels) vary from 0 to 255. Since it is necessary to find a landmark in the form of a pillar, it is advisable to average the color intensity for each column of pixels. This maps the two-dimensional color coordinates space in one-dimensional one. The average color parameters (color coordinates) for the one distance between robot and landmark is shown in Figure 2.

Visual analysis shows that the presence of a concentrated landmark (pillar) leads to falling (dip) in the intensity distribution of all three colors. Similar color distributions were obtained for different ranges from camera to landmark and show that with increasing range, the depth of falling decreases. The presence of a deep falling the curve (Figure 2) is a sign of a possible landmark in the camera's viewing area. Changing the color of a landmark against the same background may lead to a jump in color parameter.

Figure 2. Distribution of average color parameters (color coordinates) along the frame for red (bold line), green (thin line) and blue (dotted line) colors: graphics (a), (b), (c), (d) correspond to the images (a), (b), (c), (d) of the Figure 1 respectively

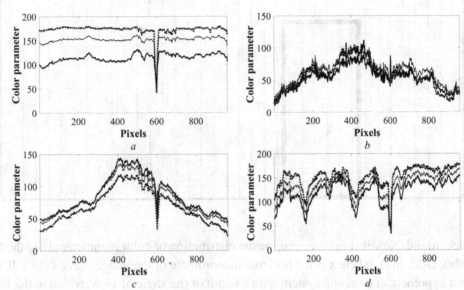

In the absence of the concentrated landmarks (pillars) at the terrain, the color parameters R,G,B dependence along the image for the first terrain type is smooth with small amplitude fluctuations. The authors can assume that for the first terrain type the color parameter distributions are homogeneous. For the second and third type, such distributions are heterogeneous. The heterogeneity of the average color parameter for the second type is due to the fact that the averaging is performed on full columns of pixels, and part of the image contains no forest. Therefore, for the second type it is necessary to perform averaging not on a full column of pixels, but only up to the level (height) of the landmark, and then the color parameter distribution will be uniform, as in the first type. It is practically impossible to find a landmark by the system of technical vision as it follows from Figure 2b. Further, we consider that the first and second types give a homogeneous distribution of colors R,G,B, and the third and fourth types – heterogeneous one.

Let's consider a scheme of an image, part of which is occupied by a narrow tall object, for example, a pillar of some color (Figure 3).

The system estimates the average color intensity in the selected area of the frame, which does not always reveal a landmark image that is different from the background color. If the position of the landmark in the photograph is clearly different in color, then this situation is the simplest and is not of practical interest. Dividing the image on the frame with width a (Figure 3) into m part creates the vertical strips with width $\Delta x = \dfrac{a}{m}$. The system can determine the average color intensity in each strip. It is desirable that the color of the separate parts of the strip is not very different. The width of each vertical strip should not exceed the width of the landmark image in the photograph, since the average color of the landmark depends partly on the color of the background. The authors believe that the width of the vertical strip is several times smaller than the width of the landmark image. After determining the average color in each strip, a search is made for adjacent strips in which this color is significantly different

Figure 3. The scheme of the landmark image and vertical dividing strips on one frame

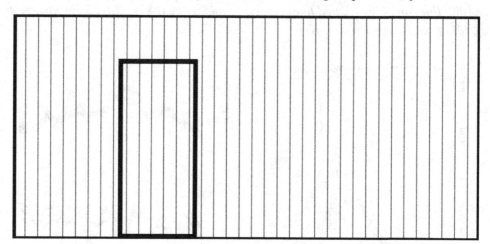

from the background. Now it is easy to imagine the distribution of color parameter along the frame with functions $R(x)$, $G(x)$, $B(x)$, where x is the horizontal coordinate of the image frame ($x \geq 0$). If we assume that there is a hypothetical scanning system with a width of the angle of view, which in the frame forms a vertical strip, then the dependencies $R(x)$, $G(x)$, $B(x)$ are transformed into time dependencies $R(t)$, $G(t)$, $B(t)$, where the time scale is determined by the scanning speed chosen by the researcher. In this case, the homogeneity of the distribution $R(x)$, $G(x)$, $B(x)$ is transformed into the stationarity of random processes with realizations $R(t)$, $G(t)$, $B(t)$, and the heterogeneity of these dependencies — into non-stationarity. Further, it will be discussed that the steady-state distributions of color intensities are the characteristics for the first and second types of the surrounding area, and the non-stationary distributions — for the third and fourth types.

OPTIMAL ESTIMATION OF THE LANDMARK COLOR PARAMETERS

Landmark color parameters on a homogeneous, for example, white background are easily determined by evaluating the color of the pixels that create a landmark. Under real conditions, the color parameters are distorted due to the effects of color noise, various interferences, and changing lighting conditions. In order to synthesize the system for optimal estimation of the color coordinates, it is necessary to first evaluate a priori capabilities of the system and noise characteristics.

Gaussian white noise $n_0(t)$ with zero mathematical expectation and one-sided spectral density N_0 is used to form a priori information about the color parameters of the landmark and background $R(t)$, $G(t)$, $B(t)$. This noise is involved in the formation of the a priori message, which is denoted as $\lambda(t)$ and described by the stochastic differential equation (Tikhonov, 1983)

$$\frac{d\lambda(t)}{dt} = g(t, \lambda) + n_0(t). \tag{1}$$

The function $g(t, \lambda)$ may be represented by functions $R(t)$, $G(t)$, $B(t)$, such as

$$\frac{dR(t)}{dt} = -\alpha \cdot R(t) + n_0(t). \tag{2}$$

The spectral density N_0 of this noise is determined from equation

$$M\{n_0(t_1) \cdot n_0(t_2)\} = \frac{N_0}{2}\delta(t_2 - t_1). \tag{3}$$

The process variance $\lambda(t)$ (or R(t), G(t), B(t)) is

$$D = \frac{N_0}{4\alpha}, \tag{4}$$

and from here

$$N_0 = D \bullet 4\alpha \tag{5}$$

We emphasize that D is a priori variance of useful message $R(t)$, $G(t)$, $B(t)$. Therefore, $n_0(t)$ it is the message noise that is required to form a differential equation describing the a priori behavior of the message $R(t)$, $G(t)$, $B(t)$. In contrast, noise $n(t)$ is the sum of internal and external noise acting on a color parameter and creating an realization that is further processed by the system for the purpose of making a decision. Noise $n_0(t)$ is created by the developer or operator of the system, and noise $n(t)$ is human independent (after the system has been developed).

As already shown, the distribution of color parameters R,G,B can be represented as realizations of random signals. Such signals must be processed together with the existing picture frame noise, which must also be transformed into time-dependence ones. The causes of image noise and their classification are detailed in (Sontakke & Kulkarni, 2015; Patro & Panda, 2016). First of all, it is necessary to study the noises that have the greatest influence on the estimation of the color parameters of landmarks against the background of random fragments of the surrounding terrain. The main causes of such noise are:

1. The relative movement between the robot and the landmark.
2. Defocusing.
3. Atmosphere turbulence.
4. Quality of lenses.

In traditional denoising algorithms, noise is considered to be white Gaussian or a mixture of Gaussian distributions, the number of which depends on the operating conditions of the cameras (Boyat & Joshi, 2015). Other colors depend on the effect of color or fractal noise, such as pink, brown, flicker, etc. Due to the possible sharp changes in the illumination of a landmark in the terrain, the pixels values that characterize the color parameter of the image element may fluctuate. Data transmission to an image processing system of the robot can distort the real color parameters, which creates pulse noise (Boyat & Joshi, 2015). As a rule, before landmark detection and recognition, as well as other possible operations (fusion, segmentation, compression), it is necessary to perform denoising of the images (Suresh, 2017).

As the chapter is exploratory, it makes no sense to justify in detail the type of noise for the landmark image. In the synthesis of the device of optimal image processing, the most common model of total noise, i.e. white Gaussian noise (WGN) was adopted, although additive WGN is not always suitable for image processing in digital cameras and camera phones (Lim, 2006). The noise variance within the frame with the image of the landmark will be automatically determined from the covariance matrix, which is estimated by the following algorithm of optimal image filtering.

The distribution of the color parameters along the horizontal coordinate of the frame x, and further in time t using degradation models can be described as:

$$R_d(t) = R_{nd}(t) \cdot \chi_R(t) + n_R(t),$$
$$G_d(t) = G_{nd}(t) \cdot \chi_G(t) + n_G(t), \tag{6}$$
$$B_d(t) = B_{nd}(t) \cdot \chi_B(t) + n_B(t),$$

where $R_{nd}(t)$, $G_{nd}(t)$, $B_{nd}(t)$ — the undistorted color parameters, respectively, red, green and blue, $R_d(t)$, $G_d(t)$, $B_d(t)$ — the distorted or the original image, $\chi_R(t)$, $\chi_G(t)$, $\chi_B(t)$ — the distortion functions of these colors due to a number of reasons, $n_R(t)$, $n_G(t)$, $n_B(t)$ — WGN for the respective colors. These functions can be used in mathematical modeling. In experimental studies, estimates of functions $\hat{R}_d(t), \hat{G}_d(t), \hat{B}_d(t)$ are known.

The one-sided spectral noise density N, included in the expression (Tikhonov, 1983), is determined from the equation

$$R_n(t_1, t_2) = M\{n(t_1) \cdot n(t_2)\} = \frac{N}{2}\delta(t_2 - t_1). \tag{7}$$

For the discrete case, where the signal and noise are sampled with a small time discrete Δt

$$n_i = \xi_i - s_i(\lambda), n_i = \frac{1}{\Delta} \int_{t_i - \Delta}^{t_i} n(t)dt. \tag{8}$$

Here ξ_i is the value of the random process realization (signal + noise) at time t_i, $s_i(\lambda)$, n_i — the value of signal and noise respectively at the same time. The parameter of the signal to be measured is designated as λ. The relations (8) can be modeled for different cases.

The noise variance

$$D_i = M\{n_i^2\} = \frac{N}{2\Delta}. \tag{9}$$

From here,

$$N = 2 \cdot \Delta \cdot D_i \tag{10}$$

The spectral noise density N can be determined from formula (10), but it depends on the magnitude of the time discrete Δt. That's why averaging across many realizations is required.

Usually, in practice, signal and noise are considered uncorrelated, and if this is not observed, there is a problem of detecting the signal or the problem of distinguishing two correlated signals, one of which is noise or interference. Therefore, one of the problems is to determine an acceptable noise model for the system. In the simulation noise is considered as white Gaussian.

A random process $\xi(t)$ is considered to be a Gaussian Markov process with a correlation function

$$R_\xi(\tau) = D_\xi \cdot e^{-\alpha \cdot |\tau|}. \tag{11}$$

It is determined by a linear stochastic differential equation

$$\frac{d\xi(t)}{dt} + \alpha \cdot \xi(t) = n(t). \tag{12}$$

The variance of this process

$$D_\xi = \frac{N}{4\alpha}. \tag{13}$$

Hence,

$$N = 4\alpha D_\xi. \tag{14}$$

Formula (8) combines all the desired characteristics α, N and D_ξ. If a dimension of the color parameter R, G, B is z, then the dimension D_ξ is z^2, the dimension α is $\frac{1}{s} = Hz$, and the dimension of the spectral density N is $\frac{z^2}{s} = z^2 \cdot Hz$. It can be seen from formula (8) that with a given variance $D\xi$, an width increasing of the process spectrum leads to an increase of N.

Otherwise, the spectral density N can be determined by setting the signal to noise ratio $q = \sqrt{\dfrac{2E}{N}}$ and signal energy E — by the known formula

$$E = \int_{-\infty}^{\infty} s^2(t)dt, \tag{15}$$

where instead of $s^2(t)$ the functions $R^2(t)$, $G^2(t)$ or $B^2(t)$ can be used. Although the color parameters are discrete, the dependencies R_i, G_i, B_i are easily transformed into continuous ones $R(t)$, $G(t)$, $B(t)$. Thus, the desired characteristics can be obtained by modeling according to the formulas above. The color coordinate R, G, B ($0 \leq R, G, B \leq 255$) are described by stochastic differential equations:

$$\frac{dR(t)}{dt} = -\alpha_R R(t) + n_{0R}, \tag{16}$$

$$\frac{dG(t)}{dt} = -\alpha_G G(t) + n_{0G}, \tag{17}$$

$$\frac{dB(t)}{dt} = -\alpha_B B(t) + n_{0B}, \tag{18}$$

where $\alpha_R, \alpha_G, \alpha_B$ are the coefficients representing the spectra width for the red, green and blue color parameters respectively, $n_{0,R}$, $n_{0,G}$, $n_{0,B}$ — independent Gaussian white noises for the same colors, having two-sided spectral densities $\dfrac{N_{0R}}{2}, \dfrac{N_{0G}}{2}, \dfrac{N_{0B}}{2}$, respectively, and creating a matrix

$$\mathbf{N}_0 = \begin{pmatrix} \dfrac{N_{0R}}{2} & 0 & 0 \\ 0 & \dfrac{N_{0G}}{2} & 0 \\ 0 & 0 & \dfrac{N_{0B}}{2} \end{pmatrix}. \tag{19}$$

The color parameters $R(t)$, $G(t)$, $B(t)$ together with noises create random processes

$$\xi_1(t) = R(t) + n_R(t), \tag{20}$$

$$\xi_2(t) = G(t) + n_G(t), \tag{21}$$

$$\xi_3(t) = B(t) + n_B(t), \tag{22}$$

where $n_R(t), n_G(t), n_B(t)$ are the additive independent white noises that create a symmetric matrix of half-divided spectral intensities

$$\mathbf{N} = \begin{pmatrix} \dfrac{N_R}{2} & 0 & 0 \\ 0 & \dfrac{N_G}{2} & 0 \\ 0 & 0 & \dfrac{N_B}{2} \end{pmatrix}. \tag{23}$$

The random process of intensities $R(t)$, $G(t)$, $B(t)$ has the form

$$\vec{\xi}(t) = \begin{pmatrix} \xi_1(t) & \xi_2(t) & \xi_3(t) \end{pmatrix}^T, \tag{24}$$

which, for given (20, 21, 22), is transformed to the next expression

$$\vec{\xi}(t) = \mathbf{H}(t)\mathbf{C}(t) + \vec{n}(t), \tag{25}$$

where

$$\mathbf{H}(t) = \begin{pmatrix} 1 & 0 & 0 \\ 0 & 1 & 0 \\ 0 & 0 & 1 \end{pmatrix} \tag{26}$$

an observation matrix which, for the simple case of a stationary random process, is independent of time. In modeling, instead of units, temporal functions may be used to describe some non-stationary process of color parameters related to the nature of the surrounding area. In relationship (25) the vectors

$$\mathbf{C}(t) = \begin{pmatrix} R(t) & G(t) & B(t) \end{pmatrix}^T, \tag{27}$$

$$\vec{n}(t) = \begin{pmatrix} n_R(t) & n_G(t) & n_B(t) \end{pmatrix}^T. \tag{28}$$

The observation matrix can be obtained from the equations (16-18)

$$\mathbf{A} = \begin{pmatrix} -\alpha_R & 0 & 0 \\ 0 & -\alpha_G & 0 \\ 0 & 0 & -\alpha_B \end{pmatrix}. \tag{29}$$

The system of optimal linear filtering color parameters according to (Tikhonov, 1983) is written in the form

$$\frac{d\hat{\mathbf{C}}(t)}{dt} = \mathbf{A}\hat{\mathbf{C}}(t) + \mathbf{K}(t)\mathbf{H}^T(t)\mathbf{N}^{-1}(\vec{\xi}(t) - \mathbf{H}(t)\hat{\mathbf{C}}(t)), \tag{30}$$

$$\frac{d\mathbf{K}(t)}{dt} = \mathbf{A}\mathbf{K}(t) + \mathbf{K}(t)\mathbf{A}^T(t) - \mathbf{K}(t)\mathbf{H}^T(t)\mathbf{N}^{-1}\mathbf{H}(t)\mathbf{K}(t) + \mathbf{N}_0, \tag{31}$$

where

$$\mathbf{K}(t) = \begin{pmatrix} K_{11}(t) & K_{12}(t) & K_{13}(t) \\ K_{21}(t) & K_{22}(t) & K_{23}(t) \\ K_{31}(t) & K_{32}(t) & K_{33}(t) \end{pmatrix} \tag{32}$$

is correlation matrix of filtering errors.

The system of two equations (30), (31) is optimal by the criterion of the generalized variance minimum. After matrix multiplication this system reduced to the form

$$\frac{d\widehat{R}(t)}{dt} = -\alpha_R \widehat{R}(t) + \frac{2K_{11}(t)}{N_R}(\xi_1(t) - \widehat{R}(t)) + \frac{2K_{12}(t)}{N_G}(\xi_2(t) - \widehat{G}(t)) + \frac{2K_{13}(t)}{N_B}(\xi_3(t) - \widehat{B}(t)), \tag{33}$$

$$\frac{d\widehat{G}(t)}{dt} = -\alpha_G \widehat{G}(t) + \frac{2K_{12}(t)}{N_R}(\xi_1(t) - \widehat{R}(t)) + \frac{2K_{22}(t)}{N_G}(\xi_2(t) - \widehat{G}(t)) + \frac{2K_{23}(t)}{N_B}(\xi_3(t) - \widehat{B}(t)), \tag{34}$$

$$\frac{d\widehat{B}(t)}{dt} = -\alpha_B \widehat{B}(t) + \frac{2K_{13}(t)}{N_R}(\xi_1(t) - \widehat{R}(t)) + \frac{2K_{23}(t)}{N_G}(\xi_2(t) - \widehat{G}(t)) + \frac{2K_{33}(t)}{N_B}(\xi_3(t) - \widehat{B}(t)), \tag{35}$$

$$\frac{dK_{11}(t)}{dt} = -2\alpha_R K_{11}(t) - \frac{2K_{11}^2(t)}{N_R} - \frac{2K_{12}^2(t)}{N_G} - \frac{2K_{13}^2(t)}{N_B} + \frac{N_{0R}}{2}, \tag{36}$$

$$\frac{dK_{12}(t)}{dt} = -\alpha_G K_{12}(t) - \alpha_R K_{12}(t) - \frac{2K_{11}(t)K_{12}(t)}{N_R} - \frac{2K_{12}(t)K_{22}(t)}{N_G} - \frac{2K_{13}(t)K_{23}(t)}{N_B}, \tag{37}$$

$$.\frac{dK_{13}(t)}{dt} = -\alpha_B K_{13}(t) - \alpha_R K_{13}(t) - \frac{2K_{11}(t)K_{13}(t)}{N_R} - \frac{2K_{12}(t)K_{23}(t)}{N_G} - \frac{2K_{13}(t)K_{33}(t)}{N_B}, . \tag{38}$$

$$\frac{dK_{22}(t)}{dt} = -2\alpha_G K_{22}(t) - \frac{2K_{12}^2(t)}{N_R} - \frac{2K_{22}^2(t)}{N_G} - \frac{2K_{23}^2(t)}{N_B} + \frac{N_{0G}}{2}, \tag{39}$$

$$\frac{dK_{23}(t)}{dt} = -\alpha_B K_{23}(t) - \alpha_G K_{23}(t) - \frac{2K_{12}(t)K_{13}(t)}{N_R} - \frac{2K_{22}(t)K_{23}(t)}{N_G} - \frac{2K_{23}(t)K_{33}(t)}{N_B}, \tag{40}$$

$$\frac{dK_{33}(t)}{dt} = -2\alpha_B K_{33}(t) - \frac{2K_{13}^2(t)}{N_R} - \frac{2K_{23}^2(t)}{N_G} - \frac{2K_{33}^2(t)}{N_B} + \frac{N_{0B}}{2}.$$ (41)

The equations for symmetrical components $K_{12}(t) = K_{21}(t)$, $K_{13}(t) = K_{31}(t)$, $K_{23}(t) = K_{32}(t)$ are omitted in the system (33-41).

Figure 2, a-d (solid line) shows the distributions of the red color parameters for the situations represented in Figure 1, a-d, on which the values were determined directly by the pixel colors. The dotted lines correspond to the estimates obtained using the system of equations (33) - (41) for the same situations.

Figure 4. Estimation of the red color parameter obtained by the system of differential equations (33)…
(41) (dotted curve) and directly from pixels (solid curve) for the situations of Figure 1, a-d, respectively

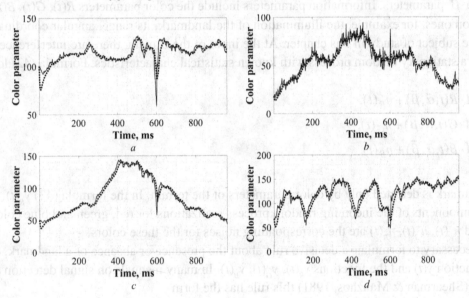

The results were obtained for $\alpha_R, \alpha_G, \alpha_B = 100$ Hz, $\frac{N_{0R}}{2}, \frac{N_{0G}}{2}, \frac{N_{0B}}{2} = 200\ z^2 \cdot$Hz, $N_R N_G N_B = 400\ z^2 \cdot$Hz.

As can be seen from Figure 4 and other figures not shown here, the system (33) … (41) keeps track well all the color parameter changes. The normalized correlation functions K_{ij} are close to one at a given observation interval, and variances are close to zero, which indicates almost perfect tracking of image color parameters. This raises doubts about the need for a developed system (33)…(41). However, in a poor light conditions, the presence of noise, etc., direct pixel color measurement can produce poor results for a robot navigation.

CLASSICAL THEORY OF LANDMARK DETECTION BY ROBOTS

There is a classical theory of signal detection (Scolnik, 1990), which will be used in the development of the method of landmark detection by mobile autonomous robots. Usually, the color parameters distribution (signals) from the camera output are obtained for further processing

$$y(t) = A \cdot x(t, \vec{\alpha}, \vec{\beta}) + n(t), \tag{42}$$

where A — a parameter that can take two values: 0 — if the echo-signal from a landmark is absent, 1 — if the echo-signal from a landmark is present at the receiver input. The signal from the landmark $x(t, \vec{\alpha}, \vec{\beta})$ (it is simply one of the signals R, G, B) depends on the time t and information $\vec{\alpha}$ and non-information $\vec{\beta}$ parameters. Information parameters include the color parameters $R(t)$, $G(t)$, $B(t)$ and non-information ones, for example, the illumination of the landmark, its range, angular coordinates, which are not the subject of study in this chapter. At the input of the camera, there are interferences $n(t)$, that represent a stationary random process with known statistical characteristics. Formula (42) looks like

$$y_R(t) = A \cdot R(t, \vec{\alpha}, \vec{\beta}) + n_R(t),$$
$$y_G(t) = A \cdot G(t, \vec{\alpha}, \vec{\beta}) + n_G(t), \tag{43}$$
$$y_B(t) = A \cdot B(t, \vec{\alpha}, \vec{\beta}) + n_B(t),$$

if the landmark is detected only by color parameters of the terrain. In the formula (43) $y_R(t)$, $y_G(t)$, $y_B(t)$ are the components of the incoming random process realizations for red, green and blue colors respectively, and $n_R(t)$, $n_G(t)$, $n_B(t)$ are the corresponding noises for the these colors.

It is necessary to formulate a decisive rule about the presence or absence of a landmark, depending on the function $y(t)$ and the functions $y_R(t)$, $y_G(t)$, $y_B(t)$. In many manuals on signal detection theory, for example, (Shearman & Manzhos, 1981) this rule has the form

$$\hat{A}_{opt} = \hat{A}_{opt}[y(t)]. \tag{44}$$

Really, the detection of a landmark is carried out by a sharp dip or jump in the color parameters distribution $\Delta y(t)$. Then the rule (44) is written as follows

$$\hat{A}_{opt} = \hat{A}_{opt}[\Delta y(t)]. \tag{44a}$$

It follows that the main task is to detect the jump or dip $\Delta y(t)$. The criterion for optimality may be the average risk minimum criterion or the weight criterion, which do not require prior information on the presence or absence of a landmark. The width of each strip can be obtained after dividing the image into vertical strips and moving to the time zone

$$\Delta t = \frac{1}{2f_{max}}, \tag{45}$$

where f_{max} is the highest maximum frequency in the signal $R(t)$, $G(t)$, $B(t)$ spectra.

The color values of the signals at discrete points in time corresponding to the previously described vertical image bands form a multidimensional vector $\vec{Y} = (y_1 \quad y_2 \quad ... \quad y_m)^T$. Multidimensional probability density in the presence in camera view only background images $p_n(\vec{Y})$ and signal from a landmark $p_{ln}(\vec{Y})$ are used in determining conditional probabilities of correct detection

$$D = P(\hat{A}_1 / A_1) = \int_{\vec{Y}} p_{ln}(\vec{Y})\hat{A}(\vec{Y})d\vec{Y}, \tag{46}$$

false alarm

$$F = P(\hat{A}_1 / A_0) = \int_{\vec{Y}} p_n(\vec{Y})\hat{A}(\vec{Y})d\vec{Y}, \tag{47}$$

missing a landmark

$$\bar{D} = P(\hat{A}_0 / A_1) = 1 - D, \tag{48}$$

correct non-detection

$$\bar{F} = P(\hat{A}_0 / A_0) = 1 - F. \tag{49}$$

In formulas (46)... (49), the integration domain $d\vec{Y} = dy_1 \cdot dy_2 \cdot ... dy_m$ describes all possible values of the value $\vec{Y}$. The situation A_1 means that there is a landmark in the camera viewing area, and A_0 means that there is no a landmark. The estimation given to these situations by the equipment are indicated as $\hat{A}_1$ and $\hat{A}_0$ respectively. The weight criterion

$$D - l_0 F = \int_{\vec{Y}} p_n(\vec{Y})[l(\vec{Y}) - l_0]\hat{A}(\vec{Y})d\vec{Y}, \tag{50}$$

where

$$l(\vec{Y}) = \frac{p_{ln}(\vec{Y})}{p_n(\vec{Y})} \tag{51}$$

is the likelihood ratio for a multidimensional random variable. Hence, the expression for the optimal decisive function

$$\hat{A}_{opt}(\vec{Y}) = \begin{cases} 1, \text{if } l(\vec{Y}) > l_0 \\ 0, \text{if } l(\vec{Y}) < l_0 \end{cases}, \tag{52}$$

The threshold l_0 depends on the ratio of the cost of the error of the robot decision-making to detect the landmark and the probabilities of the presence and absence of a landmark in the camera viewing area

$$l_0 = \frac{\rho_{10} p(A_0)}{\rho_{01} p(A_1)}, \tag{53}$$

where ρ_{10}, ρ_{01} — the cost of the wrong decision about the presence of a landmark in its actual absence and the absence of a landmark in its presence, respectively.

For optimal detection system $D_{opt} - l_0 F_{opt} \geq D - l_0 F$ or $D_{opt} \geq D$ (Neumann-Pearson optimality criterion). This criterion, together with the weight criterion, is a consequence of the general minimum risk criterion. The allowable values of the conditional probability of false alarm F and correct non-detection $\bar{F}$ are determined by the practical needs of navigating the robot. They may differ for different frame distinguishing elements in the form of vertical stripes. For the number of stripes m_l, the conditional probability of a correct non-detection of a landmark

$$\bar{F}_{m_l} = (\bar{F})^{m_l} = (1 - F)^{m_l}, \tag{54}$$

where the probability of false alarm F is small and the same for all elements. The probability of at least one false alarm for the elements that create a landmark in the image is

$$F_{m_l} = 1 - (1 - F)^{m_l}. \tag{55}$$

The probability of correct landmark detection is similarly determined as

$$D_{m_l} = 1 - (1 - D)^{m_l}. \tag{56}$$

In the general case, the likelihood ratio $l(\vec{Y})$ is different for colors *R, G, B*. Let's consider this relationship for only one color, such as red for a one-dimensional case. Suppose that the noise variance for red color is n_R^2, and the noise and the color parameters are distributed by normal law. Then the likelihood ratio for the simplest case of a known signal

$$l_R(y) = \frac{e^{-\frac{(y-R)^2}{2n_R^2}}}{e^{-\frac{y^2}{2n_R^2}}} = e^{-\frac{R^2}{2n_R^2}} e^{\frac{R \cdot y}{n_R^2}}. \tag{57}$$

For multi-dimensional fully known signal

$$l_R(\vec{Y}/\vec{\alpha}) = e^{-\frac{1}{N_0}\sum\limits_{k=1}^{m} R_k^2 \Delta t} \, e^{\frac{1}{N_0}\sum\limits_{k=1}^{m} R_k y_k \Delta t}, \tag{58}$$

where N_0 is the spectral density of the noise power. If the parameter α is, for example, the distance to the landmark, then the color intensity decreases with increasing range and then the likelihood ratio

$$l_R(y(t)/\alpha) = e^{-\frac{E(\alpha)}{N_0}} e^{\frac{2z(\alpha)}{N_0}}, \tag{59}$$

where $E(\alpha)$ is the energy of the expected signal, and

$$z(\alpha) = \int\limits_{-\infty}^{\infty} R(t,\alpha) y(t) dt = z[y(t)/\alpha] \tag{60}$$

the correlation integral.

NON-PARAMETRIC METHODS FOR DETECTING LANDMARK COLOR

The interferences for the system of detecting and evaluating color parameters are signals of all colors from the background that are slightly different from the color of the landmark. If there are no differences in color, it is impossible to find a landmark. For diverse terrain, the statistical characteristics of interference may vary considerably, and are generally unknown in advance. Therefore, the task of detecting landmarks can be considered as a statistical problem with a priori uncertainty, when the distribution functions $p_{ln}(y)$ and $p_n(y)$ can change during the observation process and may already differ from the normal distribution. In such conditions, the classical methods of signal detection are not optimal (Akimov et al, 1978). The development of adaptive signal detection algorithms is significantly complicated by the fact that the laws of change in probability densities $p_{ln}(y)$ and $p_n(y)$ are unknown.

Consequently, the classical methods cannot always be used to detect landmarks, and therefore other methods without knowledge of the functional nature of the laws of color and noise signals distribution are required. The nonparametric detector has decisive statistics that are independent of the noise distribution (Thomas, 1970; Levin, 1976).

It is impossible to construct optimal nonparametric detection algorithms for a wide range of practical problems (Akimov et al, 1984). The authors formulate two hypotheses: the first hypothesis H_0 means that the color parameters R, G, B, denoted by y, are only the color noise with distribution $p_n(y)$; the second hypothesis H_1 means that the color parameter is a signal from a landmark that also contains color noise. The law of distribution of this mixture is $p_{ln}(y)$. If the hypothesis H_0 is true, then independent observations $y_1, y_2, \ldots y_m$ are a sample of zero-median noise $n_{R_1}, \ n_{R_2}, \ \ldots \ n_{R_m}$. In the presence of a landmark, the median is greater than zero and then the statistic

$$S = \sum_{i=1}^{m} h(y_i), \tag{61}$$

where

$$h(y_i) = \begin{cases} 1, y_i > 0, \\ 0, y_i < 0. \end{cases}$$

It is clear that the statistics S in the presence of a signal from the landmark will be greater than in the presence of noise alone. After selecting the required threshold S_0 for practical reasons, the decision to detect a landmark is made when

$$S > S_0 \tag{62}$$

The probability of correct detection depends on S_0. The higher the ratio S/S_0, the greater the probability of inequality (62).

If the noises n_{R_1}, n_{R_2}, ... n_{R_m} have one sign, then

$$S = \sum_{i=1}^{m} h(y_i - n_{R_i}), \tag{63}$$

where

$$h(y_i - n_{R_i}) = \begin{cases} 1, y_i > n_{R_i}, \\ 0, y_i < n_{R_i}. \end{cases}$$

Probability of correct detection

$$P = P(S > S_0) = \sum_{i=S_0+1}^{m} C_m^i P^i (1-P)^{m-i}, \tag{64}$$

where $P = P(y > n_R)$, C_m^i is the binomial coefficients.

In the classical theory of signal detection, which includes both parametric and non-parametric methods, it is necessary to obtain the statistical laws of the color parameters distribution, in particular, the law of distribution R, G, B in the presence of a landmark, i.e. $p_{ln}(y)$. If a landmark is distributed in space, such as a building, then this law of distribution is a function of the size of the landmark and classical theory can be successfully applied to the first and second terrain types. For pillar-like landmarks, the law of distribution $p_{ln}(y)$ is weakly dependent on the presence of a landmark in the camera viewing area, and then $p_{ln}(y) \approx p_n(y)$. This complicates the application of classical signal detection theory and requires new approaches to landmark detection.

Methods for detecting concentrated landmarks by color parameters are reduced to detecting jumps or dips in the color parameters distribution. It is clear that these methods must be different for stationary signals (first and second terrain types) and non-stationary (third and fourth types). In the case of steady-state distribution R,G,B, it is necessary to determine the average value of random color parameters processes (even with a narrow jump or dip), for example $\bar{R}$, and then compare the current values of the color parameter with the threshold R_{th}, i.e.

$$|R - \bar{R}| \geq R_{th},\qquad\qquad(65)$$

as can be seen from Figure 5.

Figure 5. Illustration of the principle of detecting the red color parameter dip due to the landmark, when it exceeds the lower threshold (upper and lower thresholds are indicated by dotted line)

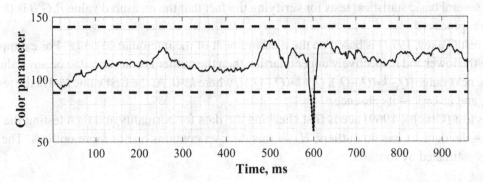

The implementation of inequality (65) is the basis for determining the moment of occurrence of a concentrated landmark or several landmarks. The probability of correct landmark detection depends on the nature of the color parameter distribution in the image. The level of the selected threshold R_{th}, and thresholds G_{th}, B_{th} for different colors may vary. Now, the task of detecting landmarks is transformed into the problem of outliers detection. In the statistical processing of random processes, their outliers are considered anomalous values and eliminated from processing. The main causes of outliers are usually (Zimek & Schubert, 2017): human, executing, dummy errors, unintended mutations of data set and others.

The outliers are informative in this problem, as they indicate to the presence of concentrated objects in the camera's area of view, which can be taken as landmarks. Detecting a random process outlier R,G,B means either detecting a landmark (correct detection) or obtaining abnormal values (false alarm). Such a result can be verified using other robot systems.

The outliers detection can be performed by parametric (probabilistic and statistical modeling, extreme or Z-score value analysis, various regression models) and non-parametric (high dimensional outlier detection, proximity-based models) methods (Hodge & Austin, 2004). To detect landmarks, the authors choose the well-known and relatively simple method (Zimek & Schubert, 2017), which is based on the use of quartiles and is illustrated in Figure 6.

Figure 6. Explanation of the principle of using quartiles for a landmark detection

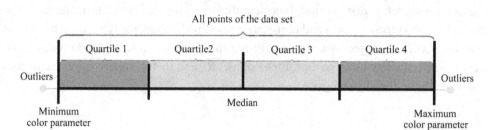

Quartiles divides all points of the data set (color parameter values) into four parts (Figure 6). The average number between the lowest color parameter value (possibly a data dip) and the median of this parameter is the first quartile Q_1. The median for the data set is the second quartile Q_2, and the third quartile Q_3 is the average between the median and the maximum color parameter value (possibly a jump). There are several basic statistical tests for verifying the fact that the measured value R,G,B is the outlier of a random process.

Tukey test (Tukey, 1977) is based on the measurement of the interquartile range. For example, if Q_1 and Q_3 are the lower and respectively upper quartile, then the outlier is considered to be any value R,G,B beyond the next range $[Q_1-k\bullet(Q_3-Q_1), Q_3+k\bullet(Q_3-Q_1)]$, where $k>0$. At the first value $k=1.5$ the test shows an outlier, and at $k=3$ — its absence.

Grubbs test (Grubbs, 1969) needs first checking the data for normality and then testing the hypotheses: H_0 — the data set has no outliers, H_1 — this data set contains one or more outliers. The Grabbs criterion is estimated by formula

$$G = \frac{\max|Y_i - \bar{Y}|}{s}, \text{ where } i=1,2,\ldots,N \tag{66}$$

$\bar{Y}$ and s mean selective mean color parameter and standard deviation. The no-outliers hypothesis is rejected at significance level α if

$$G > \frac{N-1}{\sqrt{N}}\sqrt{\frac{t^2_{\alpha/(2N),N-2}}{N-2+t^2_{\alpha/(2N),N-2}}}, \tag{67}$$

where $t^2_{\alpha/(2N),N-2}$ is the critical value of the Student's distribution with $N-2$ degrees of freedom and the level of significance $\alpha/(2N)$.

Before modeling, the distribution law (it is Gaussian one) was determined according to the Kolmogorov criterion, and then the Grabbs criterion with the critical value of the t-distribution 0.01 & 0.1 and Tukey test were applied. According to the Grabbs criterion, no outliers of color parameters for situations (Figure 1, a-d) were found. An illustration of the application of the Tukey test for the same situations (for red color) is shown in Figure 7, where the corresponding quartiles with $k=1.5$ and $k=3$ are indicated by the horizontal thin and bold dotted lines.

Figure 7. Illustration of the Tukey test application to the landmark detection

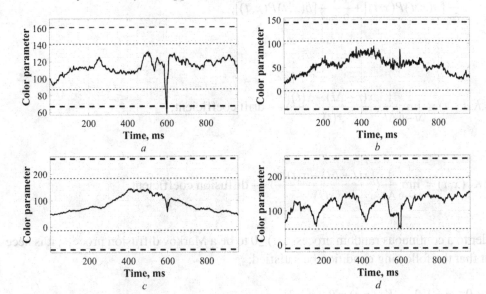

Figure 7 shows that detecting landmark on the red color is possible only for the image (Figure 1, a), and for the blue color — also for the image (1, d) (not shown here). This demonstrates the limited capabilities of traditional methods of detecting the landmarks.

Consequently, the classical signal detection theory can be used to detect the landmarks that have not small horizontal dimension. For concentrated landmarks (such as pillars), it is advisable to use statistical threshold crossing theory or the theory of random process outliers.

OPTIMAL DETECTION OF LANDMARKS BY THE FAST COLOR CHANGE CRITERION

Another approach is based on using the optimal signal detection system (Poliarus et al., 2018; Poliarus & Poliakov, 2020b). The term "jump" will be used not only for the color parameter jump, but also for the dip of this parameter distribution. Inverting color dependence R,G,B from time turns the dip into a jump.

The optimal jump detection system uses prior scientific knowledge. First, it is the Marian Smoluchowski equation (68), which describes the evolution of the function $P(z,v,t)$ for the Brownian particle with the mass m for one-dimensional coordinate z and velocity v under the influence of an external force $F(z)$ in a viscous medium with a viscosity parameter γ at temperature T:

$$\frac{\partial P(z,v,t)}{\partial t} = [-\frac{\partial v}{\partial z} + \frac{\partial}{\partial v}(\gamma v - \frac{F(z)}{m}) + \frac{\gamma k_B T}{m}\frac{\partial^2}{\partial^2 v}]P(z,v,t), \qquad (68)$$

where k_B — Boltzmann's constant.

Secondly, it is Fokker-Planck equation for a random variable x with transition probability density $P(x,t)$, which is a consequence of the first equation and is written as follows (Tikhonov & Mironov, 1977)

$$\frac{\partial P(x,t)}{\partial t} = -\frac{\partial}{\partial x}[a(x,t)P(x,t)] + \frac{1}{2}\frac{\partial^2}{\partial x^2}[b(x,t)P(x,t)], \tag{69}$$

where

$$a(x,t) = K_1(x,t) = \lim_{\Delta t \to 0}\frac{1}{\Delta t}\langle\frac{x(t+\Delta t)-x(t)}{x(t)}\rangle - \text{drift coefficient}, \tag{70}$$

$$b(x,t) = K_2(x,t) = \lim_{\Delta t \to 0}\frac{1}{\Delta t}\langle\frac{[x(t+\Delta t)-x(t)]^2}{x(t)}\rangle - \text{diffusion coefficient}. \tag{71}$$

In order to a continuous random process $x(t) \geq 0$ to be a Markov diffusion process, it is necessary and sufficient that the following condition be satisfied:

$$K_n(x,t) \neq 0, \quad n = 1, 2; \quad K_n(x,t) \equiv 0, \quad n \geq 3.$$

Third, it is the Stratonovich integral-differential equation for a vector function $\vec{x}(t)$ (Stratonovich, 1968)

$$\frac{\partial P(\vec{x},t)}{\partial t} = L\{P(\vec{x},t)\} + [F(\vec{x},t) - \int F(\vec{x},t)P(\vec{x},t)d\vec{x}]P(\vec{x},t), \tag{72}$$

where

$$F(\vec{x},t) = -\frac{1}{2}[\vec{\xi}(t) - \vec{s}(\vec{x},t)]^T \mathbf{N}^{-1}[\vec{\xi}(t) - \vec{s}(\vec{x},t)] \tag{73}$$

with the initial conditions

$$P(\vec{x},0) = P_{pr}(\vec{x},0) = P_{pr_0}(\vec{x}). \tag{74}$$

Here $P_{pr_0}(\vec{x})$ is a priori probability density of the initial color parameter and $\mathbf{N}^{-1}$ — the matrix, which is the inverse matrix of the spectral intensity of the observations noise in the model.

The Stratonovich equation (72) is written for a random process $x(t)$, which is Markovian. It describes the evolution in time of a posteriori probability density $P(x,t)$. The maximum value of this density is achieved for the conditional mathematical expectation $\hat{x}(t)$, which is also the optimal estimation of the function $x(t)$ by the criterion of the minimum mean square error on the interval $(0,t)$. The first term in (72) is an a priori part of the equation (72) and is described by an operator L that affects a posteriori probability density of the process $\vec{x}(t)$. This process is extracted from a vector of observations $\vec{\xi}(t)$

that includes a useful vector signal $\vec{s}(\vec{x}, t)$ and vector of white Gaussian noise $\vec{n}(t)$ with zero mathematical expectation and a matrix of spectral densities $\mathbf{N}(t)$, i.e.

$$\vec{\xi}(t) = \vec{s}(\vec{x}, t) + \vec{n}(t). \tag{75}$$

In accordance with (Tikhonov, 1983) the description of the parameters $\vec{x}(t)$ can be carried out by a differential equation of type (1)

$$\frac{d\vec{x}(t)}{dt} = g(t, \vec{x}(t)) + \vec{n}_0(t), \tag{76}$$

where $\vec{g}(t, \vec{x}(t))$ — vector-function, $\vec{n}_0(t)$ — a column vector of white Gaussian noises and a matrix of spectral densities $\mathbf{N}_0(t, \vec{x}(t))$. In (Poliarus & Poliakov, 2020b), a system of stochastic differential equations was synthesized on the basis of the Stratonovich equation to estimate the jumps of random processes parameters for echo — signals generated by radar. With small corrections, such a system can be used to estimate the jumps of color parameters due to a landmark presence in the camera's viewing area.

$$\frac{dp_1(t)}{dt} = P_{\tau_{jump}}(t) \cdot e^{-z(t)} + \frac{1}{N} \cdot p_1(t) \cdot (1 - p_1(t)) \cdot \left\{ A(t) \cdot \Delta\hat{A}_1(t) + \frac{1}{2} \cdot (A_1^2(t)) - \sigma_{\Delta A_1}^2(t) + 2 \cdot n(t) \cdot \Delta\hat{A}_1(t) \right\}, \tag{77}$$

$$\frac{dz(t)}{dt} = \frac{p_1(t)}{N} \cdot \left\{ A(t) \cdot \Delta\hat{A}_1(t) + \frac{1}{2}(\Delta\hat{A}_1^2(t) - \sigma_{\Delta A_1}^2(t)) + 2 \cdot n(t) \cdot A_1(t) \right\}, \tag{78}$$

$$\frac{d\Delta\hat{A}_1(t)}{dt} = \frac{1}{p_1(t)} \cdot P_{\tau_{jump}}(t) \cdot e^{-z(t)} \cdot \left(\Delta\hat{A}_0(t) - \Delta\hat{A}_1(t) \right) + V_1(t) \cdot \frac{1}{N} \cdot \left[2 \cdot y(t) - A(t) - \Delta\hat{A}_1(t) \right], \tag{79}$$

$$\frac{dV_1(t)}{dt} = \frac{1}{p_1(t)} \cdot P_{\tau_{jump}}(t) \cdot e^{-z(t)} \cdot \left[\left(\Delta\hat{A}_0(t) - \Delta\hat{A}_1(t) \right)^2 + V_0(t) - V_1(t) \right] - \frac{1}{N} \cdot V_1^2(t), \tag{80}$$

$t > 0$,

where p_1 is the a posteriori probability of detecting a jump of one of the colors R, G, B, z is the operation speed of the system, ΔA — the amplitude of a jump of some color, $V_1(t)$ — the variance of the posterior distribution of a random color, $V_0(t)$ — the variance of its a priori distribution, τ_{jump} — the moment of the jump of the color amplitude from the value $A_0(\tau_{jump})$ to $A_1(\tau_{jump})$. The simulation is performed under initial conditions:

$$\Delta \hat{A}_1(t)\Big|_{t=0} = \Delta \hat{A}_0, \ V_1(t)\Big|_{t=0} = V_0,$$ (81)

$$P_1(t)\Big|_{t=0} = \int_{-\infty}^{0} P_\tau(t)dt.$$ (82)

For a qualitative estimation of the jump parameters, it is advisable to have a priori information, for example τ_{jump}, that can be formed in advance if there is information about the terrain and the trajectory of the robot. In general, there is no such information, so it should be formed immediately before estimating the jump parameters by other methods.

One such method is wavelet analysis, which is essentially a "mathematical microscope". Figure 8 shows a wavelet image obtained for the conditions of Figure 1, a, b, c, d using a basic function of type wavelet "haar".

Figure 8. Wavelet image of red color parameter obtained using Figure 2

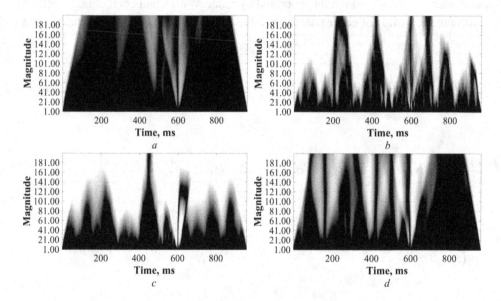

From the Figure 8 it follows that the wavelet image can serve as an approximate indicator of the presence of a landmark, i.e. it uses for creating a priori information about it. The wavelet transforms of the signal shown in Figure 8 indicates that there are several "jumps" in the analyzed color parameter. In particular, the jump is clearly registered at a time of 600 ms, which corresponds to the real time of the landmark detection. The wavelet transform can select the specified area of the image more clearly and has the tools to improve the quality of the jump detection, however, the areas filtering that are not associated with the appearance of a landmark (for example, at a time of 700 ms) for more complex cases, will be difficult. An example of modeling the system of equations (77) ... (80) for the first model is shown in (Figure 9) for the red color of the image.

Figure 9. The results of modeling the system of differential equations (77)… (80) for the first image type: a — estimation of a random color parameter with the jump; b is the probability of detecting the color parameter jump; c is a parameter that characterizes the speed of the system operation; d is the variance of the color parameter determined by the system

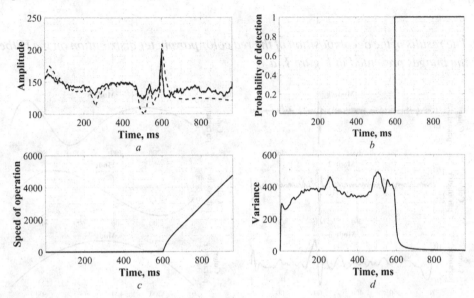

From Figure 9 it follows that the estimation of the red color jump (Figure 9, a) in the first model is well performed by the system with a high probability of its detection (Figure 9, b), high speed operation (Figure 9, c) and low variance of the jump (Figure 9, d) at the time of its appearance. For the second and third image models, these parameters are worse, which limits the color jump method application. For radar, the scope of this method is wider (Poliarus et al., 2018; Poliarus & Poliakov, 2020b).

The biggest problems with the detection of landmarks occur with non-stationary color parameter changes of the images, that is, for the conditions of third and fourth terrain type.

DETECTION OF LANDMARKS WITH NON-STATIONARY COLOR PARAMETERS CHANGING

In the case of non-stationary change of the color parameters (the third and fourth types of the background), these methods of landmarks detection are not always effective. Hence the task of eliminating the non-stationarity of the signals arises. The use of smoothing techniques is unacceptable, as jumps or dips of color parameters, which are the features of a landmark, may be lost. A trade-off approach is to decompose $R(t)$, $G(t)$, $B(t)$ in the basis of orthogonal functions, which would be acceptable for any distribution of color parameters, especially those with sharp variations due to the presence of terrestrial landmarks. Such a basis should include the Hilbert-Huang modes (Huang et al, 1998) which are suitable for decomposition of non-stationary distributions $R(t)$, $G(t)$, $B(t)$, moreover the number of modes is small and most of them is stationary or close to stationary ones. It should be noted that the residuals in the decomposition algorithm are removed and therefore the color coordinate values may be negative.

This is not a problem for estimating the possibility of a landmark detection, because the main task is the search of a color parameter jump (dip) due to the presence of a landmark. The results of the decomposition of the red color parameter (Figure 2, a-d) on the Hilbert-Huang modes are respectively shown in Figures 10-13.

Figure 10. The results of the decomposition of the red color parameter distribution on the Hilbert-Huang modes for the images presented in Figure 1, a

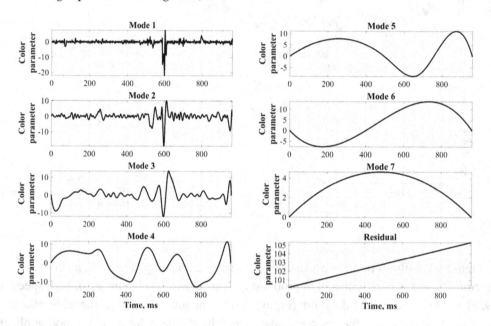

Figure 11. The results of the decomposition of the red color parameter distribution on the Hilbert-Huang modes for the images presented in Figure 1, b

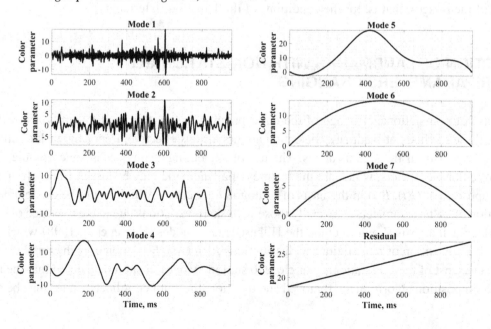

Figure 12. The results of the decomposition of the red color parameter distribution on the Hilbert-Huang modes for the images presented in Figure 1, c

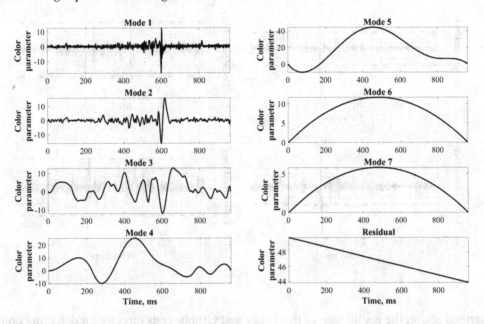

Figure 13. The results of the decomposition of the red color parameter distribution on the Hilbert-Huang modes for the images presented in Figure 1, d

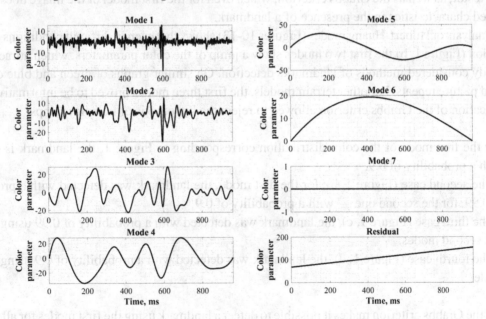

Figure 14. Illustration of the landmark detection using the first Hilbert-Huang mode for red color

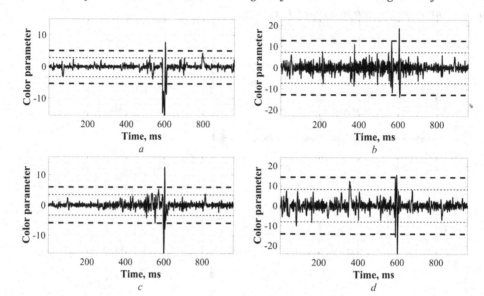

As described above, the application of the Tukey and Grubbs tests directly for detecting color jumps (Figure 2) is ineffective, since the Tukey test allowed detecting the color jump (i.e., landmark) only for the image (Figure 1, a), where the landmark is well recognized visually; for other images — it gave a negative result, as well as the Grabbs criterion, which even for the first model of the image does not give the desired characteristics in the presence of a landmark.

The analysis of Hilbert-Huang modes (Figures 10-13) shows that even in difficult conditions of visual observation (Figure 1, b) the first two modes show a jump of the color parameters, which attracts to use the already considered methods of landmarks detection. On similar graphs of green and blue color the described picture repeats. For other terrain models, the first three modes proved to be informative. Thus, the application of the Grabbs criterion allowed to reject the hypothesis of no color jumps:

- For the first mode of the color distribution corresponding to Figure 1, the landmark is detected with a probability of 0.9.
- In the second case (Figure 1, b): for the first mode, the landmark was detected with a probability of 0.99, for the second one — with a probability of 0.9.
- In the third case (Figure 1, c), the landmark was detected with a probability of 0.99 using the first and second modes.
- In the fourth case (Figure 1, d), the landmark was detected with a probability of 0.9 using the first mode.

Thus, the Grabbs criterion makes it possible to detect a landmark using the first modes for all the considered situations. The results of landmark detection for these modes of red color parameter distribution for the cases (Figure 1) are shown in Figure 14. The landmark is considered to be detected if the curve crosses the bold dotted line and this is observed for all four cases. Similar dependences were obtained for the second modes (Figure 15).

Figure 15. Illustration of the landmark detection using the second Hilbert-Huang mode for red color parameter

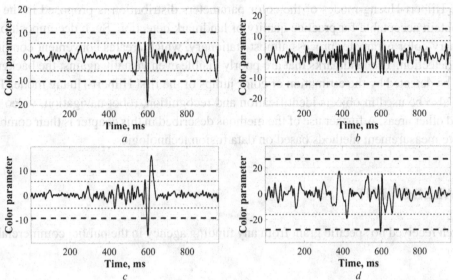

In this case, the landmark is detected for the cases a, c, d, i.e. everywhere where it can be seen visually. The case (Figure 1, b) is difficult to detect a landmark, but the proposed method is effective even in this situation, however, only using the first mode. Thus, the use of the Hilbert-Huang decomposition makes it possible to detect a landmark, even in the cases with a complex background, when the classical methods are ineffective.

CONCLUSION

Navigation of mobile autonomous robots in the absence of GPS is an urgent problem for society. In such conditions, there is a problem of detecting landmarks on the robot path. The most accurate "binding" of a robot is provided to the landmarks like a pillar with simultaneous use of various remote measuring instruments. The chapter analyzes in detail the possibilities of detecting landmarks based on jumps (dips) of color parameters estimation. Unlike radars, a landmark detection is performed after direct measurement of the color parameters R, G, B. The developed system of stochastic differential equations is used to obtain optimal values of color parameters R, G, B by the criterion of generalized variance minimum in the conditions of interference. On the basis of phenomenological analysis, the four surrounding terrain types with a landmark have been proposed. The first and second types characterize the color parameters of background, which vary little. The third and fourth types describe the background with parameters that change significantly and unpredictably. The spatial distribution of the color parameters is mapped into a time domain and then the first two types give a stationary process of random color parameters, and the third and fourth ones — a non-stationary one. The classical methods of detecting landmarks in these conditions are unacceptable. Non-parametric detection methods by different statistical criteria can be used most effectively for the first terrain type. The developed system for optimal detection of the color parameters jumps is quite effective for the first terrain type. Non-stationary color parameters of background almost exclude the detection of landmarks by traditional methods. The method of processing the Hilbert-Huang modes of the color parameters distribution is proposed in the chapter and allows to obtain acceptable for practice quality of landmark detection. Such decomposition makes the non-stationary color parameter process almost stationary, which creates convenient conditions for the landmarks detection. The landmarks that are poorly visible even visually against the background of the terrain can be detected by the estimation of color jumps of the first Hilbert-Huang mode. The obtained results can also be used in objects identification and recognition, robot navigation, video surveillance systems and other areas. A further use of the methods described in this chapter is their combination with other remote measurement methods based on data fusion technologies.

ACKNOWLEDGMENT

This research received no specific grant from any funding agency in the public, commercial, or not-for-profit sectors.

REFERENCES

Abdellatif, M. (2013, November). Color-Based Object Tracking and Following for Mobile Service Robots. *International Journal of Innovative Research in Science. Engineering and Technology*, 2(11), 5921–5928.

Akimov, P. S., Bakut, P. A., & Bogdanovich, V. A. (1984). *Theory of signal detection* [Lexical characteristics of Russian language]. Radio and Communications.

Akimov, P. S., Efremov, V. S., & Kubasov, A. N. (1978). On the stability of the sequential rank detection procedure [Lexical characteristics of Russian language]. *Radio Engineering and Electronics*, 23(7), 1427–1431.

Bajscy, R., Lee, S. W., & Leonardis, A. (1990). Color image segmentation with detection of highlights and local illumination induced by inter-reflections. In *Proceedings of the International Conference on Pattern Recognition* (pp. 785-790). Academic Press.

Bouman, C. A., & Shapiro, M. (1994, March). A Multiscale Random Field Model for Bayesian Image Segmentation. *IEEE Transactions on Image Processing*, 3(2), 162–177. doi:10.1109/83.277898 PMID:18291917

Boyat, A. K., & Joshi, B. K. (2015, April). A review paper: Noise model in digital image processing. *Signal and Image Processing: an International Journal*, 6(2), 63–75. doi:10.5121ipij.2015.6206

Brainard, D. H., & Wandell, B. A. (1986). Analysis of the retinex theory of color vision. Optical Society of America, 3(10), 1651-1660.

Ellison, S. L. R., & Williams, A. (Eds.). (2012). Quantifying Uncertainty in Analytical Measurement. EURACHEM/CITAC Guide CG 4.

Elmogy, M., & Zhang, J. (2009). Robust Real-time Landmark Recognition for Humanoid Robot Navigation. In *Proceedings of the 2008 IEEE International Conference on Robotics and Biomimetics* (pp. 572-577. 10.1109/ROBIO.2009.4913065

Gasvik, K. (2002). *Optical Metrology* (3rd ed.). John Wiley & Sons, Ltd. doi:10.1002/0470855606

Grubbs, F. (1969). Procedures for Detecting Outlying Observations in Samples. *Technometrics*, *11*(1), 1–21.

Hodge, V. J., & Austin, J. (2004). A Survey of Outlier Detection Methodologies. *Artificial Intelligence Review*, *22*(2), 85–126. doi:10.1023/B:AIRE.0000045502.10941.a9

Huang, N. E., Shen, Z., Long, S. R., Wu, M. C., Shih, H. H., Zheng, Q., Yen, N.-C., Tung, C. C., & Liu, H. H. (1998). The empirical mode decomposition and the Hilbert spectrum for nonlinear and non-stationary time series analysis. In Proceedings of R. Soc. London (pp. 903-995). doi:10.1098/rspa.1998.0193

Johnson, G., Song, X., Montag, E. D., & Fairchild, M. D. (2010, December). Derivation of a Color Space for Image Color Difference Measurement. *Color Research and Application*, *35*(6), 387–400. doi:10.1002/col.20561

Krantz, D. (1975). Color Measurement and Color Theory: Opponents Colors Theory. *Journal of Mathematical Psychology*, *12*(3), 304–327. doi:10.1016/0022-2496(75)90027-9

Levin, B. R. (1976). Theoretical foundations of statistical radio engineering [Lexical characteristics of Russian language]. Sov. radio, book.

Lim, S. H. (2006). Characterization of Noise in Digital Photographs for Image Processing. *Proceedings of SPIE Digital Photography II*, *6069*, 1–11. doi:10.1117/12.655915

Mori, H., Kobayashi, K., Ohtuki, N., & Kotani, S. (1997). Color impression factor: an image understanding method for outdoor mobile robots. In *Proceedings of the IEEE/RSJ International Conference of Intelligent Robots and Systems* (pp. 380-387). 10.1109/IROS.1997.649088

Ningombam, D., Rai, P., & Chingntham, T. S. (2018, December). Autonomous Mobile Robot Navigation and Localisation using Cloud, based on Landmark. *International Journal of Management. Technology And Engineering*, *8*(12), 1265–1278.

Nishad, P. M., & Chezian, R. M. (2013, January). Various colour spaces and colour space conversion algorithms. *Journal of Global Research in Computer Science*, *4*(1), 44–48.

Onta, Y., Kanade, T., & Sakai, T. (1980). Color Information for Region Segmentation. *Computer Graphics and Image Processing*, *13*(3), 222–241. doi:10.1016/0146-664X(80)90047-7

Patro, P. P. & Panda, C. S. (2016). A review on: noise model in digital image processing. *International Journal of Engineering Sciences & Research Technology*, 891-897.

Poliarus, O., & Poliakov, Y. (2020a) The Methods of Radar Detection of Landmarks by Mobile Autonomous Robots. In O. Sergiyenko, W. Flores-Fuentes & P. Mercorelli (Eds.), Machine Vision and Navigation (pp. 171-196). Springer. doi:10.1007/978-3-030-22587-2_6

Poliarus, O., & Poliakov, Ye. (2020b) The Problem of Using Landmarks for the Navigation of Mobile Autonomous Robots on Unknown Terrain. In O. Sergiyenko, M. Rivas-Lopez, W. Flores-Fuentes, Ju. Rodríguez-Quiñonez, & L. Lindner (Eds.), Control and Signal Processing Applications for Mobile and Aerial Robotic Systems. IGI Global. doi:10.4018/978-1-5225-9924-1.ch008

Poliarus, O., Poliakov, Ye., & Lindner, L. (2018). Determination of landmarks by mobile robot's vision system based on detecting abrupt changes of echo signals parameters. In *The 44th Annual Conference of the IEEE Industrial Electronics Society* (pp. 3165-3170). 10.1109/IECON.2018.8591362

Poliarus, O., Poliakov, Ye., Sergiyenko, O., Tyrsa, V., & Hernandez, W. (2020). Azimuth estimation of landmarks by mobile autonomous robots using one scanning antenna. *Proceedings of IEEE 28th International Symposium on Industrial Electronics (ISIE 2019)*.

Ramírez-Hernández, L. R., Rodríguez-Quiñonez, J. C., Castro-Toscano, M. J., Hernández-Balbuena, D., Flores-Fuentes, W., Rascón-Carmona, R., Lindner, L., & Sergiyenko, O. (2020). Improve three-dimensional point localization accuracy in stereo vision systems using a novel camera calibration method. *International Journal of Advanced Robotic Systems*, *17*(1), 1–15. doi:10.1177/1729881419896717

Romero, J., Hernández-Andrés, J., Nieves, J. L., & García, J. A. (2003, February). Color Coordinates of Objects with Daylight Changes. *Color Research and Application*, *28*(1), 25–35. doi:10.1002/col.10111

Scolnik, M. I. (Ed.). (1990). *Radar handbook*. McGraw-Hill.

Sergiyenko, O. Yu. (2010). Optoelectronic system for mobile robot navigation. *Optoelectronics, Instrumentation and Data Processing*, *46*(5), 414–428. doi:10.3103/S8756699011050037

Sharma, R. R., & Ingole, P. V. (2015, June). Review of Vision Based Robot Navigation System in Dynamic Environment. *International Journal for Research in Applied Science and Engineering Technology*, *3*(VI), 393–397.

Shearman, J. D., & Manzhos, V. N. (1981). Theory and techniques of processing radar information on the background noise [Lexical characteristics of Russian language]. Radio and Communication.

Shih, C., & Ku, Y. (2016). Image-Based Mobile Robot Guidance System by Using Artificial Ceiling Landmarks. *Journal of Computer and Communications*, *4*(11), 1–14. doi:10.4236/jcc.2016.411001

Sontakke, M. D., & Kulkarni, M. S. (2015, January). Different types of noises in image and noise removing technique. *International Journal of Advanced Technology in Engineering and Science*, *3*(1), 102–115.

Souto, L., Castro, A., Gonçalves, L., & Nascimento, T. P. (2017). Stairs and Doors Recognition as Natural Landmarks Based on Clouds of 3D Edge-Points from RGB-D Sensors for Mobile Robot Localization. Sensors, 17(8), 1824.

Stockman, A., & Sharpe, L. T. (2006). Physiologically based color matching functions. In *ISCC/CIE Expert Symposium '06: 75 Years of the CIE Standard Colorimetric Observer* (pp. 13-20) Vienna: CIE Central Bureau.

Stratonovich, R. L. (1968). *Conditional Markov Process and Their Application to the Theory of Optimal Control*. Elsevier.

Suresh, S. (2017). A Survey on Types of Noise Model, Noise and Denoising Technique in Digital Image Processing. *International Journal of Innovative Research in Computer and Communication Engineering*, *5*(2), 50-56.

Thomas, J. B. (1970). Nonparametric signal detection methods. *TIER*, *5*, 23-31.

Tikhonov, V. I. (1983). Optimal reception of signals [Lexical characteristics of Russian language]. Radio and Communication.

Tikhonov, V. I., & Mironov, M. A. (1977). Markov's processes [Lexical characteristics of Russian language]. Sov. radio.

Tukey, J. W. (1977). *Exploratory Data Analysis*. Addison-Wesley.

Varghese, D., Wanat, R., & Mantiuk, R. K. (2014) Colorimetric calibration of high dynamic range images with a ColorChecker chart. In *HDRi 2014 - Second International Conference and SME Workshop on HDR imaging* (pp. 1-6). Academic Press.

Zhengt, J. Y., Barth, M., & Tsuji, S. (1991). Autonomous Landmark Selection for Route Recognition by A Mobile Robot. In *Proceedings of the 1991 IEEE International Conference on Robotics and Automation* (pp. 2004-2009). 10.1109/ROBOT.1991.131922

Zhong, X., Zhou, Y., & Liu, H. (2017). Design and recognition of artificial landmarks for reliable indoor self-localization of mobile robots. *International Journal of Advanced Robotic Systems*, *14*(January-February), 1–13. doi:10.1177/1729881417693489

Zimek, A., & Schubert, E. (2017). *Outlier Detection. Encyclopedia of Database Systems*. Springer. doi:10.1007/978-1-4899-7993-3_80719-1

Chapter 9
Recognition System by Using Machine Vision Tools and Machine Learning Techniques for Mobile Robots

Jesús Elias Miranda Vega

 https://orcid.org/0000-0003-0618-0455

Polytechnic University of Baja California, Mexico

Anastacio González Chaidez

Polytechnic University of Baja California, Mexico

Cuauhtémoc Mariscal García

Faculty of Engineering, Autonomous University of Baja California, Mexico

Moisés Rivas López

Polytechnic University of Baja California, Mexico

Wendy Flores Fuentes

 https://orcid.org/0000-0002-1477-7449

Faculty of Engineering, Autonomous University of Baja California, Mexico

Oleg Sergiyenko

 https://orcid.org/0000-0003-4270-6872

Engineering Institute, Autonomous University of Baja California, Mexico

ABSTRACT

The systems based on image recognition play an important role in many cases where inspection methods are critical for industrial processes. Machine vision is required in industry for monitoring and detecting objects in real-time applications. Industrial robots are increasingly used in a variety of industries and applications such as manipulators and mobile robots. These devices are necessary for dangerous work conditions and tasks that humans cannot do; however, these industrial robots can also do some human activities such as transport material, select, and identify objects. In order to substitute human capabilities is no easy task. Nevertheless, the use of artificial intelligence has been replacing some human activities for increased productivity. By supporting machine learning technologies, this chapter presents a k nearest neighbors' algorithm for image classification of mobile robots to detect and recognize objects.

DOI: 10.4018/978-1-7998-6522-3.ch009

INTRODUCTION

Throughout history the technological advances have increased the human potential, however, there are 4 key moments that have presented a turning point and represent relevant moments in the societies. The very first of these key events was 18 centuries AD, a period of time during which the artisan production mode prevailed, and the force of motion was provided by natural means like water, beasts, or human beings. With the introduction of hydraulic and steam machine, a great change was made because it revolutionizes the manufacturing industry, mainly in England and it spreads to Occidentalist world, this was known as The First Industrial Revolution as well as the transportation means like the train and ships which also which helped by these advances (Li, 2017).

The second industrial revolution started a short time later within the birth of electricity and the incorporation of the electromotive force in the industry changes, which revolutionized again the life of the humankind by creating the electric motor and artificial illumination. According to (Mokyr, 1998) the second Industrial Revolution is usually dated between 1870 and 1914.

The Third Industrial Revolution initiates with the invention of computers, through binary logic, perforated cards at the beginning and with the advance in the transistors, integrated circuits, and its miniaturization, where the keystone for the emergence of Analysis and Control elements, for example, the personal computer (PC) and Programmable Logic Controls (PLC), which in the industry helped to increase the automation level and the quality of the processes, controlling those which are whether dangerous or repetitive. In 1969, the charge-coupled device (CCD), was invented by George Smith and the late Willard Boyle. One year later, computer vision first received serious attention (Zeuch, 2000) The use of computer vision in an industrial application is known as machine vision (MV). The widespread use of MV in industry arose in the 1980s. In the mid-1980s, smart cameras for industrial applications were introduced (Belbachir, 2010).These devices can be developed with two types of sensors such as CCD and Complementary Metal Oxide Semiconductor (CMOS). Techniques such as time-of-flight (ToF), laser scanning range finders became available during this era for outdoor mobile robot navigation (Kanade, 2012). These techniques are based on robust computer vision algorithms that allowed the development of the solution of the reliable problem of object recognition in the industry. Other types of algorithms that had a significant impact and gained wide popularity on image recognition and image segmentation in the 1990s were Generic Algorithms (GA) and artificial neural networks (ANNs) (Davies, 2004). In the mid-1990s, the communication and energy created a powerful new infrastructure for his epoch (Rifkin, 2011).

In 2011 appears in Hannover´s Technology Fair the 4.0 Industry concept, this means that a Fourth Industrial Revolution has been foreseen and despite it's focused on the industry, undoubtedly, impacts the modern society´s daily life and the advances generated by the innovations of direct consuming emerging companies, provided the base to integrate all this technological advances and created an environment, which when it was applied to the manufacturing systems, this industry evolved. Some of the recent advancements in Artificial Intelligence (AI), big data analytics, automation, additive manufacturing, the Internet of Things (IoT), Industrial Internet of Things (IIoT) and evolution of 5G networks are accelerating the change towards a fourth industrial revolution (Singh, 2019).

On a day-to-day basis, we could see elements as IoT, where intelligent sensors of our daily life elements (e.g. coffee makers, cars, smartphones, refrigerators, etc) interact between them and humans through the internet. The capacity to analyze the data obtained by sensors is processed by AI which makes decisions that can be corrected in case of mistakes with machine learning (ML), as well as enhanced by the strength of computerized vision, sharing this information with other systems and creating an integrating environment.

On the other hand, inspection and industrial difficult tasks such as welding, assembly, spray-painting, material handling, and integral now form part of most real-time. Currently, these requirements for real-life environments are changing and it is important to reduce the risk of accidents. One solution to reduce the risk is using a robot programmed to perform these industrial tasks. There are other reasons for using robots. First, they can execute repetitive tasks under varying environmental conditions. Second, robots can work 27/7. Third, they can improve their productivity.

The purpose this chapter is to introduce to the readers a basic video system based on machine learning tools for mobile robots and analyze the state of the art and the different approaches. And main direction of this work is for academics and engineers in order to be used as a reference work. It is also important to mention that a code for a mobile robot is developed by using an open source and inexpensive platform like a Psoc 4 Pioneer Kit. This chapter gives an overview of currently available and emerging optical technologies for sensing and communication applications by using mobile robots and smart cameras. Moreover, this work reviews its possible application in the context of the IoT for realizing smart systems for industry.

The organization of the present chapter is detailed as follows: First section introduces to the visual human activities realized in industry in this part it deals with the critical tasks that humans cannot do. Section 2 corresponds to the machine vision for industry 4.0 where manipulator and mobile robot are showed. Section 3, a brief description of the IIoT is provided. Section 4 describes important open resources that can be used for MV applications for industry such as Field-programmable gate array (FPGA), smart cameras and Raspberry Pi. Machine learning for machine vision application are discussed in section 5. Section 6 is discussed machine learning techniques and tools for machine vision applications. Section 7 shows a basic example for image pre-processing and test-training dataset for object detection. Section 8 the results of the experiments carried out in this chapter are discussed.

Visual Human Activities Replaced by Machine Vision

Modern robotics technologies in conjunction with machine vision have been introduced in industry 4.0 by replacing some visual human activities with the purpose of improving the quality of the products and reduce the headcount of the industry. Robotics systems use cameras to detect objects during their navigation. The modern cameras used for machine vision can adapt to different lighting conditions founded inline inspections of the industry.

Depending on light conditions the visual human activities can be affected. Industry illuminance level recommendations, for spaces intended mostly for use by older adults, are substantially higher than those for similar spaces intended for the "general population" (Rea, 2000).

Several norms regulate the light conditions in the industry, for example, according to the norm Mexican NOM-025-STPS-2008 the inspectors or workers have to be satisfied with the amount of light for reviewing parts or products, assembly, packing and other tasks. This norm, the minimum level of illumination incident light that should be on the workstation for every visual task or area work is detailed as follows in table 1.

The illuminance is measured in Lux (lm/m^2)

Table 1. Illumination Levels

Visual task of the workstation	Work area	Minimum levels of illumination (lux)
In outdoors: distinguish the area of transit, move around on foot, vehicle movement,	General outdoors: courtyards and parking lots	20
Moderate distinction of details: simple assembly, simple inspection, packing, office work.	Packing area and assembly	300
High accuracy for distinguish details: assembly, process and inspection of small complex pieces, fine polished finishing	Process: assembly and inspection of small complex pieces, fine polished finishing.	1000
High degree of specialization to distinguish details	Process of high accurate: Execution of visual task: • Of low contrast and small for long periods • Exact task and prolonged, and • Very special of extremely contrast and small size	2000

Another norm is the European standard EN 12464-1 that regulates and stipulates the light conditions of workplaces. In the context of the United States, the occupational safety and health administration also regulates the light conditions of work by using the standard 1926.56 – illumination. One example of this American standard can be described the area of operation such as first aid stations, infirmaries, and offices are standardized with 30 foot-candles or 322.917 lux.

A robot equipped with machine vision technology can adapt to a new industrial task with the light conditions shown in Table 1, and preserve the original requirements of its design. Human beings respond to wavelengths from about 380 nm to 740 nm, however, some machine vision systems are also able to inspect the objects in light invisible to humans, such as ultraviolet (UV), near-infrared (NIR), and infrared (IR) (Chen, 2002). A vision system can be based on spectral response to detecting light that humans cannot do. This is an advantage for robots with machine vision technology.

On the other hand, there are difficult tasks that require doing a job under conditions of ultraviolet light (UV), however, this natural radiation can be the cause of cancer. In these cases, it is better using a robot or manipulator to do the task than a human. At the other extreme of the wavelength range of the visible light, appears the infrared (IR) light, and this is located about 750nm–1mm. In the industry, this radiation is used for curtain beam detectors, although this radiation does not affect human health and is still invisible for the human eye. IR light is commonly used in mobile robots for avoiding and detecting obstacles.

The use of the MV for critical processes such as pharmaceutical and food packaging have been received by these industries to enhance their competitiveness. Critical defects such as a label or product mix-ups, wrong product, or mixed in a container, defective container and loose labeled they can be reduced significantly if a vision system is installed to filter these defects. According to (Hubbard, 2012), critical defects are those which are certain to cause the failure of a product to function as designed and the most serious are those which can endanger the health of consumers. Figure 1 shows a typical vision system developed to detect different labels for the pharmaceutical industry. When something wrong labeled occurs the machine vision software sends a signal to PLC in order to control the entire robotic arm by using the robot controller. Once the camera detects the label correctly, the process is ready to continue again.

Figure 1. A typical Machine vision applied on pharmaceutical industry

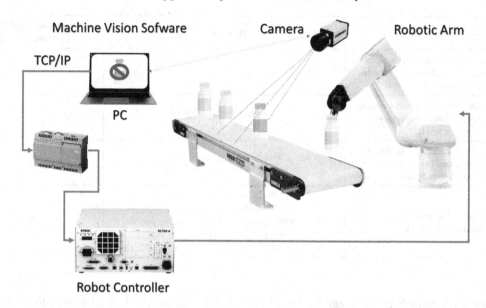

The typical response of the system

- An image is captured by the camera
- The camera sends the image captured
- Computer Software processes the image with the wrong label.
- A digital signal is sent from PC to PLC to control the robotic arm to collect the container and remove the conveyor.

Machine Vision for Industry 4.0

With the advance of technology in computer science, Industry 4.0 has been beneficial for monitoring processes in real-time. Today, Industry 4.0 requires MV applications for enhancing the manufacturing process or making inspection at higher speed, without defects, and at a lower cost. In any manufacturing process, the inspection of the products or services is indispensable. In this context, manipulator and mobile robot satisfies these necessaries.

Manipulator Robot and Mobile Robot

Industrial applications robots can be classified such as semi-autonomous and autonomous. These types of robots have been used in industry to increase effectiveness. The semiautonomous can be controlled by humans and the fully automated robots are already indispensable for industries, for example, Figure 2 illustrates a Fanuc LR mate 200i that is mounted on a CNC machine. This model is available with reaches up to 911 mm, a max load capacity of 7 kg, and a calculated TCP speed of 11 meters/second maximum.

Figure 2. Robotic manipulator taking a piece

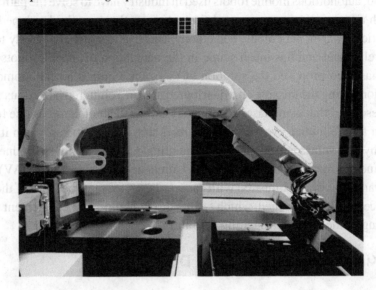

The teach pendant is a control box or hand-held device for programming the motions of this Fanuc robot, also this device is commonly used for control and manage any other industrial robot. This device is usually called a teaching box and has many advantages because is ease of learning to program and can direct programming. Figure 3 is illustrated a ''teach pendant'' of Fanuc. Although there are other functions of the Teach pendant, only 6 main functions will be detailed as follows.

1. Emergency stop button: brings the robot to an emergency stop.
2. Mode Selector switch: Switches between the operation modes such as auto, manual, and teach check modes.
3. Cursor keys: moves the cursor on the display and entry screens.
4. Hand strap: Secures the handle.
5. LCD screen: provides a display and a touch panel.
6. Function Keys: perform function assigned.

Figure 3. Teach pendant Fanuc

On the other hand, autonomous mobile robots used in industry have to solve the path planning problem and challenges such as precise measures in real-time. Among the recently developed in path planning techniques like particle swarm, ant colony, bee colony, bacteria forging, and firefly techniques, firefly technique is relatively novel and has much scope in the navigation of mobile robots (Zafar, 2018). In (Ivanov, 2020), it has been proposed the solution of tasks set required for autonomous robotic group behavior optimization during the mission on a distributed area in a cluttered hazardous terrain. In (Reyes-García, 2019), is described that using laser scanning system instead of cameras due to these reproduce the natural function for the human eye, they have remarkable disadvantages, when it comes to precise measurements of physical values and capturing surrounding data in real-time. (Lindner, 2017) proposes the use of laser scanners as machine vision systems in Unmanned Aerial Vehicle (UAV) navigation offers a wide range of advantages, when compared with camera-based systems because of the measurement of real physical distances can be mentioned, which results in a reduction of measurement times and thereby fast image processing of the UAV surrounding medium.

Mobile Robot Kinematics of Differential Drive

In order to design controllers for mobile robots, it is important to create models of how the robots actually behave. A differential drive mobile robot has two independently driven wheels and one or more caster wheels to keep it horizontal. Differential drive wheeled mobile robot has two wheels and the wheels can turn at different rates and by turning the, the wheels at different rates, you can make the robot move around.

For instance, if the two wheels turning at the same rate, the mobile robot is moving straight ahead. If one wheel is turning slower than another, then the mobile robot is going to be turning towards the direction in which the slower wheel is. The following paper shows the kinematic model of a mobile differential drive robot for trajectory tracking and control along a predefined regular geometrical path made up of these primitives (Mathew, 2016). Figure 4 shows the kinematic model for the mobile robot used for this chapter.

Figure 4. Kinematic model of a differential drive mobile robot for the

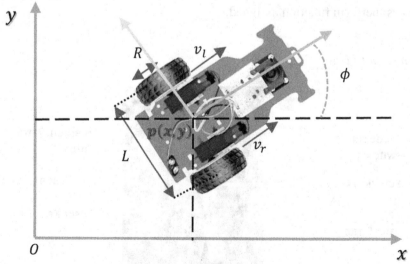

The first to take in account is to determine what are the dimensions of the robot. It is needed know the radius of the wheel. But these parameters play a little bit of a role when we're trying to design controllers for these robots.

Phi represents the heading or the orientation of the robot. The forward kinematic model of the mobile robot position and orientation can be defined by Eq.(1).

$$
\dot{q} = \begin{bmatrix} \dot{x} \\ \dot{y} \\ \dot{\phi} \end{bmatrix} = \begin{bmatrix} \dfrac{R}{2}\cos\phi & \dfrac{R}{2}\cos\phi \\ \dfrac{R}{2}\sin\phi & \dfrac{R}{2}\sin\phi \\ \dfrac{R}{L} & -\dfrac{R}{L} \end{bmatrix} \begin{bmatrix} v_r \\ v_l \end{bmatrix}.
\tag{1}
$$

where

$\dot{\phi}$ represents the rolling angles of the left and right wheels.

v_l and v_r correspond to the left and right wheel velocities, respectively.

L Specifies the track with of the mobile robot.

R represents the radius of the wheel.

Industrial Internet of Things (IIoT)

The application of IoT to the industry for the purpose of add value to the supply chains, products and service is widely known as IIoT. In other words, is all about getting the information that every single consumer need in their hand when they want it. IIoT is a subcategory of the Internet of things based on key concepts of operational efficiency, preventive maintenance, and cost optimization (Veneri & Capasso, 2018). Nowadays, the requirements for access to a network by any device anywhere are indispensable to improve the productivity of an industry. There are several advances in science computers that enhance technologies such as artificial intelligence, big data, and the internet of things. Devices such as Arduino and Raspberry are available to replace existing options for programmer logic controllers (PLCs), human-man interface (HMI), and supervisory control and data acquisition (SCADA) due to these devices can work by using open source software to solve industrial automation problems.

In (Ray, 2016), they introduced Internet of Robotic Things (IoRT) architecture considering conjugation between recently grown IoT and robotics together allows robots or robotic systems to connect, share, and disseminate the distributed computation resources, business activities, context information, and environmental data.

IIoT Comunication Procols

Industrial MV applications help to detect and identify failures in an industrial process and communicate with SCADA systems in order to monitor specific tasks or behaviors through IIoT devices (sensors and actuators placed in the field). SCADA architecture typically consists of four subsystems that correspond to I/O Network supervisory control, Control Network and Corporate Control. I/O network integrates all IIoT devices from a specific industrial application. Supervisory control is the second subsystem and the elements that form part of this are engineering workstations such as computer and human-machine Interface HMI to provide a graphical user interface to enable the operators to interact with the system or IIoT devices. The third subsystem of the SCADA corresponds to the Control Network integrated by PLC's that can control and connect with the IIoT devices such as sensors or actuators. The Corporate Network is the last subsystem that manages general services such as Web hosting, mail servers, computers, etc. The communication protocols allow managing the Corporate Network. IIoT data transmission protocols are an indispensable part of the SCADA to allow communication with the general system. The objective of these protocols is to implement data exchange between actuators, controllers, sensors and other systems in real-time control of local or remote processes.

One of the fundamental challenges for system engineers is to choose the appropriate communication protocol for their specific IoT system requirements (Dizdarević, 2019). The most common SCADA protocols typically implemented are: Modbus (Modbus is an open control that allows serial communication for devices IIoT by tradicional master-slave model, which master and slave requires and supply information, respectively), Ethernet/IP, DNP3, Profinet, DCOM, etc. The Modbus communication protocols most widely used in the SCADA systems.

Currently, SCADA networks are being connected to corporate networks with the goal of increasing productivity and efficiency, consequently, problems related to security can appear. An effective security strategy is required as any vulnerability of the SCADA system could generate severe financial and/or safety implications (Upadhyay, 2020). Lack of security can be make IIoT systems unattractive, as a consequence, authentication mechanisms have been proposed in order to ensure secure integration of IIoT solutions in the future production systems (Esfahani, 2017 Aug 9).The following author (Skwarek, 2017 Dec 4), explain Blockchain mechanisms can secure the wireless communication of Internet-of-Things-devices in a lightweight and scalable manner.

A classification of IoT connectivity technologies which can be used in industrial applications can be grouped into low power or battery-operated solutions (Wang, 2019 Jul 8). The classification for LP consists of three parts such as short range (BLE, WiSun and ZigBee), Medium Range (Dash7,Ingenu, PRMA, Weightless (N/P/W)) and Long Range (LoRa,NB-IoT, NB-Fi, Sigfox and Telensa). The classification of the connectivity solution for HP showed by the authors ided into two groups such as Short/Medium Range (WiFis: 802.11af, 802.11ac, 802.11ah and 802.11ax) and Long Range (LTE,LTE-A, LTE-A-Pro and 5G).

Summary of various advanced optical communication technologies and approaches (DWDM: Dense Wavelength-Division Multiplexing, CDC-ROADM: Colorless Directionless Contentionless Reconfigurable Optical Add-Drop Multiplexer, OTN: Optical Transport Networks, SDH: Synchronous Digital Hierarchy, SONET: Synchronous Optical Networks, EOS: Elastic Optical Networking, WSS: Wavelength Selective Switch, NG-PON: Next-Generation Passive Optical Network, NG-EPON: Next-Generation-Ethernet Passive Optical Network, P2P: Point-to-Point, OWC: Optical Wireless Communications, LiFi: Light Fidelity, VLC: Visual Light Communication, LAN: Local Area Network) these optical technologies have the potential to greatly help in realizing future smart infrastructures and systems (Aleksic, 2019 Sep).

The next authors (Sun, 2019 Nov 22) have designed a Modbus to MQTT gateway for Industrial IoT cloud applications using inexpensive Raspberry Pi single-board computer and a RS485 add-on board with the necessary integrated circuits for data communication. The following article (Hussain, 2019 Aug 27), presents a light-weight convolutional neural network (CNN) and IIoT-based computationally intelligent (CI)Multiview video summarization (MVS) framework by using an IIoT network containing smart devices, Raspberry Pi (RPi) (clients and master) with embedded cameras to capture multiview video data.

Resources for Machine Vision

The resources that can be used for machine vision applied to a mobile robot are integrated by smart cameras based on RGB-D and Raspberry Pi.

Raspberry Pi Hardware

Raspberry Pi is a minicomputer of small dimensions and price and has been developed for small prototypes and to stimulate the teaching of computer science in educational centers. Developed in free hardware, it has GNU / Linux operating systems such as Raspbian although we can find other operating systems optimized for the Raspberry Pi hardware. Acorn RISC Machine (ARM) processors are commonly used in electronic devices such as tablets, smartphones, and other mobile devices. The advantages of using these processors are they have reduced instruction set as a result they require fewer transistors, which enables a smaller die size for the integrated circuitry (IC). According to (Eben Upton, 2014) the ARM-based BCM2835 is the secret to the Raspberry Pi's capacity to operate on just the 5V 1 A power supply, this means that the device does not require heat sinks.

Raspberry Pi hardware for machine vision supplies the elementary tools required for install and program the project based on Raspberry Pi 3. Raspberry has a massive community around the world, the availability and the cost are the advantages of using this device for industrial environments. The cost of other industrial equipment could cost thousands, even hundreds, of dollars. Industrial automation can be benefited by using open-source devices. One of the most important goals in the industry is to reduce manufacturing costs by preserving the quality of the products. There are indispensable tasks that require devices such as PLC's, however, Raspberry can execute difficult tasks that PLC can't do such as higher processing, pattern recognition (creates models, supervision, and classification), machine learning. The authors (Gundecha, 2020) proposes the use of Revolution pi due to this is an open, modular and inexpensive industrial PC based on Raspberry pi and it can be used as a small controller unit.

Raspberry Pi Software

Python is a high-level programming language and object-oriented with dynamic semantics according to python developers. This software facilitates the design projects for machine vision by using python libraries such as Scikit-Image and OpenCV. These libraries collected algorithms for image processing by using Python software. The term image processing has been known as a method or technique to extract useful information from an image through an computatational algorithm.

Scikit-Image is used for segmentation, geometric transformations, color space manipulation, analysis, filtering, morphology, feature detection. OpenCV is a highly optimized library of Python developed to solve computer visión problems focused on real-time applications. All the OpenCV array structures are converted to and from NumPy arrays, this makes it easier to integrate it with other libraries that use Numpy. For example, a basic operation that can be performed in OpenCV is presented as follows.

```
Import cv2 (import the OpenCV Module)
Img = cv2.imread("Penguins.jpg",1) # read the image in RGB / colored format
Img = cv2.imread("Penguins.jpg",0) # read the image as a gray scale image or
black and White image, where Penguins is the path of the image. Python stores
the image as a Numpy array / matrix of numbers
print(type(img)) # numpy n-dimensional array
```

RGB-D Cameras

The irruption of the RGB-D cameras in several research fields such as robotics, 3D reconstruction, human-machine interaction, automotive and other industrial applications. Commercial RGB-D cameras have become one of the main sensors for indoor scene 3D perception and robot navigation and localization (Zhong, 2020). It is well known that Microsoft Kinect, Asus Xtion Pro, Intel RealSense cameras and PrimeSense carmine are very popular RGB-D cameras used to develop machine vision for industry based on techniques for depth measurements.

The commonly types of technique or working principles for depth measurements used for these cameras are the following: ToF (measures the round trip time of a light source to determine the distance between the ToF camera and the object observed), structured light (consists in projecting different light patterns on to a scene or object under study). The type of structure light used for the RGB-D cameras is infrared light. According to the author (Moreno, 2012 Oct 13), the use of the patterns projected on the scene can be generated by a projector or other devices. The other type of working principle of the stereo cameras are based on the human binocular vision to capture 3D images by two or more lenses. The next types are examples embedded stereo camera, FPGA stereo camera, active stereo camera, Active IR Stereo using Global Shutter Sensors. These authors show an example of an active stereo camera applied to a mobile robot (Kijima, 2017 Oct 18). Applications based on FPGA stereo cameras to develop real-time 3D stereo vision for parallel computation can be consulted in (Schauwecker, 2018) The camera Intel RealSense D415 Depth is based on active stereo and (Carfagni, 2019) characterized and provided metrological considerations for this camera.

With the techniques mentioned before, it is possible to determine depth values and are widely used in industrial applications to develop machine vision systems. However, something important to consider is to implement the MV system under difficult environments like outdoors. Time-of-flight, laser scanning range finders techniques became available for outdoor mobile robot navigation (Kanade, 2012). These techniques are based on robust computer vision algorithms that allow the development of the solution of the reliable problem of object recognition in the industry for mobile robots. Other types of algorithms that had a significant impact and gained wide popularity on image recognition and image segmentation in the 1990s were Generic Algorithms (GA) and artificial neural networks (ANNs) (Davies, 2004).

An increasing number of publications of object detection has gained attention in the past two decades . Current challenges in object detection are the following aspects: object rotation and scale changes (e.g., small objects), accurate object localization, automatic detection of multi-class objects, speed for real-time object detection or Visual object tracking and background subtraction .

Currently, robot vision has successfully applied in industrial testing detection, such as product packaging and printing quality detection . Mobile robots depend on the development of video systems to realize localization and navigation. They can use three different of cameras: monocular, binocular and multi-camera vision system . These cameras are usually mounted on a mobile robot and can be applied for mapping, localization, and obstacle detection especially for scene-related tasks .

FPGA for Measuring Pulses of the Encoder

It is common knowledge that FPGA integrates the elements and resources for solving specific requirements of the design. It was decided that the best equipment for this investigation was an FPGA. In this research work, an experiment of a data acquisition system using the FPGA platform is presented, data acquisition systems are used in a wide variety of fields one of them is the industry whose use is for control and automation process, in the field of research we have that is used to perform monitoring, testing, and signal measurements.

One of the characteristics of signal acquisition systems is to measure the physical phenomena whose signals are analog and convert it to a digital signal understandable by the system and then manipulate this signal and generate data that will be used in another subsequent process.

This research focuses on the design and implementation of a tachometer system which consists of activating a direct current motor, this being activated by means of an external source to the Cyclone II EP2C5T144C8 plate of the Altera family.

This experiment is performed on a Cyclone II EP2C5T1448 platform from Intel's Altera family. The tachometer is a device that is used to measure the rotation of an axis, it is measured by revolutions per minute (RPM), currently the most common are digital tachometers which are found in vehicles and motorcycles, applications such as turbine engines also to carry out a study on the life of a machine and perform predictive maintenance on it.

The card used in this experiment is a cyclone II EP2C5T144C8 that contains a two-dimensional architecture whose row and column arrangement allows logical designs to be implemented in a personalized way. This architecture allows signal interconnections between LABs and integrated memory blocks. A logical matrix consists of LABs, with 10 LEs in each LAB. An LE is a small logical unit for logical user functions. The cyclone devices contain between 2,910 and 20,060 LE. The following Figure 5 shows a LE from cyclone II.

The Figure 6 shows the basic equipment of the development board Cyclone II Card EP2C5T144C8.

Quartus II software provides parameterizable mega functions ranging from units of simple arithmetic, such as adders and counters, up to advanced phase blocks blocked (PLL), multipliers, and memory structures. These mega functions have a performance-optimized for Altera devices and therefore provide a synthesis logic and more efficient device implementation. These mega functions automate the coding process and save you valuable design time. You must use these functions during design implementation to be able to consistently meet its design goals.

The experiment is carried out using block diagram design. This is another option that facilitates the development of projects that are not very complex to carry out.

First a new file is created and then select block diagram as shown in the following Figure 7:

To select the component to be used in this case, it is a 7447 common anode decoder to later display the result on a 7-segment display that will show when the motor makes a complete turn, the appropriate connections are made to work, that is, the outputs are connected as can be seen in the diagram that will go to the inputs of the Display 7 segments, in this case, the output pins of the Cyclone card of Pin_112, Pin_113, Pin_114, Pin_115, Pin_118, Pin_111, Pin_121, were selected for the connection of the Display.

Subsequently, the counter 74193 was selected, which a four-bit is descending up counter, in Figure 8 is illustrated the block diagram selected and inserted:

Figure 5. Diagram blocks of the register LE from Cyclone II..

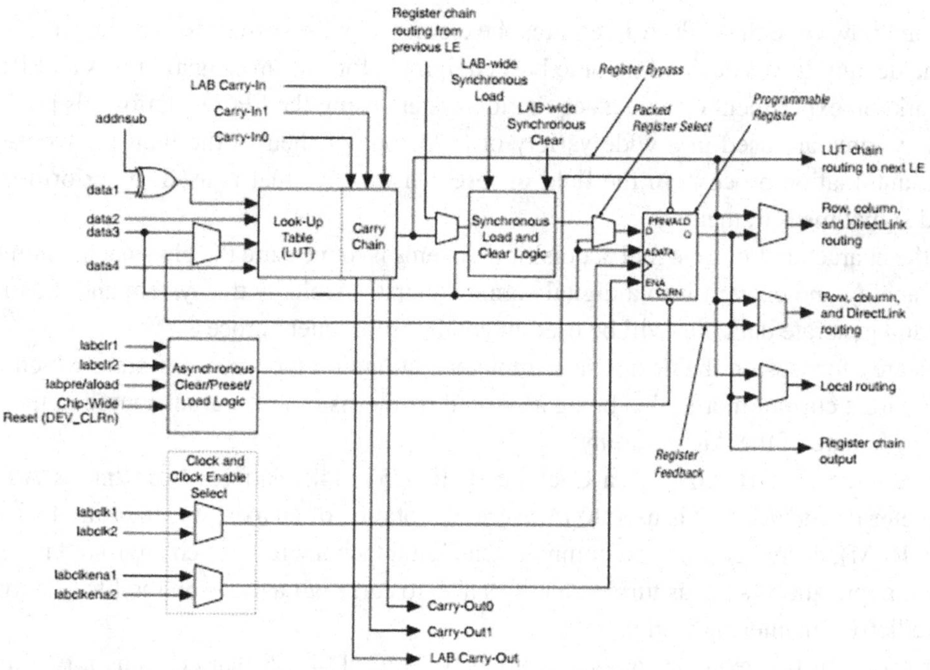

Figure 6. Basic equipment of the Cyclone II.

Figure 7. Procedure to create a new design file (Block diagram).

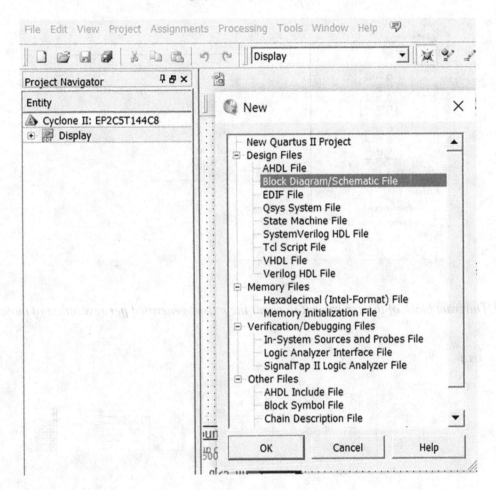

Connections are made from the counter to the 7447 decoder and an and gate is implemented with two inputs whose output goes to the counter reset this in order to count from 0 to 9, later it can be expanded to two or more displays to make it a Two-digit account, or it is recommended to use a 16x2 LCD display of the Hitachi Family to see the number of revolutions of the engine. To generate the changes in the counter, the outputs of the Hall sensor were taken, which gives us a high-level signal when detecting a complete turn of the motor, which facilitated the experiment of detecting turns. Figure 9 shows this process.

The result of the experiment can be seen in the following photo illustrated in Figure 10 that was taken to demonstrate the experiment in a practical way.

It is important to know that cards for embedded systems such as the Cypress PSOC 4 Pioneer Kit CY8CKIT-042 family microcontroller are used for signal acquisition; this microcontroller is also used for embedded systems. Psoc Creator contains a workspace explorer in which files and folders are added. Figure 11 shows Schematic Editor displays the top-level schematic file known as TopDesign.cysch, and the Component Catalog opens to display a list of Components to use in the present design.

Figure 8. Diagram block of counter 74193.

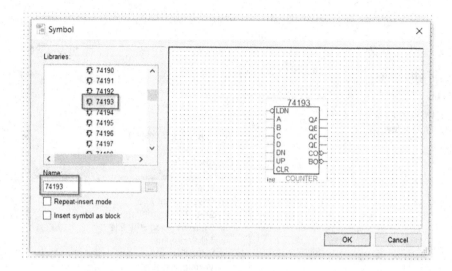

Figure 9. Diagram block of the experiment to count the pulses generated per revolution of motor.

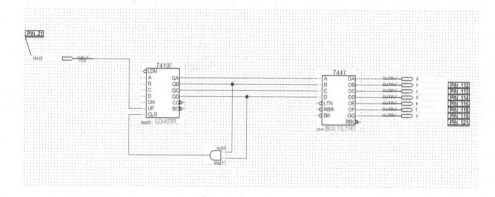

Figure 10. Experimental test for measuring the revolution of motor with cyclone II.

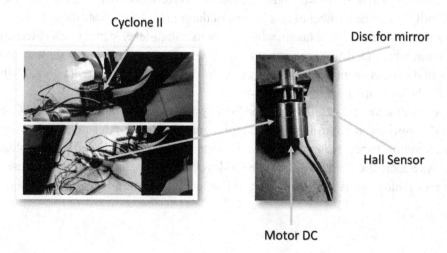

Figure 11. Top design of the experimental test for detecting obstacles by using Sharp sensor.

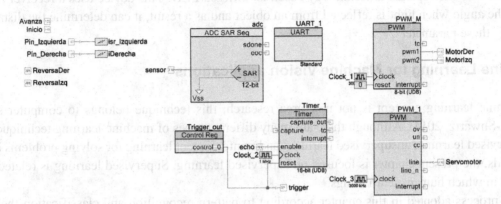

Optical Sensors for Measuring Distance

There are several technologies and methods for measuring distance such as infrared triangulation, Laser, Time-of-flight based on optical sensors.

For example, the infrared proximity sensor made by Sharp with model number GP2Y0A21YK can be used to measure the distance to an object by applying the triangulation method. The output of this sensor is inversely proportional because when the measurement of the distance is growing the output is decreasing. The typical measuring range of this sensor is 10 cm – 80 cm and the output voltage can reach up to 3v.

In Figure 12 is illustrated as an example of a time-of-flight system based on the active stereo. Figure 12a) shows a point laser that emits a pulses train that is reflected from the object under study, as a result, the difference between the initial time when the pulse was transmitted and the time that it returns to the detector correlates with the object's distance. An example of this application is a LIDAR which is an acronym of light detection and ranging. This device used a pulsed laser to measure range and generate 3D data.

Figure 12. Proximity sensors based on LASER and LED devices.

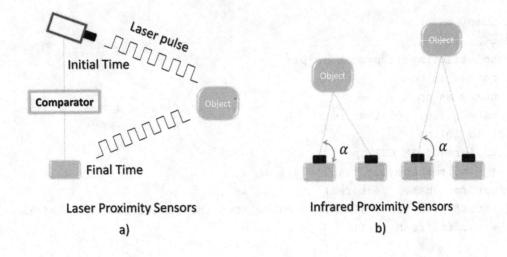

Figure 12b corresponds to technology based on infrared LED. This device uses a receiver that can detect the angle when light is reflected from an object and as a result, it can determine the distance by knowing these parameters.

Machine Learning for Machine Vision applications

A machine learning concept is not new area research, this technique belongs to computer science (Shalev-Shwartz, 2014). Although there are many different types of machine learning techniques such as supervised learning, unsupervised learning, and reinforcement learning for solving problems in various fields, the present chapter is focused on supervised learning. Supervised learning is related to the process in which humans learn things.

The process adopted in this chapter according to pattern recognition and classification theory for applying the techniques and methods of machine learning to machine vision is detailed as follows.

1. **Sensing**: This step deals with raw data captured by the sensors.
2. **Segmentation process**: This step of the process consists in segregating any aggregated entity into separate parts or groups (Sarkar, 2018).
3. **Feature extraction**: the extraction of features is an important step for recognition systems that describe the main characteristics of the phenomenon under study. The process of feature extraction can be carried out by using wavelet transform for pattern recognition as is presented in (Hubeli, 2001). Another method for feature extraction is based on the fast Fourier transform (FFT) (Tachaphetpiboon, 2006). These transformation techniques are used to enhance the characteristics of the data set.
4. **Classification process:** The patterns are related or classified into specific class or category by using classifiers like Bayesian Classifiers, Hidden Model Markov (HMM), K-Nearest Neighbor (KNN), Artificial Neural Network (ANN), Support Vector Machines (SVM) and others.
5. **Performance of the classifiers:** In this process is measured the performance of the model selected of the before step.

Feature extraction is an important part of determining a model for machine learning. When the problem deals with image or video it will be needed to adequate the dataset before estimate the model.

For example, the following code is used for vectorizing an image by using python software.

```python
import numpy as np
import cv2
from matplotlib import pyplot as plt
import pandas as pd
import numpy as np
import matplotlib.pyplot as plt
#matplotlib inline
img = cv2.imread('IMAGE1.jpg',0)
features = np.reshape(img, (452*678))
print(features.shape, features)
f = np.fft.fft(features) #Then, is calculated the Fast fourier transform
fshift = np.fft.fftshift(f)
```

```
magnitude_spectrum = 20*np.log(np.abs(fshift)) #it is extracted, the magnitude
spectrum or real part.
```

#With the magnitude spectrum obtained it is possible to create a corresponding model.

In (Ramík, 2014), it was presented as an intelligent machine vision system able to autonomously learn individual objects presented in the real environment.

If there is a need for using an algorithm easy-to-implement for machine vision application, the *k*-nearest neighbors (*k*-NN) is an appropriate algorithm that can be used for classification problems with the advantage that there is no need to build a model. This algorithm takes the *k*-NN of the new data point according to their Euclidean distance. Figure 13 illustrates the algorithm where the red triangle represents a new instance to classify and circles correspond to a certain type of class.

Figure 13. k-NN algorithm

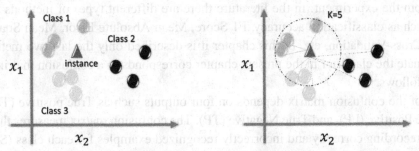

k-Nearest Neighbor Algorithm

Visualizing dataset in two dimensions. Three classes are illustrated and an observation or instance.

Euclidean distance is calculated for finding closets similar object.

This machine learning technique is based on a distance metric between two patterns x and y. In the case of this chapter the Euclidean distance is used for the experiments, this distance metric is calculated with Eq(2).

$$d(x,y) = \sum_{i=1}^{n} (x_i - y_i)^2 .$$ (2)

Although, there are other methods to calculate the distance metric in the present chapter are described the following: Tchebyshev Metric, Minkowski Metric, and Hamming Metric in Table 2.

Table 2. Different metrics to calculate the distance between two patterns.

Distance	Formula	Equation
Hamming	$d(x,y) = \sqrt{\sum_{i=1}^{n} |x_i - y_i|}$	(3)
Minkowski	$d(x,y) = \sqrt[p]{\sum_{i=1}^{n} (x_i - y_i)^p}$	(4)
Tchebyshev	$d(x,y) = \max_{i=1,2,...,n} |x_i - y_i|$	(5)

Performance of the Classifiers

Once a model is implemented and developed, the next step is to determine the performance of the model. This part evaluates the machine learning algorithm used in order to know if satisfying results have been gotten from the experiment. In the literature there are different types of methods to evaluate the classifiers such as classification accuracy, F1 Score, Mean Absolute Error, Mean Sqare Error, Confusion Matrix, Cross-Validation, etc. In this chapter it is described only the last two metrics. The metric used for evaluate the classifier in the present chapter corresponds to a confusion matrix method that is explained as follows.

The parameters of the confusion matrix depends on four outputs such as True positive (TP), False Negative (FN), False Positive (FP), and True Negative (TP). The confusion matrix measures the quality of classification by recording correctly and incorrectly recognized examples for each class (Sokolova, 2006). Figure 14 illustrates the confusion matrix table.

These metrics are described as follows.

1. **True Positive**: corresponds to the correct predictions, for example when a sample is positive and the model classifies correctly as positive.
2. **False Negative**: A false negative error occurs when the classifier has been incorrectly predicted a class negatively, given class is positive.
3. **False Positive**: This output occurs when the classifier has been incorrectly predicted a class positively, given class is negative.
4. **True Negative**: Given class is negative and the classifier has been correctly predicted is as the negative class is known a true negative.

Another method to evaluate the classifier corresponds to Cross-Validation (CV). The method of CV splits the data set into two parts: The first part has the data used for creating a model. The second part contains the data points for probing the model, in (Shao, 1993) it can be consulted this method. One of the configurations of CV is called leave-one-out that consist in removing one sample a at time for testing. *k*-fold CV is another method where *k* is equal sized subsamples from a given data is to be split into.

Figure 14. Confusion Matrix

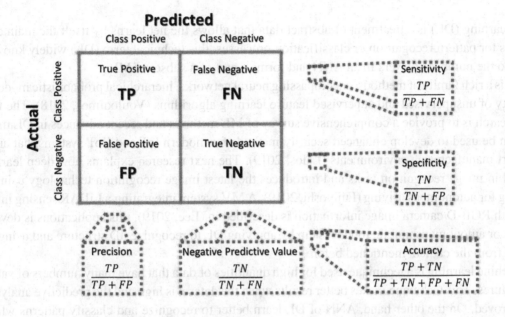

Figure 15 illustrates the leave-one-out cross-validation method, this configuration of CV means that one point is extracted for testing the model and the rest of the data points are used for fitting the model. This process is repeated for all elements of the dataset, obtaining performance for everyone. This method can be modified for leave-two-out, leave-there-out, leave-four-out, or many configurations required for calculating the goodness-of-fit.

Figure 15. Cross Validation

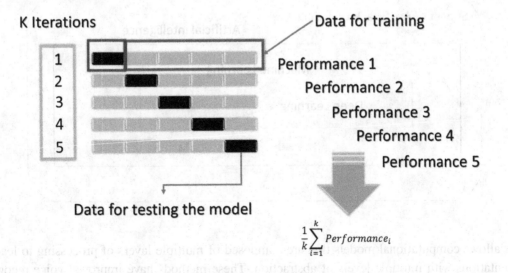

Deep Learning

Deep learning (DL) is a treatment of abstract data that allows the net learns by itself the main characteristics for pattern recognition or classification processes, this technical term (DL) widely known that refers to the number of layers in an ANN and forms a part of a subset of ML.

DL is a rich family of methods, encompassing neural networks, hierarchical probabilistic models, and a variety of unsupervised and supervised feature learning algorithms (Voulodimos, 2018). The following research is to provide a comprehensive survey of ML methods and recent advances in DL methods that can be used to develop enhanced security methods for modern IoT and IIoT systems that are used in smart manufacturing environments (Lalos, 2019). The next research explains how deep learning is applied in image recognition tasks and introduces the latest image recognition technology using deep learning for autonomous driving (Fujiyoshi, 2019). A MV system integrating a LiDAR sensing information with RGB-D camera image information is described in (Lee, 2019), this applications is developed for indoor autonomous mobile robot system by applying DL to recognize 3D structure and using depth images from the camera mentioned before.

Machine learning is a technique used for high quantities of data that have many numbers of variables and features. This technique makes better results when the dataset is high and the predictive analysis can be improved. On the other hand, ANN of DL learn better to recognize and classify patterns when the data set is high. AI can be defined as a system that interacts with its environment: machines of artificial intelligence are dotted with sensors that allow knowing the environment which are tools where they can relate to them. AI contributes to informatic with intelligent behaviors and collaborates with data abstraction for programming algorithms to treat data for their behavior and can do the task programmed.

Figure 16. Relation among Artificial intelligence, Machine Learning and Deep Learning.

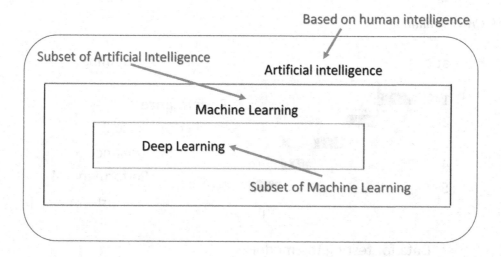

DL allows computational models that are composed of multiple layers of processing to learn data representations with multiple levels of abstraction. These methods have improved voice recognition, audio and visual object recognition.

The following characteristics are related to Figure 164.

- Artificial intelligence:
  - Can interact with natural language with humans.
  - Do specific tasks.
  - Solve problems faster than humans.
- Machine Learning:
  - Given a dataset, it learns by itself.
  - Recognizes patterns of data and can predict.
  - Do not require to be programmed before.
- Deep Learning:
  - Imitates the connectivity of the neurons of the human brain.
  - If the dataset is big, the prediction can be improved.
  - Classify a dataset and calculate the correlation.

Artificial Neural Networks

ANN have been tried to be defined with mathematical models during the information process. Some mathematical models are integrated by many numbers of elements organized in levels. These levels are known as layers of the network. The neural networks can be interconnected with themselves and organized hierarchically. When it has tried to define ANN it always has been compared to the biological nervous system. The vision system of the brain has the function of recognizing objects such as persons. Then, artificial neuron networks try to imitate a vision system based on pattern recognition. Since the human brain is not a linear system and can work in parallel to execute many tasks at the same time. By contrast, the computers only can do one task at a time. In this context, artificial neural networks can process information distributed in parallel through units called neurons, see Figure 17.

Figure 17. Graphic representation of a neuron.

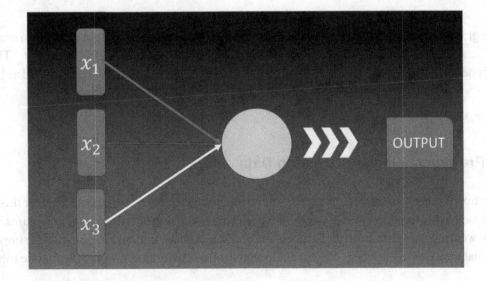

Types of neural networks

- Simple perceptron
- Multilayer perceptron
- Hopfield
- Convolutional

Simple perceptron: Simple perceptron networks contain only a single layer so neurons of the output are not linear as shown in Figure 18. The input layer is not often considered. This network was designed for pattern recognition, however, but at the same time has a single layer that only can be used for linear separability patterns.

Figure 18. Graphic representation of a neuronal network perceptron.

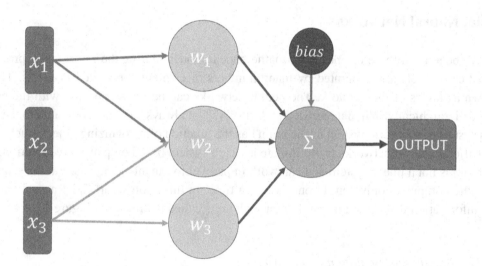

In the graphic representation shown before in Figure 18, the inputs of simple perceptron are denoted by $x_1, x_2, ..., x_m$ and the synaptic weight of the simple perceptron corresponds to $w_1, w_2, ..., w_m$. Therefore, the bias is denoted by b. In this graphic, induced local field of the neuron can be determined as by Eq(6).

$$v = \sum_{i=1}^{m} w_i x_i + b .$$

(6)

Image Pre-processing and Training Data

In this section is presented the pre-processing of the image and training-testing data used for this chapter. The purpose of this experiment is to study the performance of k-NN with minimun features, in other words, by working with data from high dimension and select a low-dimensional space. It is important to mention that the experiments were carried out in a controlled environment where the mobile robot could move freely until detect an object. The mobile robot takes pictures and sends to Matlab software in order to process and build the *k*-NN model and the shows the prediction calculated from dataset selected. The procedure of those are described as follows.

First, it was taken 20 images test the *k*-NN and separed by two groups of 10 images each one. The first group has 10 images that correspond to Class 1 (which corresponds to a person in front of mobile robot) and the rest of the images represent Class 2 (this class corresponds an object infront mobile robot). The second step of the pre-processing is for resizing the images 1/64 times. The original size of 3-dimensional image array (3492, 4656, 3; where the values correspond to the rows, columns, and numbers of color channels). The image was resized to 55,73,3.

Due to the original images containing lots of data, according to the image array, it was necessary to convert it to the grayscale image. The dimension of the grayscale image resulted in 55,73 for each image. Then, for the third step it was calculated the two-dimensional Fourier transform of a matrix (grayscale image) with a dimension of 55,73. However, due to computational cost that represents for the experiment to use these dimensions, only it was taken the two first columns. And the final dimension selected was 55,2 for the image array.

In Figure 19 is illustrated the two first features of the grayscale image.

Figure 19. Features of the two first features of the two-dimensional Fourier transform.

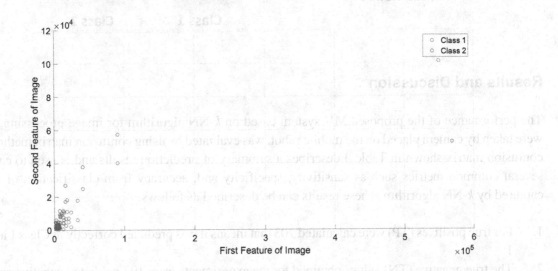

Finnaly, based on these vectors the dataset is separated into two groups that correspond to training and test datasets. However, 5 images were used for training the dataset and the rest of the images were separated to be used to test the model based on the *k*-NN algorithm. The dimension of the training dataset was *275×2* for Class 1 and Class 2. The test dataset has the same dimensions that train dataset.

Figure 20 illustrates the sequence of the experimental test carried out in this chapter. The following detailed procedure specifyes the sequence adopted for the pre-processing and applying *k*-NN algorithm stages.

1. Psoc 4 Pionner Kit controls the servo by sending an electric pulse Infrared proximity Sharp sensor GP2Y0A21YK.
2. Two objects are detected by a Sharp sensor.
3. Raspberry Pi 3 takes pictures of objects detected.
4. k-NN algorithm written in Matlab Code is used to create a model depending on the class.
5. Finally, the features are classified according to the k-NN algorithm.

Figure 20. Experimental test by using Psoc 4 and Raspberry Pi 3.

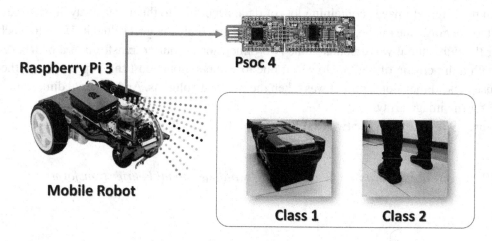

Results and Discussion

The performance of the proposed MV system based on k-NN algorithm for image processing, which were taken by camera placed on the mobile robot, was evaluated by using confusion matrix method. The confusion matrix shown in Table 3 describes a summary of prediction results and is used to calculate several common metrics such as sensitivity, specificity and, accuracy from classification of images captured by k-NN algorithm. These results can be described as follows.

1. For true positives (TP) were calculated 103 that means it was predicted correctly in Class 1 as Class 1.
2. The true negative (TN) values obtained for the experiments were 197 predicted negative correctly the class 2 as a class 2.
3. False positives (FP) are known as a type 1 Error, in this table is illustrated the 78 instances labeled as a class 1 when they are class 2.
4. The type 2 error or false negative (FN) calculated were 172, that means it was classified incorrectly as class 2 when it was class 1.
5. The precision of the classification was 0.5691 while sensitivity, specificity and accuracy were 0.3745, 0.7164, and 0.5455 respectively.

It is important to mention that sensitivity is low compared to specificity. That reveals that 2-class classification problems presented in this work have a considerable prediction from true negatives in the comparison between true positives rates.

Table 3.Confusion matrix of the experiments carried out with two objects detected by a mobile robot.

	Positive	Negative	
Positive	TP= 103	FN=172	Sensitivity=0.3745
Negative	FP= 78	TN=197	Specificity= 0.7164
	Precision = 0.5691	Negative Predictive Value= 0.5339	Accuracy=0.5455

FUTURE RESEARCH DIRECTIONS

The future improvements of the present project include the integration of an aperture to replace sharp sensors and implement in real-time the classification of the obstacles detected by using Raspberry Pi. Improve the dataset by incrementing the dataset of images and comparing several ML methods such as ANN, AdaBoost, HMM or SVM. Those are the next challenges and integration for the present work described in this chapter.

CONCLUSION

The purpose of this work was to create and train an image recognition device to be used in a mobile application, this limits the kind of device that can be used, a laptop is both too big and power-hungry, after thorough research it was decided that a Raspberry Pi should be used, the processing power on it is limited in comparison with a computer so the idea of using it to train an algorithm to recognize images was discarded, instead, a computer will be used to train the algorithm and then the trained file will be exported to a Raspberry pi.

The most important findings in this work that using *k*-NN and matrix confusion showed that the system can classify the Class correctly with a specificity of 0.7164. the classes are not separated completely based on the results of Sensitivity.

Although the k-NN algorithm is a non-parametric method used for classification in this chapter, this method has advantages that it is easy to implement in comparison to neural networks.

REFERENCES

Aleksic, S. (2019, September). A Survey on Optical Technologies for IoT, Smart Industry, and Smart Infrastructures. *Journal of Sensor and Actuator Networks*, 8(3), 47. doi:10.3390/jsan8030047

Belbachir, A. N. (2010). *Smart cameras*. Springer. doi:10.1007/978-1-4419-0953-4

Carfagni, M. R., Furferi, R., Governi, L., Santarelli, C., Servi, M., Uccheddu, F., & Volpe, Y. (2019). Metrological and critical characterization of the Intel D415 stereo depth camera. *Sensors (Basel)*, *19*(3), 489. doi:10.339019030489 PMID:30691011

Chen, Y.-R., Chao, K., & Kim, M. S. (2002). Machine vision technology for agricultural applications. *Computers and Electronics in Agriculture*, *36*(2-3), 173–191. doi:10.1016/S0168-1699(02)00100-X

Davies, E. R. (2004). *Machine vision: theory, algorithms, practicalities*. Elsevier.

Deng, Z. H., Sun, H., Zhou, S., Zhao, J., Lei, L., & Zou, H. (2018). Multi-scale object detection in remote sensing imagery with convolutional neural networks. *ISPRS Journal of Photogrammetry and Remote Sensing*, *145*, 3–22. doi:10.1016/j.isprsjprs.2018.04.003

Dizdarević, J. F.-B., Carpio, F., Jukan, A., & Masip-Bruin, X. (2019). A survey of communication protocols for internet of things and related challenges of fog and cloud computing integration. *ACM Computing Surveys*, *51*(6), 1–29. doi:10.1145/3292674

Eben Upton, G. H. (2014). *Raspberry Pi User Guide*. John Wiley & Sons.

Esfahani, A. G., Mantas, G., Matischek, R., Saghezchi, F. B., Rodriguez, J., Bicaku, A., Maksuti, S., Tauber, M. G., Schmittner, C., & Bastos, J. (2017, August 9). A lightweight authentication mechanism for M2M communications in industrial IoT environment. *IEEE Internet of Things Journal*, *6*(1), 288–296. doi:10.1109/JIOT.2017.2737630

Fujiyoshi, H. T., Hirakawa, T., & Yamashita, T. (2019). Deep learning-based image recognition for autonomous driving. *IATSS Research*, *43*(4), 244–252. doi:10.1016/j.iatssr.2019.11.008

Gundecha, T. J. (2020). *Automation of Mechanical Press Machine Using Revolution Pi and PLC*. Academic Press.

He, A. C. (2018). A twofold siamese network for real-time object tracking. In *Proceedings of the IEEE Conference on Computer Vision and Pattern Recognition* (pp. 4834-4843). Salt Lake City: CVF. 10.1109/CVPR.2018.00508

Hubbard, M. R. (2012). *Statistical quality control for the food industry*. Springer Science & Business Media.

Hubeli, A. a. (2001). *Multiresolution feature extraction for unstructured meshes. In Proceedings Visualization, 2001. VIS'01*. IEEE.

Hussain, T., Muhammad, K., Ser, J. D., Baik, S. W., & de Albuquerque, V. H. C. (2019, August 27). Intelligent Embedded Vision for Summarization of Multiview Videos in IIoT. *IEEE Transactions on Industrial Informatics*, *16*(4), 2592–2602. doi:10.1109/TII.2019.2937905

Ivanov, M. O.-F.-Q. (2020). Influence of data clouds fusion from 3D real-time vision system on robotic group dead reckoning in unknown terrain. *IEEE/CAA Journal of Automatica Sinica*, *7*(2), 368-385.

Kanade, T. (2012). Three-dimensional machine vision (Vol. 21). Boston: Springer Science & Business Media.

Kijima, T. N.-o. (2017, Oct. 18). Study on object recognition by active stereo camera for clean-up robot. In *2017 17th International Conference on Control, Automation and Systems (ICCAS)* (pp. 27-31). Jeju, South Korea: IEEE.

Lalos, A. S. (2019). Secure and safe IIoT systems via machine and deep learning approaches. In W. F. Gilb (Ed.), Security and Quality in Cyber-Physical Systems Engineering (pp. 443-470). Springer. doi:10.1007/978-3-030-25312-7_16

Lee, S.-Y. J.-Y.-H.-y. (2019). Educational Indoor Autonomous Mobile Robot System Using a LiDAR and a RGB-D Camera. *Journal of IKEEE*, *23*(1), 44–52.

Li, G. Y., Hou, Y., & Wu, A. (2017). Fourth Industrial Revolution: Technological drivers, impacts and coping methods. *Chinese Geographical Science*, *27*(4), 626–637. doi:10.100711769-017-0890-x

Lindner, L. O.-L.-Q.-B.-F.-R. (2017). Machine vision system errors for unmanned aerial vehicle navigation. In *2017 IEEE 26th international symposium on industrial electronics (ISIE)* (pp. 1615-1620). Edinburgh, UK: IEEE.

Long, Y. Y., Gong, Y., Xiao, Z., & Liu, Q. (2017). Accurate object localization in remote sensing images based on convolutional neural networks. *IEEE Transactions on Geoscience and Remote Sensing*, *55*(5), 2486–2498. doi:10.1109/TGRS.2016.2645610

Mathew, R. a. (2016). Trajectory tracking and control of differential drive robot for predefined regular geometrical path. *Procedia Technology*, 1273-1280.

Mokyr, J. (1998). The second industrial revolution, 1870-1914. *Storia dell'economia Mondiale*, 1-18.

Moreno, D. a. (2012 Oct 13). Simple, accurate, and robust projector-camera calibration. In *2012 Second International Conference on 3D Imaging, Modeling, Processing, Visualization & Transmission* (pp. 464-471). Zurich, Switzerland: IEEE. 10.1109/3DIMPVT.2012.77

Patel, K. K., Kar, A., Jha, S. N., & Khan, M. A. (2012). Machine vision system: A tool for quality inspection of food and agricultural products. *Journal of Food Science and Technology*, *49*(2), 123–141. doi:10.100713197-011-0321-4 PMID:23572836

Pérez, L. Í., Rodríguez, Í., Rodríguez, N., Usamentiaga, R., & García, D. (2016). Robot guidance using machine vision techniques in industrial environments: A comparative review. *Sensors (Basel)*, *16*(3), 335. doi:10.339016030335 PMID:26959030

Prasad, D. K. (2016). *Challenges in video based object detection in maritime scenario using computer visio.* arXiv preprint arXiv:1608.01079

Ramík, D. M., Sabourin, C., Moreno, R., & Madani, K. (2014). A machine learning based intelligent vision system for autonomous object detection and recognition. *Applied Intelligence*, *40*(2), 358–375. doi:10.100710489-013-0461-5

Ray, P. P. (2016). Internet of robotic things: Concept, technologies, and challenges. *IEEE Access: Practical Innovations, Open Solutions*, *4*, 9489–9500. doi:10.1109/ACCESS.2017.2647747

Rea, M. S. (2000). *The IESNA lighting handbook: reference & application*. Illuminating Engineering Society of North America.

Reyes-García, M. O.-Q.-B.-F.-A.-R. (2019). Defining the Final Angular Position of DC Motor shaft using a Trapezoidal Trajectory Profile. In *2019 IEEE 28th International Symposium on Industrial Electronics (ISIE)* (pp. 1694-1699). Vancouver, Canada: IEEE.

Rifkin, J. (2011). *The third industrial revolution: how lateral power is transforming energy, the economy, and the world*. Macmillan.

Sarkar, D. a. (2018). Practical Machine Learning with Python. In *A Problem-Solvers Guide To Building Real-World Intelligent Systems*. Berkeley: Apress.

Schauwecker, K. (2018). Real-time stereo vision on FPGAs with SceneScan. In T. P. Längle (Ed.), Forum Bildverarbeitung (p. 339). Academic Press.

Shalev-Shwartz, S. D. (2014). *Understanding machine learning: From theory to algorithms*. Cambridge University Press. doi:10.1017/CBO9781107298019

Shao, J. (1993). Linear model selection by cross-validation. *Journal of the American Statistical Association*, *88*(422), 486–494. doi:10.1080/01621459.1993.10476299

Singh, I., Centea, D., & Elbestawi, M. (2019). IoT, IIoT and Cyber-Physical Systems Integration in the SEPT Learning Factory. *Procedia Manufacturing*, *31*, 116–122. doi:10.1016/j.promfg.2019.03.019

Skwarek, V. (2017 Dec 4). Blockchains as security-enabler for industrial IoT-applications. *Asia Pacific Journal of Innovation and Entrepreneurship*.

Sokolova, M. a. (2006). Beyond accuracy, F-score and ROC: a family of discriminant measures for performance evaluation. In *Australasian joint conference on artificial intelligence* (pp. 1015-1021). Berlin: Springer.

Sun, C. a. (2019 Nov 22). Design and Development of Modbus/MQTT Gateway for Industrial IoT Cloud Applications Using Raspberry Pi. In *2019 Chinese Automation Congress (CAC)* (pp. 2267-2271). Hangzhou, China: IEEE. 10.1109/CAC48633.2019.8997492

Tachaphetpiboon, S. a. (2006). Applying FFT features for fingerprint matching. In *1st International Symposium on Wireless Pervasive Computing* (pp. 1-5). Phuket, Thailand: IEEE.

Upadhyay, D., & Sampalli, S. (2020). SCADA (Supervisory Control and Data Acquisition) systems: Vulnerability assessment and security recommendations. *Computers & Security*, *89*, 1–18. doi:10.1016/j.cose.2019.101666

Veneri, G., & Capasso, A. (2018). *Hands-On Industrial Internet of Things: Create a powerful Industrial IoT infrastructure using Industry 4.0*. Packt Publishing Ltd.

Voulodimos, A., Doulamis, N., Doulamis, A., & Protopapadakis, E. (2018). Deep learning for computer vision: A brief review. *Computational Intelligence and Neuroscience*, *2018*, 1–14. doi:10.1155/2018/7068349 PMID:29487619

Wang, W. S.-S., Capitaneanu, S. L., Marinca, D., & Lohan, E.-S. (2019, July 8). Comparative analysis of channel models for industrial IoT wireless communication. *IEEE Access: Practical Innovations, Open Solutions, 7*, 91627–91640. doi:10.1109/ACCESS.2019.2927217

Xu, L.-Y. Z.-Q., Cao, Z.-Q., Zhao, P., & Zhou, C. (2017). A new monocular vision measurement method to estimate 3D positions of objects on floor. *International Journal of Automation and Computing, 14*(2), 159–168. doi:10.100711633-016-1047-6

Zafar, M. N., & Mohanta, J. C. (2018). Methodology for path planning and optimization of mobile robots: A review. *Procedia Computer Science, 133*, 141–152. doi:10.1016/j.procs.2018.07.018

Zeuch, N. (2000). *Understanding and applying machine vision, revised and expanded.* CRC Press. doi:10.1201/b16927

Zhong, J. M., Li, M., Liao, X., & Qin, J. (2020). A Real-Time Infrared Stereo Matching Algorithm for RGB-D Cameras' Indoor 3D Perception. *SPRS International Journal of Geo-Information, 9*(8), 472. doi:10.3390/ijgi9080472

Zou, Z. a. (2019). *Object detection in 20 years: A survey.* arXiv preprint arXiv:1905.05055

Chapter 10
Technical Vision Model of the Visual Systems for Industry Application

Oleg Sytnik

A. Ya. Usikov Institute for Radiophysics and Electronics of the National Acadademy of Sciences of Ukraine, Ukraine

Vladimir Kartashov

Kharkiv National University of Radio Electronics, Ukraine

ABSTRACT

The problems of highlighting the main informational aspects of images and creating their adequate models are discussed in the chapter. Vision systems can receive information about an object in different frequency ranges and in a form that is not accessible to the human visual system. Vision systems distort the information contained in the image. Therefore, to create effective image processing and transmission systems, it is necessary to formulate mathematical models of signals and interference. The chapter discusses the features of perception by the human visual system and the issues of harmonizing the technical characteristics of industrial systems for receiving and transmitting images. Methods and algorithms of pattern recognition are discussed. The problem of conjugation of the characteristics of the technical vision system with the consumer of information is considered.

INTRODUCTION

Optoelectronics is one of the most dynamically developing scientific and technical areas due to the sharp expansion of the range of applications and the ability to solve emerging problems with non-traditional methods. The key place of optoelectronics in information systems is because more than 90% (Andrews, 1990) of the information a person receives is video information.

DOI: 10.4018/978-1-7998-6522-3.ch010

The optoelectronic vision system forms the object model and the surrounding scene based on the signals received by the sensor system. Obviously, as a rule, such kind of the model is presented in the form of analog or digital signals that are transmitted to the systems for processing and reproduction. At each stage of the formation of such an image model, as well as in its transmission and processing, distortions and noise appear. The result is an image of an object that is different from the original. Therefore, the designer of the technical vision system should choose a criterion of image correction using the models of distortions and noise. An important point in system design is the question of who is the consumer of the image. If the consumer is a human, then the system of technical vision should be compatible with the visual system of a person. According to the received visual information, the human brain restores not only the image on the plane, but also the spatial "volumetric" arrangement of objects on an image. In different technical systems its preferably to form flat image, because it is easier, cheaper and faster than produce 3D holographic scene. Thus, digital image processing is usually performed in the form of some mathematical operations on data represented by two-dimensional arrays of numbers.

Image processing includes procedures that allows the consumer to efficiently recognize images and make adequate decisions, and can be divided into rectification, enhancement, restoration and reconstruction (Bates, 1989). The rectification includes some procedures of spatial image transformation to exclude geometric distortion. The enhancement processing are noise suppression procedures, algorithms to correct nonlinear distortion, contrast to background corrections, methods for emphasizing the boundaries of objects, etc. Problems of restoration and reconstruction are associated with improving visual image quality. Particularly, image restoration includes estimation of image distortion parameters and used them for the correction of prior data. The reconstruction procedure consists in the extraction of image details and using them for pattern recognition.

In general, most important image distortions are caused by three types of technical visual systems blocks - linear, nonlinear and complex (Castleman, 1979). Linear blocks are completely described by amplitude - frequency and phase - frequency characteristics. Nonlinear blocks have nonlinear amplitude characteristics that caused by wrong functional dependencies of image brightness between input and output video signals. Complex blocks create both amplitude-phase and nonlinear distortions, which are difficult to take into account when developing image correction and restoration algorithms.

In this chapter, the features of image models formed by optoelectronic systems of technical vision is considered, taking into account distortions and noise. Mathematical description of different types of distortions and methods of its compensation at digital signal processing is discussed. Particular attention is paid to models of the human visual system and pattern recognition algorithms. The problem of the conjugation of the characteristics of the technical vision system with the human visual system is discussed too.

MAIN FEATURES OF HUMAN VISUAL SYSTEM

Technical vision systems, as a rule, form an image for a person. Therefore, it is important to conjugate technical features of a system with properties of human eyesight. Thereby, it's necessary to form the mathematical models of the human being eyesight system. For more than half of a century, much attention has been paid to this issue in the scientific literature, for example (Bates, 1989) or (Castleman, 1979).

The main areas of application of the functional model of the human visual system are image rectification, enhancement, restoration, coding, etc. The basis of the human visual system is a huge number of interconnected brain neurons that have the unique ability to generalize the observed images, highlight characteristic details on them, form abstract images and perform object recognition. The human eye is able to work in an extremely wide range of illuminations. This range is nine orders of magnitude, and the difference in brightness within a single image can reach 30-40 dB. The wide range of the human eyesight system bases on two kinds of adaptive photoreceptors. Each of these systems is able to adapt to changes over a wide range of illumination values. These are night and day vision systems. Optoelectronic technical vision systems, as a rule, forms images for human day eyesight.

Therefore, the main focus of this chapter will be on the main characteristics of the human day eyesight model. The human day eyesight contains three types of light sensitive elements, which differs to sensitiveness to red (r), blue (b) and green (g) light wavelength λ. In the construction of a functional model of the visual system, this property will be taken into account by three items whose signals are calculated by the formulas:

$$e_r(x,y) = \int_\Delta L(x,y,\lambda)\varepsilon_r(\lambda)d\lambda, \tag{1}$$

$$e_g(x,y) = \int_\Delta L(x,y,\lambda)\varepsilon_g(\lambda)d\lambda, \tag{2}$$

$$e_b(x,y) = \int_\Delta L(x,y,\lambda)\varepsilon_b(\lambda)d\lambda, \tag{3}$$

where $L(x,y,\lambda)$ – is the distribution of image brightness according to scene coordinates and light wavelengths, $\varepsilon_r(\lambda)$, $\varepsilon_g(\lambda)$, $\varepsilon_b(\lambda)$ – are the spectral sensitivities of "red", "green" and "blue" light sensitive elements, Δ – is the visible wavelength range.

Primary image processing takes place in the retina. There, achromatic and color signals are formed, which are then filtered in the retina. When filtering, the lower spatial frequencies of the image are suppressed. Moreover, if the image is motionless, then the attenuation of the lower frequencies is maximum. In the case of flickering images, the attenuation of the lower ones is often the less, the higher the flicker frequency. In a human eye high spatial frequencies are filtered too.

Thus, a complete functional model of the human eyesight system at the retina level, which would reflect all it's currently known characteristics, is too complex, although it can be built.

A simplified model of a visual system was used for technical applications, discussed by (Li, 2011), (Faugeras, 1979) and (Wang, 2010). Figure 1 shows a generalized functional model of human eyesight system.

The model contains three channels, which generate the signals, calculated by equations (1) - (3). Then these signals are converted using the three spatial frequency filters by factors $G_r(\omega_x,\omega_y)$, $G_g(\omega_x,\omega_y)$, $G_b(\omega_x,\omega_y)$ respectively. In the next step, the output signals are amplified by logarithmic amplifiers and combined to obtain achromatic and color different images according to following formulas

$$s_a(x,y) = a\left[\alpha \log(e_{r1}(x,y)) + \beta \log(e_{g1}(x,y)) + \gamma \log(e_{b1}(x,y))\right], \tag{4}$$

Figure 1. Functional model of human eyesight system

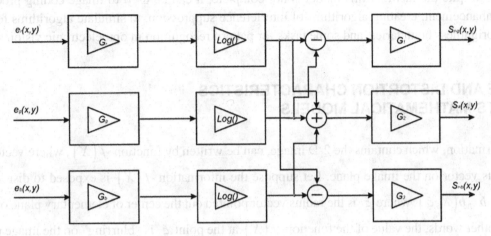

$$s_{r-g}(x, y) = b_1 \left[\log\big(e_{r1}(x, y)\big) - \log\big(e_{g1}(x, y)\big) \right],$$ (5)

$$s_{r-b}(x, y) = b_2 \left[\log\big(e_{r1}(x, y)\big) - \log\big(e_{b1}(x, y)\big) \right],$$ (6)

where $\alpha=0.612$, $\beta=0.369$, $\gamma=0.019$, and coefficients a, b_1 and b_2 are chosen so that the differences in the perception of light or color are represented by a sphere of unit radius in the space of signals

$$s_a(x, y), s_{r-g}(x, y), s_{r-b}(x, y) \text{ (Keelan, 2002).}$$

Achromatic and color signals

$$s_{r-g}(x, y), s_a(x, y), s_{r-b}(x, y)$$

are followed by the input of spatial frequency filters with coefficients $G_1(\omega_x, \omega_y)$, $G_1(\omega_x, \omega_y)$, $G_3(\omega_x, \omega_y)$ and, after filtering, goes to the brain for further processing. It is important to see, that in this model the scene brightness $s_a(x, y)$ changing do not cause any change in spectral components $s_{r-g}(x, y)$, $s_{r-b}(x, y)$, which is in good agreement with experimental data. Obviously, if a scene is in a black- and white mode, the components $s_{r-g}(x, y)$, $s_{r-b}(x, y)$ become equal to zero and the model is simplified to one channel (Liu, 2012). The main objective of the considered human eyesight system is to estimate the visibility of interference and distortion evenly distributed over the surface of the entire image. A quantitative measure of the visibility of interference, in accordance with experiments (Tang, 2011) and (Lin, 2011), is the average square of the interference at the output of the functional model of the human eyesight system. So, this model can be used for process modeling, which concerned with boundaries of brightness detection, some kind of tasks for pattern recognition etc.

Thus, despite the fact that this model is not complete, it can be used to image coding procedures, image enhancement, creating algorithms of interference suppression, to simulate algorithms for determining brightness boundaries, and some tasks for pattern recognition in optoelectronic vision systems.

IMAGE AND DISTORTION CHARACTERISTICS AND ITS MATHEMATICAL MODELS

The information, which contains the 2-D image, can be written by function $f(\vec{X})$, where vector $\vec{X}$ is the radius vector on the image plane. Let suppose the information $f(\vec{X})$ is exposed to distortion by function $b = b(\vec{X}, \vec{\xi})$, where $\vec{\xi}$ is the radius vector pointed out the center of elementary plane of distortion. In other words, the value of the function $f(\vec{X})$ at the point $\vec{\xi}$ is "blurring" on the image plane in accordance with the function. The distorted image in general can be written as

$$g(\vec{X}) = \iint_\infty f(\vec{\xi}) b(\vec{X}, \vec{\xi}) d\sigma(\vec{\xi}), \tag{7}$$

where $d\sigma(\vec{\xi})$ is an element of the plane centered at a point defined by a radius – vector $\vec{\xi}$.

The double integral in (7) is taken because the image has 2-D size, and the infinity symbol within the integration limits indicates that the entire image is used in the processing. Well-known processing methods for restoration and reconstruction image were designed under assumption of spatial invariance of image distortion (Andrews, 1990). The invariance means that blurring is the same for all points $\vec{X}$ of an image. In mathematical model it can be written by following way

$$b(\vec{X}, \vec{\xi}) = b(\vec{X} - \vec{\xi}). \tag{8}$$

If (8) is true, equation (7) is a convolution procedure and (7) can be rewritten in compact form

$$g(\vec{X}) = f(\vec{X}) \otimes b(\vec{X}), \tag{9}$$

where $\otimes$ is a convolution symbol.

Consider some typical kinds of distortion function $b(\vec{X}, \vec{\xi})$. First, it is a smudge image effect. This effect occurs when the speed performance of the sensors of the optoelectronic system is insufficient to fix a moving object. The second reason for image distortion is defocusing. The distortion function has a circle view. When an object is viewed through a turbulent medium using a high-resolution optoelectronic system, the function $b(\vec{X}, \vec{\xi})$ is well described by a set of random short pulses. If the sensors of the optoelectronic system are sufficiently inertial, then the form of the function $b(\vec{X}, \vec{\xi})$ approaches to Gaussian.

Consider the process of image formation in an optoelectronic system separated from the object by a distorting medium. The light signal spread along the line from arbitrary point ζ to point $\vec{Z}$ at the image. The suppression and delay of the signal can be described by complex numbers $\Re(\zeta, \vec{Z})$. If distortion is isoplanatic, the complex number is independent from ζ. This statement will be true, if the source of light would have a small size (point light source). Let denote the radiation field at an arbitrary point in time t at the point $\vec{Z}$ by $H(\vec{Z}, t)$, and its Fourier image by $F(\vec{X}, t)$. Suppose that the point $\vec{Z}$ lies in the plane of the aperture diaphragm of the imaging device. Then $F(\vec{X}, t)$ will be the instant image formed by this device. The properties of the image formed by the optoelectronic device depend on the degree of spatial coherence of the radiation source. When assessing the degree of spatial coherence, the individual spectral components of the images are usually considered. For example, the model of image $r(\vec{X})$ can be written as

$$r(x) = \begin{cases} F(\vec{X}, t)\exp(j\omega t) & \text{in case of complete coherence} \\ E\left\{\left|F(\vec{X}, t)\right|^2\right\} & \text{in case of complete incoherence} \end{cases} \tag{10}$$

where $E\{\bullet\}$ is a symbol of mathematical expectation on time.

It's important to note, that calculation of the mathematical expectation in time domain should be done on a large amount of periods of a light field central frequency. This is not difficult for most optoelectronic devices, because the time interval of expectation, as usual, is a very small part of the recording interval. If to neglect the noise, that is always present in vision systems, and consider distortions as isoplanatic, then the function $r(\vec{X})$ coincides with the function $g(\vec{X})$ in formula (9). However, the image $r(\vec{X})$ is a diffraction-limited image. This is because the aperture D of the real technical visual system is always limited. So, if λ is the central light wavelength, then the imaging device cannot solve the details of the scene smaller than λ/D.

Image distortions are not always constant over time in the real vision systems. Neither the distortions themselves, nor the process of converting light into electrical signals in optoelectronic systems of technical vision is perfectly linear and spatially invariant. In addition, noise is always present in the real image. When constructing a mathematical model of a real image, all image defects are conveniently denoted by the function $C = C(\vec{X})$. Then the generalized mathematical model of the distorted image can be a function of the following form:

$$r(\vec{X}) = g(\vec{X}) + C(\vec{X}) = f(\vec{X}) \otimes b(\vec{X}) + C(\vec{X}). \tag{11}$$

Of course, formula (11) is not comprehensive, but for most practically important cases, it can be used without restrictions. Therefore, the main attention in the analysis of vision systems and the synthesis of digital image processing methods will be given to issues within this model.

PATTERN RECOGNITION PROBLEMS UNDER NON-CORRELATED JAMMERS

One of the most widely used image processing methods is the deconvolution method (Lizhe, 2020). Image processing is necessary to perform various measurements in areas such as physics, chemistry, biology, medicine, as well as technology, industry, etc. The choice of processing method when solving the deconvolution problem depends on many factors. In particular, the main factors are the shape and extent of the point blur function, the nature of the original image, the degree of distortion of the signal spectrum in the image transmission path, etc. However, no matter what processing method is selected, it is necessary to pre-process the distorted image to convert it into a form that is convenient for the de-convolution procedure.

According to the properties of the Fourier transform (Campbell, 1978), formula (9) in the spectral region takes the form:

$$G(\omega) = F(\omega) \bullet B(\omega), \tag{12}$$

where $G(\omega)$, $F(\omega)$, $B(\omega)$ – are the Fourier transforms of functions $g(\vec{X}), f(\vec{X}), b(\vec{X})$ respectively.

The ideal deconvolution problem is: there are known functions $g(\vec{X})$ and $b(\vec{X})$, it is required to restore the function $f(\vec{X})$, if all functions are finite-dimensional. So, the problem of deconvolution solution from (12) can be found by

$$f(\vec{X}) = \Im^{-1}\{G(\omega)/B(\omega)\}, \tag{13}$$

where $\Im^{-1}\{\cdot\}$ – is inverse Fourier transform.

The division operation inside curly brackets in formula (13) is a simple inverse filtering. Since the processed images are usually stored in the computer memory in the form of quantized values of continuous signals, digital filters are used for filtering. The characteristic of a digital filter is represented by an array of complex numbers. Therefore, both functions $G(\omega)$ and $B(\omega)$ in formula (13) can be considered as filters. The theory of digital filters originally arose in the field of signal processing as a function of time. This theory is also suitable for image processing with the only difference being that a spatial variable is used instead of time. The term "sample" in the theory of signal processing goes into the term "image element" or "pixel" in the theory of image processing. According to these statements, it will be considered the idea of image processing based on the analysis of the eigenvalues of the spatial correlation function. For simplicity, processing a one-dimensional string is considered, although the results can easily be generalized to the two-dimensional case.

Let it be done the sequence of estimators of correlation function Bk in the discrete set m *(k= 1,…,m)* (Hamming, 1962)

$$B_k = \sum_{j=1}^{r} \rho_j \cdot \exp(i \cdot \omega_j \cdot k), \, r \le m, \tag{14}$$

where i – is an imaginary unit; ρ_j, ω_j – is the amplitude and phase of spectral components respectively.

The correlation function that is calculated by the ensemble of receiving signal plus noise realizations can be written in the following way

$$R_k = A \cdot \sum_{j=1}^{r} \rho_j \cdot \exp\left(i \cdot \omega_j \cdot k\right) + n_k, \qquad (15)$$

where A – is an unknown factor; n_k – is the noise discrete set (particularly, that may be a normal Gaussian noise set having average of distribution equal null and dispersion equal σ^2).

The problem is finding some spectral decomposition $F(\omega)$ associated with correlation function R as

$$R_k = (2\pi)^{-0,5} \int_{-\pi}^{\pi} \exp\left(i \cdot \omega \cdot k\right) \cdot dF(\omega), \qquad (16)$$

that the vectors $\vec{B} = \left(B_1, B_2, ..., B_m\right)$ and $\vec{N} = \left(n_1, n_2, ..., n_m\right)$ are orthogonal.

Orthogonal Decomposition

Let's examine the same, in generically, non-orthogonal and non-normed function system $\left\{\vec{\xi}_1, \vec{\xi}_2, ..., \vec{\xi}_m\right\} = \left\{\vec{\xi}\right\}$ and then vector $\vec{B}$ expands by basic set $\left\{\vec{\xi}\right\}$

$$\vec{B} = B_1 \cdot \vec{\xi}_1 + B_2 \cdot \vec{\xi}_2 + \cdots + B_m \cdot \vec{\xi}_m. \qquad (17)$$

As it follow from problem definition, the vector $\vec{N}$ must satisfy to orthogonal condition of all vectors: $\vec{\xi}_1, \vec{\xi}_2, ..., \vec{\xi}_m$, i.e. $\left(\vec{N}, \vec{\xi}\right) = 0$, where $(\bullet, \bullet)$ is the symbol of scalar product. Then, in view of (4), we can write the system of equations

$$
\begin{aligned}
\left(\vec{N}, \vec{\xi}_1\right) &= \left(\vec{R} - A \cdot \vec{B}, \vec{\xi}_1\right) \\
\left(\vec{N}, \vec{\xi}_2\right) &= \left(\vec{R} - A \cdot \vec{B}, \vec{\xi}_2\right) \\
\vdots \quad &= \quad \vdots \qquad \vdots \\
\left(\vec{N}, \vec{\xi}_m\right) &= \left(\vec{R} - A \cdot \vec{B}, \vec{\xi}_m\right)
\end{aligned}
\qquad (18)
$$

The determinant of the system of equations (18) can be written in follow way

$$\det = \begin{vmatrix} \left(\vec{\xi}_1, \vec{\xi}_1\right) & \left(\vec{\xi}_2, \vec{\xi}_1\right) & \cdots & \left(\vec{\xi}_m, \vec{\xi}_1\right) \\ \left(\vec{\xi}_1, \vec{\xi}_2\right) & \left(\vec{\xi}_2, \vec{\xi}_2\right) & \cdots & \left(\vec{\xi}_m, \vec{\xi}_2\right) \\ \vdots & \vdots & \vdots & \vdots \\ \left(\vec{\xi}_1, \vec{\xi}_m\right) & \left(\vec{\xi}_2, \vec{\xi}_m\right) & \cdots & \left(\vec{\xi}_m, \vec{\xi}_m\right) \end{vmatrix}. \qquad (19)$$

The scalar product, under definition (Halmos, 2019), is symmetrical, positively defined bilinear form. So the determinant (19) is nonzero. To find factor set B_j, where $j = 0,\ldots,m$, lets solve the system of equations (18) using (Halmos, 2019)

$$B_j = \frac{1}{\det} \cdot \begin{vmatrix} (\vec{\xi}_1,\vec{\xi}_1) & \cdots & (\vec{\xi}_{j-1},\vec{\xi}_1) & (\vec{R},\vec{\xi}_1) & (\vec{\xi}_{j+1},\vec{\xi}_1) & \cdots & (\vec{\xi}_{m-1},\vec{\xi}_1) & (\vec{\xi}_m,\vec{\xi}_1) \\ (\vec{\xi}_1,\vec{\xi}_2) & \cdots & (\vec{\xi}_{j-1},\vec{\xi}_2) & (\vec{R},\vec{\xi}_2) & (\vec{\xi}_{j+1},\vec{\xi}_2) & \cdots & (\vec{\xi}_{m-1},\vec{\xi}_2) & (\vec{\xi}_m,\vec{\xi}_2) \\ \vdots & \vdots & \vdots & \vdots & \vdots & \vdots & \vdots & \vdots \\ (\vec{\xi}_1,\vec{\xi}_m) & \cdots & (\vec{\xi}_{j-1},\vec{\xi}_m) & (\vec{R},\vec{\xi}_m) & (\vec{\xi}_{j+1},\vec{\xi}_m) & \cdots & (\vec{\xi}_{m-1},\vec{\xi}_k) & (\vec{\xi}_m,\vec{\xi}_m) \end{vmatrix}. \tag{20}$$

Spectral Estimate

To calculate the estimation of spectral function $F(\omega)$ using estimates B_k, that had been calculated by the equation (20), it is necessary to follow to the Pisarenko method (Pisarenko, 1973) and keep in mind the validation of the model (14). The model (14) in terms of Caratheodory's theorem (Hoang, 2020) can be written as follow

$$B_k = \rho_0 \cdot \delta_k + \sum_{j=1}^{r} \rho_j \cdot \exp\left(i \cdot \omega_j \cdot k\right), \tag{21}$$

where $\delta_0 = 1$, $\delta_k = 0$ when $k \neq 0$ and denotes Kroneker's delta.

The unknown factors r, ρ_j, ω_j in equation (21) can be calculated by the follow way.

First, it is necessary to form the correlation matrix

$$\vec{B} = \begin{bmatrix} B_0 & B_1 & \cdots & B_m \\ B_{-1} & B_0 & \cdots & B_{m-1} \\ \vdots & \vdots & \vdots & \vdots \\ B_{-m} & B_{-m+1} & \cdots & B_0 \end{bmatrix}. \tag{22}$$

The next step is find the minimal eigenvalue μ_0 of matrix (22) equal to amplitude factor $\rho_0 = \mu_0$. The order of eigenvalue μ_0 denotes as v. Difference in m and v defines parameter r

$$r = m - v. \tag{23}$$

At the second step, we need to build the matrix $\vec{Z} = \vec{B} - \mu_0 \cdot \vec{I}$, where $\vec{I}$ is an identity matrix and calculates the eigenvector $\vec{P} = \left(p_0, p_1, \ldots, p_r\right)$ that corresponds to eigenvalue equal zero. Roots of polynomial

$$p_0 + p_1 \bullet \alpha + \ldots + \mathrm{p}, \alpha^r = 0 \tag{24}$$

defines the frequencies of spectral function $F(\omega)$

$$\alpha_j = \exp(i \cdot \omega_j).$$ (25)

The system of equations $\sum_{j=1}^{r} \rho_j \cdot \sin\left(\omega_j \cdot k\right) = \text{Im}\left\{B_k\right\}$, that in matrix mode can be written

$$
\begin{bmatrix}
\sin(\omega_1) & \sin(\omega_2) & \cdots & \sin(\omega_r) \\
\sin(2 \cdot \omega_1) & \sin(2 \cdot \omega_2) & \cdots & \sin(2 \cdot \omega_r) \\
\vdots & \vdots & \vdots & \vdots \\
\sin(r \cdot \omega_1) & \sin(r \cdot \omega_2) & \cdots & \sin(r \cdot \omega_r)
\end{bmatrix}
\begin{bmatrix}
\rho_1 \\ \rho_2 \\ \vdots \\ \rho_r
\end{bmatrix}
=
\begin{bmatrix}
\text{Im}\{B_1\} \\ \text{Im}\{B_2\} \\ \vdots \\ \text{Im}\{B_r\}
\end{bmatrix},
$$ (26)

the model parameters can be calculated: r, $\vec{\rho} = \left(\rho_1,...,\rho_r\right)$, $\vec{\omega} = \left(\omega_1,...,\omega_r\right)$ and restore the function $F(\omega)$.

Numerical Experimental Data

The considered algorithm of signal processing is used for detection and identification of some signal sources was four narrowband Gaussian with average of distribution equal to zero and carrier frequencies: $f_1 = 1.27$ Hz, $f_2 = 1.5$ Hz, $f_3 = 2.7$ Hz, $f_4 = 3.4$ Hz respectively. Observation interval was 10 s. As a noise factor a wideband white noise with average of distribution equal zero and dispersion from 0,1 to 1,0 was defined. Figure 2 shows the spectral function under noise dispersion equal to 0,1

Figure 2. The spectral function under signal to noise ratio equal 40 dB.

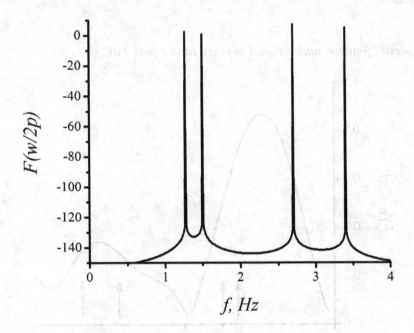

As you can see in Figure 3 the first and second ($f_1 = 1.27$ Hz, $f_2 = 1.5$Hz) signals, so as the third and fourth ($f_3 = 2.7$ Hz, $f_4 = 3.4$ Hz) signals are irresoluble.

Figure 3. The traditional Fourier analyses under signal to noise ratio equal 40 dB.

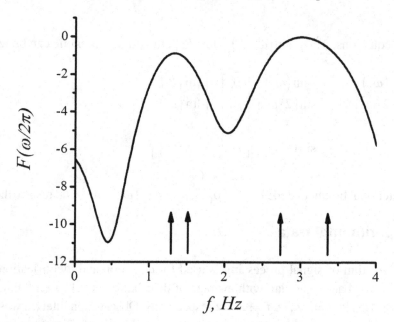

The errors of estimates signal's amplitude and frequencies increases when the signal to noise ratio decreases (see Figure 4). So the suggested signal processing algorithm has not been efficient under low signal to noise ratio.

Figure 4. The spectral function under signal to noise ratio equal 3 dB

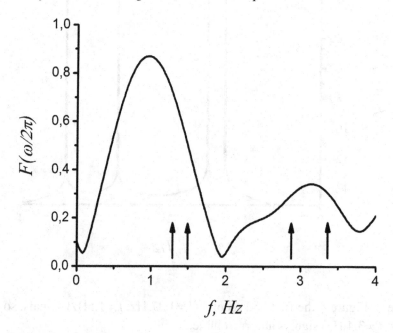

The errors of frequencies estimations can appear in the signal processing algorithm as a result of a non-adequate choice of quantity discrete set base function or if a number of signal sources are more than the quantity discrete set of correlation function in model.

CONJUGATION OF THE CHARACTERISTICS OF THE TECHNICAL VISION SYSTEM WITH THE CONSUMER OF INFORMATION

A characteristic feature of real objects images is that they consist of areas separated by more or less sharp light boundaries, inside which the brightness and color change relatively slowly. These light boundaries convey the shape of the object and are the basis for its recognition. The human brain easily recognizes a familiar face only by contour. In this regard, it can be argued that information about color, gradations of brightness is somewhat redundant for recognition. Therefore, it's important to find out how the characteristics of different boundaries can affect on the problem of pattern recognition. Let's determine the relationship between the structure of images of real objects and their spatial frequency spectra, obtained as a result of the Fourier transform. To do this, it is needed to take in consideration the spectra of three typical boundaries oriented perpendicular to the x axis. This arrangement of boundaries is convenient since in this case, the brightness of the images does not depend on the y coordinate and the two-dimensional problem can be reduced to one-dimensional. The mathematical models of boundaries can be written in the following way

$$B_1(x) = \begin{cases} 0 & when\ x \le x_0 \\ B & when\ x > x_0 \end{cases}, \tag{27}$$

$$B_2(x) = \begin{cases} \dfrac{B}{2} - \dfrac{B}{2}\exp\left[\alpha(x - x_0)\right] & when\ x \le x_0 \\ \dfrac{B}{2} + \dfrac{B}{2}\exp\left[-\alpha(x - x_0)\right] & when\ x > x_0 \end{cases}, \tag{28}$$

$$B_3(x) = \begin{cases} \dfrac{1}{2}B\exp\left[\alpha(x - x_0)\right] & when\ x \le x_0 \\ B\left\{1 - \dfrac{1}{2}\exp\left[-\alpha(x - x_0)\right]\right\} & when\ x > x_0 \end{cases}, \tag{29}$$

where $B(x)$ — is the brightness function, B, α — are some constants.

The form of boundaries (27) - (29) (curves (1) - (3), respectively) is shown in Figure 5.

The spectra of functions (27) - (29) have the following form:

$$F_1(\omega_x) = (B / \omega_x)\exp\left[-j(\pi / 2 + \omega_x x_0)\right], \tag{30}$$

Figure 5. The view of boundaries (27) – solid line 1, (28) – dash line 2, (29) – dash dot line 3.

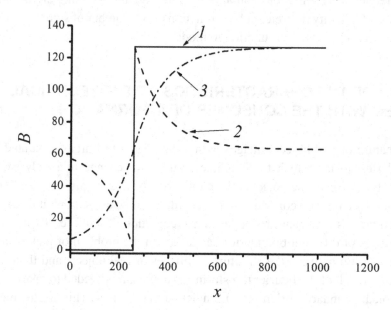

x

$$F_2(\omega_x) = B\pi\delta(\omega_x) + \left[(B\omega_x)/(\alpha^2 + \omega_x^2)\right]\exp\left[-j(\pi/2 + \omega_x x_0)\right],\tag{31}$$

$$F_3(\omega_x) = \left\{(B\alpha^2)/\left[\omega_x(\alpha^2 + \omega_x^2)\right]\right\}\exp\left[-j(\pi/2 + \omega_x x_0)\right],\tag{32}$$

where $\delta(\bullet)$ – is the delta function, ω_x – is the spatial frequency along the coordinate x.

An analysis of the spectral functions (30) - (32) shows the identity of the phase spectra for all three boundaries (27) - (29). This indicates that the phase spectrum contains information about the position of the brightness boundary in the image. In contrast, the amplitude spectrum contains information about the sharpness of the image and is invariant with respect to information about the position of the brightness boundary.

Based on this, it can be argued that in the optoelectronic system of technical vision, the image can have both linear and nonlinear distortions. However, it is important that in this case the distortions of the phase spectrum would be as small as possible. With small phase distortions, there will be a low probability of the appearance of new or the disappearance of existing brightness boundaries in the image, and as a result, the pattern recognition system will work with minimal errors. Therefore, the problem of conjugate the characteristics of an optoelectronic technical vision system with an information consumer is the accuracy of preserving the phase spectrum during image processing and transmission. Obviously, the most destructive factor for the image phase spectrum is white noise. Experimentally established (Hassen, 2010), that in the presence of white noise, the human visual system is capable of do pattern recognition with approximately the same probability as the optimal receiver, based on the maximum likelihood criterion. It should be noted that the probability of correct pattern recognition on images in noise depends both on the noise probability density and on the degree of nonlinearity of the amplitude characteristic of the image transmission and processing system. This dependence is stronger, than is greater the contrast of the image and the greater the variance of the noise. Nonlinear image transformations reduce the likelihood of correct recognition of images by a person in comparison with the optimal receiver by about 10%. Therefore, to harmonize the human visual system with the characteristics of vision systems, pre distortions can be introduced into the image.

It will be useful to discuss some typical methods of noise suppression and phase spectra restoration.

IMAGE PREPROCESSING FOR SPATIAL SAMPLING NOISE SUPPRESSION

To reduce the noise level of spatial sampling, it is necessary to limit the width of the spectrum of spatial frequencies of images within half the spatial sampling band. This can be done either by optical methods or by filtering the signal with the aperture of the optoelectronic system. Let the transmission coefficient of the sensor of the optoelectronic system satisfy the condition

$$K\left(\omega_x,\omega_y\right) = \begin{cases} 1 & at \left|\omega_x\right| < \dfrac{\omega_d}{2}, \ \left|\omega_y\right| < \dfrac{\omega_y}{2} \\ 0 \text{ if this condition is not met} \end{cases} \tag{33}$$

where ω_d is sampling frequency.

Function (33) is separable by variables ω_x and ω_y:

$$K\left(\omega_x,\omega_y\right) = K\left(\omega_x\right)K\left(\omega_y\right). \tag{34}$$

The transmission coefficient $K(\omega_x,\omega_y)$ and the transparency function $h(x,y)$ of the image sensor are related by the Fourier transform.

$$h\left(x,y\right) = \frac{1}{4\pi^2} \int\limits_{-\infty}^{\infty} \int\limits_{-\infty}^{\infty} K\left(\omega_x,\omega_y\right) \exp\left[j\left(\omega_x x + \omega_y y\right)\right] d\omega_x d\omega_y. \tag{35}$$

Substituting (33) into (35) and performing integration, will obtain

$$h\left(x,y\right) = \frac{1}{\Delta_d} \frac{\sin\left(\pi x/\Delta_d\right)}{\pi x/\Delta_d} \frac{1}{\Delta_d} \frac{\sin\left(\pi y/\Delta_d\right)}{\pi y/\Delta_d}, \tag{36}$$

where Δ_d – is a sampling step by coordinates x and y, $\omega_d = 2\pi/\Delta_d$.

The function (36) has the following features. First of all, function $h(x,y)$ can have both positive and negative value. Secondly, it is separable by variables x and y. In a discrete form, the function $h(x,y)$ can be written as follows

$$h_d\left(\omega_x,\omega_y\right) = \sum_{n=-\infty}^{\infty} \sum_{k=-\infty}^{\infty} h_s\left(x-n\Delta_s, y-k\Delta_s\right), \tag{37}$$

where $h_d(x,y)$ – is the transparency function of the image sensor in discrete form, $h_s(\bullet)$ – is the function of distribution of transparency in the aperture of a real image signal sensor, Δ_s – is a shift between the centers of elementary apertures.

Using (37), the transmission coefficient $K_d(\omega_x,\omega_y)$ of the sensor in discrete form can be written as

$$K_d\left(\omega_x,\omega_y\right)=K_s\left(\omega_x,\omega_y\right)\sum_{n=-\infty}^{\infty}\sum_{k=-\infty}^{\infty}b_{nk}\left[\exp\left(jn\Delta_s\omega_x\right)\right]\exp\left(jk\Delta_s\omega_y\right). \tag{38}$$

In order for function (38) to satisfy condition (33), it is necessary that the sum in formula (38) be equal to $1/K_s(\omega_x,\omega_y)$ within $|\omega_x|<\omega_d/2$, and zero outside this interval. Theoretically, this is possible in the case of $\Delta s\to 0$. Unfortunately, the requirement of an infinitesimal step of decomposition is not feasible in practice. However, this requirement can be significantly reduced while maintaining the necessary noise reduction. For this, it is necessary to require that the sum in formula (38) represent a double trigonometric series expansion with period $2\omega_d$ of some function $\varphi(\omega_x,\omega_y)$, which is defined as follows:

$$\varphi\left(\omega_x,\omega_y\right)=\begin{cases}1/K_s\left(\omega_x,\omega_y\right) & at\ |\omega_x|<\dfrac{\omega_d}{2},\ |\omega_y|<\dfrac{\omega_y}{2}.\\0 & \text{if this condition is not met}\end{cases} \tag{39}$$

Thus, the preliminary image processing that is formed by the optoelectronic vision system consists in decomposing the original signal with a discrete step Δ_s. After the image is recorded in the memory, signal samples are formed for transmission over the communication channel with step $\Delta_d=2\Delta_s$.

The formula (38) can be rewritten with (39) taken into account

$$K_d\left(\omega_x,\omega_y\right)=K_s\left(\omega_x,\omega_y\right)\varphi\left(\omega_x,\omega_y\right), \tag{40}$$

where the function

$$\varphi\left(\omega_x,\omega_y\right)=\sum_{n=-\infty}^{\infty}\sum_{k=-\infty}^{\infty}b_{nk}\left[\exp\left(jn\Delta_s\omega_x\right)\right]\exp\left(jk\Delta_s\omega_y\right)$$

can be considered as a function that corrects the transmission coefficient of the elementary aperture. It is the same that the transmission coefficient of real sensor. In order to use relation (40) in practice, it is necessary to know the coefficients b_{nk}. To calculate the coefficients b_{nk}, it can be assumed that the form of the function $K_s(\omega_x,\omega_y)$ is known and it is separable on its arguments

$$K_s(\omega_x,\omega_y)=K(\omega_x)K(\omega_y). \tag{41}$$

In this case, the function $\varphi(\omega_x,\omega_y)$ will also be separable, i.e.

$$\varphi(\omega_x,\omega_y)=\varphi(\omega_x)\varphi(\omega_y). \tag{42}$$

where

$$\varphi(\omega_x) = \begin{cases} 1/K_s(\omega_x) \ at \ |\omega_x| < \omega_d/2 \\ 0 \ \ at \ (\omega_d/2) < |\omega_x| < \omega_d \end{cases}, \qquad (43)$$

$$\varphi(\omega_y) = \begin{cases} 1/K_s(\omega_y) \ at \ |\omega_y| < \omega_d/2 \\ 0 \ \ at \ (\omega_d/2) < |\omega_y| < \omega_d \end{cases}. \qquad (44)$$

Wherein

$$\varphi(\omega_x) = \sum_{n=-\infty}^{\infty} b_n \exp(jn\Delta_s\omega_x), \qquad (45)$$

$$\varphi(\omega_y) = \sum_{k=-\infty}^{\infty} b_k \exp(jk\Delta_s\omega_y). \qquad (46)$$

The functions (45), (46) are even; therefore, for the coefficients b_{nk}, the relations $b_n = b_{-n}, b_k = b_{-k}$ are true. Given this, formulas (45) and (46) can be written as

$$\varphi(\omega_x) = \sum_{n=0}^{\infty} b_n \cos(n\Delta_s\omega_x), \qquad (47)$$

$$\varphi(\omega_y) = \sum_{k=0}^{\infty} b_k \cos(k\Delta_s\omega_y). \qquad (48)$$

The expansion coefficients b_{nk} for the correction function can be calculated from the equations taking into account formulas (47), (48).

$$b_{0x} = \frac{1}{2\pi} \int_0^{2\pi} \varphi(\omega_x) d(\Delta_s\omega_x), \qquad (49)$$

$$b_{nx} = \frac{1}{\pi} \int_0^{2\pi} \varphi(\omega_x) \cos(n\Delta_s\omega_x) d(\Delta_s\omega_x), \qquad (50)$$

$$b_{0y} = \frac{1}{2\pi} \int_0^{2\pi} \varphi(\omega_y) d(\Delta_s\omega_y), \qquad (51)$$

$$b_{ky} = \frac{1}{\pi} \int\limits_{0}^{2\pi} \varphi(\omega_y) \cos(k\Delta_s \omega_y) d(\Delta_s \omega_y). \tag{52}$$

The $K_d(\omega_x, \omega_y)$ dependencies calculated by the above procedure and ideal $K_d(\omega_x, \omega_y)$ have shown at Figure 6. They are calculated under assumption that $\omega_y = 0$ sensor transmission coefficient $K_s(\omega_x, \omega_y)$ was approximated by function

$$K_s(\omega_x, \omega_y) = \exp\left[-(\omega_x^2 + \omega_y^2)\Delta_d^2 / \pi^2\right]. \tag{53}$$

Figure 6. The ideal (curve 1) and calculated dependencies (curve 2) from ωx ρωd.

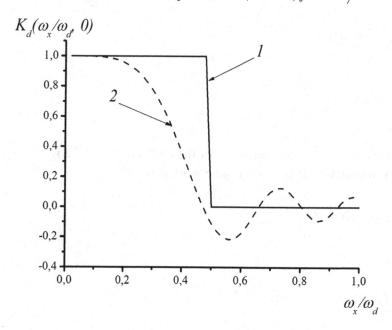

It is interesting to note that curve 2 in Figure 6 was calculated with a total number of sum elements equal to 20 in formulas (47), (48).

So, it is advisable to use the image processing described above to eliminate spatial sampling noise when a matrix is used as an image signal sensor.

THE PROBLEM OF IMAGE SPECTRUM RECOVERY

In many practical problems image signals are represented in the frequency domain. In such cases, the true image must be restored from its spectrum. But as noted earlier, the phase spectrum plays an important role in pattern recognition in vision systems. In optoelectronic systems operating in the optical range, direct measurement of the phase of the signal is difficult. Therefore, the focus will be on algorithms for reconstructing the phases of the spectra from records of their intensities.

Let $f(x,y)$ is a scene (or cadre) of monochrome distribution of brightness on a receiving matrix. Where x and y are coordinates. The image spectrum $F(\omega x, \omega y)$ connected with function $f(x,y)$ by Fourier transform

$$F(\omega_x,\omega_y) = \int\limits_{-\infty}^{\infty}\int\limits_{-\infty}^{\infty} f(x,y)\exp\left[-j(\omega_x x + \omega_y y)\right]dxdy, \tag{54}$$

$$f(x,y) = \frac{1}{4\pi^2}\int\limits_{-\infty}^{\infty}\int\limits_{-\infty}^{\infty} F(\omega_x,\omega_y)\exp\left[-j(\omega_x x + \omega_y y)\right]d\omega_x d\omega_y, \tag{55}$$

where ω_x, ω_y – are the circular spatial frequencies of the spectrum in the direction of the x *and* y *a*xes, respectively.

Let similarly define the spectrum of the difference of two images

$$F_{ik}(\omega_x,\omega_y) = \int\limits_{-\infty}^{\infty}\int\limits_{-\infty}^{\infty}\left[f_i(x,y) - f_k(x,y)\right]\exp\left[-j(\omega_x x + \omega_y y)\right]dxdy, \tag{56}$$

$$f_i(x,y) - f_k(x,y) = \frac{1}{4\pi^2}\int\limits_{-\infty}^{\infty}\int\limits_{-\infty}^{\infty} F_{ik}(\omega_x,\omega_y)\exp\left[-j(\omega_x x + \omega_y y)\right]d\omega_x d\omega_y, \tag{57}$$

where i,k – are indexes of scenes which offset relative to each other by $i\Delta_x - k\Delta_y$, Δ_x, Δ_y – are the step of spatial image discretization on x and y axis respectively.

The spectra defined in this way contain complete information about both the amplitudes and the phases of the frequency components. When the image translated to customer through some system it got distortions. The brightness distribution in a scene $f_s(x,y)$ reproduced by a linear system can be found using convolution integral

$$f_s(x,y) = \int\limits_{-\infty}^{x}\int\limits_{-\infty}^{y} h(x-\xi, y-\eta)f(\xi,\eta)d\xi d\eta, \tag{58}$$

where ξ, η – are integration variables, $h(x,y)$ – is the impulse response of the system, which, up to a constant factor, coincides with the brightness distribution in the image.

The function $h(x,y)$ fully characterizes the distortions introduced by the system into the signal.

The spectrum of the original image $F(\omega_x, \omega_y)$ and the spectrum $F_s(\omega_x, \omega_y)$ of image for the consumer are related by the following relation

$$F_s(\omega_x,\omega_y) = F(\omega_x,\omega_y)K(\omega_x,\omega_y), \tag{59}$$

where $F_s(\omega_x,\omega_y)$ – is the spectrum of $f_s(x,y)$, $K(\omega_x,\omega_y)$ – is the transmission coefficient (frequency response) of the system.

The spectrum $F_{sik}(\omega_x,\omega_y)$ of the difference in images received by the consumer can be found similarly to (59). The spectral intensity function of the image can be found as

$$S(\omega_x,\omega_y) = C|F(\omega_x,\omega_y)|^2, \tag{60}$$

where $C-$ is a normalization constant that is inversely proportional to the image area.

In the same way, it can be determined the spectral intensity of the image difference

$$S_{ik}(\omega_x,\omega_y) = C|F_{ik}(\omega_x,\omega_y)|^2. \tag{61}$$

The spectral intensities of images and their differences make it possible to calculate the corresponding energies

$$E_i = \frac{C}{4\pi^2} \int_{-\infty}^{\infty}\int_{-\infty}^{\infty} S_i\left(\omega_x,\omega_y\right)d\omega_x d\omega_y , \tag{62}$$

$$E_{ik} = \frac{C}{4\pi^2} \int_{-\infty}^{\infty}\int_{-\infty}^{\infty} S_{ik}\left(\omega_x,\omega_y\right)d\omega_x d\omega_y . \tag{63}$$

As analysis shows, the bulk of the energy of images is concentrated in the region of low spatial frequencies. At the same time, about half of it falls on the constant component. This directly follows from the analysis of the probability density formula for the brightness distribution in the image

$$W\left(f_i,f_{ik}\right) = \frac{1}{E\{f_i^2\}\left[1-\rho_i\left(i,k\right)\right]}\left\{\exp\left[-\frac{f_i+f_{ik}}{E\{f_i^2\}\left[1-\rho_i\left(i,k\right)\right]}\right]\right\}I_0\left\{\frac{2\sqrt{f_if_{ik}}\rho\left(i,k\right)}{E\{f_i^2\}\left[1-\rho_i\left(i,k\right)\right]}\right\}, \tag{64}$$

where f_i and f_{ik} – are the brightness values in the image at two points displaced one relative to the other along the x and y axis respectively on i and k elements; $E\{f_i\}$ – is an average image brightness; $E\{\bullet\}$ – is the operator of mathematical expectation; $\rho(i,k)$ – is the image autocorrelation coefficient; $I_0\{\bullet\}$ – is the zero-order Bessel function of the first kind of imaginary argument.

The spectral intensity of the difference of the images at their best combination has a maximum in the region of low spatial frequencies, but in contrast to the spectral intensity of the image itself, it decreases more slowly with increasing frequency. In addition, on average, the entire energy of the difference of the images falls on the variable components of the spectrum. Despite the fact that the energy of high spatial frequencies in the image is small, these components of the spectrum are important, because they provide image sharpness. It is important to note that the spectral intensities of the images are anisotropic. This leads to a relatively slow decrease in the spectral components along the x, y axes with respect to the directions located to these axes under 450. Therefore, as a rule, vertical and horizontal contours prevail in the images.

The spectral intensities (60) and (61) can be approximated by the following formulas:

$$S_i\left(\omega_x,\omega_y\right)=\frac{2a_iE_i}{\left(a_i^2+\omega_x^2\right)\left(a_i^2+\omega_y^2\right)C}+\frac{2\pi^2}{C}E_i\delta\left(\omega_x\right)\delta\left(\omega_y\right),\tag{65}$$

$$S_{ik}\left(\omega_x,\omega_y\right)=\frac{4a_{ik}E_{ik}}{\left(a_{ik}^2+\omega_x^2\right)\left(a_{ik}^2+\omega_y^2\right)C},\tag{66}$$

where $\delta(\bullet)$ – is delta function, a_i and a_{ik} – are constants determined by the nature and size of the image.

To describe the statistical relationships between image elements, it can be used the autocorrelation function too.

$$R_i\left(\xi,\eta\right)=C\int_0^{x_0}\int_0^{y_0}f_i\left(x,y\right)f_i\left(x+\xi,y+\eta\right)dxdy,\tag{67}$$

where ξ and η – are displacements of image $f_i(x+\xi, y+\eta)$ from the same image $f_i(x,y)$ along axis x and y respectively.

Integration within infinite limits in (67) does not make sense, since the brightness function is identically zero outside the image. It was experimentally established that the autocorrelation (67) is well approximated by an exponential function

$$R_i\left(\xi,\eta\right)=\frac{E_i}{2C}\left\{1+\exp\left[-a_i\left(|\xi|+|\eta|\right)\right]\right\}.\tag{68}$$

As Figure 7 has shown the autocorrelation functions of an image from an offset in the x-axis for two images. The solid line corresponds to an image that does not contain a large number of small details. Therefore, the statistical relationships between the elements are strong enough and the autocorrelation function decreases slowly. An example of such an image can be a person's face close-up. The dashed line shows the autocorrelation function of an image with a large number of small details, for example, a stadium tribune.

Due to the fact that the autocorrelation functions of images are approximated fairly accurately by exponential functions, two-dimensional Markov processes can be used as stochastic image models.

CONCLUSION

Optoelectronic vision systems are very diverse and widely used in various fields of science and technology. It is impossible to cover all aspects and features of image processing and harmonizing the characteristics of the technical system and the consumer of information within one chapter. However, there are important technical applications in which it is known, that the point blur function or impulse response of a system is known in advance. An example is the image blurred due to the movement of the receiving matrix relative to the object or due to the defocusing of the optical system. In this case, to restore the image and its coordination with the visual system of the consumer, it is not enough to know that the im-

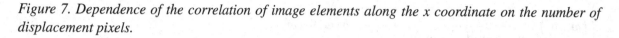

Figure 7. Dependence of the correlation of image elements along the x coordinate on the number of displacement pixels.

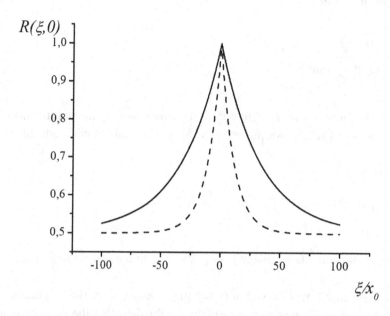

pulse response of a system has a simple form. It is important to know the specific form of this function. In particular, if the point blur function is simple, then the optical transfer function usually has explicit real zeros that are easy to detect in the spectrum of the distorted image. Since the form of the optical transfer function is completely determined by its zeros, the analysis of the spectra of distorted images is a convenient tool for solving the problem of matching the characteristics of the optical system and the image consumer. The image blurring, associated with the linear movement of the receiving matrix, is shown in the fact that the transfer function vanishes on parallel lines in the frequency domain. If the point blur function has elliptic symmetry, then its spectrum will have zeros on ellipses in the frequency domain. The identification of such zeros makes it possible to establish the symmetry of this function and restore its shape by selecting the appropriate mathematical model.

In complex and expensive systems, it is often more profitable to plan digital image processing in advance than to perform precise system tuning. In this case, mathematical models of corrective functions are obtained using test (reference) images and solving the convolution equation. Among the huge number of different methods for recovering distorted images, the calibration method is undoubtedly the simplest and most effective. If for any reason calibration is not possible, it would rather be used the method of detecting zeros in the spectral domain and the border detection and convolution methods.

An important aspect of the problem of matching the characteristics of optoelectronic systems of technical vision with the characteristics of the visual system of the consumer are the issues of image sampling. In particular, these are the issues of choosing the distance between samples, the spectral frequency band of the device forming the image, the level and statistical characteristics of the noise, the dynamic range of amplitude quantization, image recording parameters, etc. Some of these characteristics can be determined by the image. For example, it is the point blur function. Some types of distortion can be caused by imperfections in the recording equipment, others by digitalization errors, for example, an insufficient number of samples in space and amplitude. In this chapter, the main attention was paid to the methods of analysis and reconstruction of distorted images to match them with the visual system of the consumer under the assumption that the distortions are linear. Nonlinear distortions introduced by the equipment into the images should be corrected in the processing algorithms in the reverse order to the one in which they are introduced. Within this chapter, it was assumed that the fluctuation interference is an additive Gaussian noise of low intensity, which had a negligible effect on the quality of the images transmitted to the consumer. However, in some cases it should be remembered that multiplicative speckle noise might also be present in the image signal. In practice, the quality of image restoration is much more limited by edge effects, the choice of deconvolution method, and other effects than the inaccuracies of the fluctuation interference model.

REFERENCES

Andrews, H. C. (1990). *Computer Techniques for Image Processing*. Academic Press.

Bates, R.H.T. & McDonnel, M.J. (1989). *Image Restoration and Reconstruction*. Clarendon Press.

Campbell, G. A., & Foster, R. M. (1978). *Fourier Integrals for Practical Application*. Von Nostrand.

Castleman, K. R. (1979). *Digital Image Processing*. N.J. Prentice Hill.

Faugeras, O. D. (1979). Digital Color Image Processing within the Framework of Human Visual Model. *IEEE Transactions on Acoustics, Speech, and Signal Processing, ASSP-27*(Aug), 380–393. doi:10.1109/TASSP.1979.1163262

Halmos, P. R. (2019). *Finite-Dimensional Vector Spaces* (2nd ed.). Dover Publications Inc.

Hamming, R. W. (1962). *Numerical Methods for Scientists and Engineers*. MC Gaw-Hill Book Company.

Hassen, R., Wang, Z., & Salama, M. (2010). No-reference image sharpness assessment based on local phase coherence measurement. *Proceedings of the IEEE International Conference on Acoustics, Speech, and Signal Processing (ICASSP '10)*. 10.1109/ICASSP.2010.5496297

Keelan, B. (2002). *Handbook of Image Quality: Characterization and Prediction, Optical engineering*. Taylor & Francis. doi:10.1201/9780203910825

Li, S., Zhang, F., Ma, L., & Ngan, K. N. (2011). Image quality assessment by separately evaluating detail losses and additive impairments. *Proceedings of the IEEE Transactions on Multimedia*, 13(5). 10.1109/TMM.2011.2152382

Lin, W., & Kuo, C. C. J. (2011). Perceptual visual quality metrics: A survey. *Journal of Visual Communication and Image Representation*, 22(4), 297–312. doi:10.1016/j.jvcir.2011.01.005

Liu, A., Lin, W., & Narwaria, M. (2012). Image quality assessment based on gradient similarity. *Proceedings of the IEEE Transactions on Image Processing*, 21(4).

Pisarenko, V. F., (1973), The Retrieval of Harmonics from a Covariance Function. *Geophys. J.R. astr. Soc., 33*.

Saad, M. A., Bovik, A. C., & Charrier, C. (2012). Blind image quality assessment: A natural scene statistics approach in the DCT domain. *IEEE Transactions on Image Processing*, *21*(8), 3339–3352. doi:10.1109/TIP.2012.2191563 PMID:22453635

Tan & Jiang. (2020). Digital Signal Processing. 3rd Edition. Fundamentals and Applications. Academic Press.

Tang, H., Joshi, N., & Kapoor, A. (2011). Learning a blind measure of perceptual image quality. *Proceedings of the International Conference on Computer Vision and Pattern Recognition*. 10.1109/CVPR.2011.5995446

Tuy, H. (2020). *Convex Analysis and Global Optimization*. Springer.

Wang, J., Chandler, D. M., & Le Callet, P. (2010), Quantifying the relationship between visual salience and visual importance. In Human Vision and Electronic Imaging, (vol. 7527). doi:10.1117/12.845231

KEY TERMS AND DEFINITIONS

Brightness: Is an attribute of visual perception in which a source appears to be radiating or reflecting light. It is not necessarily proportional to luminance.

Deconvolution: Is an algorithm-based process used to reverse the effects of convolution on recorded data.

Human Visual System: Is a mathematical model of image processing which make the image processing like the human eye and brain.

Image Processing: Is a procedure for rectification, enhancement, reconstruction, restoration, coding, and pattern recognition.

Sampling: Is a procedure of digitizing an image according to spatial coordinates and amplitude.

Spectrum: Is the Fourier transform of the original image.

Compilation of References

Abdellatif, M. (2013, November). Color-Based Object Tracking and Following for Mobile Service Robots. *International Journal of Innovative Research in Science. Engineering and Technology, 2*(11), 5921–5928.

Abidi, M. A. (1992). *Data fusion in robotics and machine intelligence.* Academic Press Professional, Inc.

Ackland, Resnikoff, & Bourne. (2017). World Blindness and Visual Impairment: Despite Many Successes, the Problem Is Growing. *Community Eye Health, 30*(100).

Advani, S., Zientara, P., Shukla, N., Okafor, I., Irick, K., Sampson, J., Datta, S., & Narayanan, V. (2016). A Multitask Grocery Assist System for the Visually Impaired: Smart Glasses, Gloves, and Shopping Carts Provide Auditory and Tactile Feedback. IEEE Consumer Electronics Magazine, 6(1), 73–81.

Akimov, P. S., Bakut, P. A., & Bogdanovich, V. A. (1984). *Theory of signal detection* [Lexical characteristics of Russian language]. Radio and Communications.

Akimov, P. S., Efremov, V. S., & Kubasov, A. N. (1978). On the stability of the sequential rank detection procedure [Lexical characteristics of Russian language]. *Radio Engineering and Electronics, 23*(7), 1427–1431.

Aleksic, S. (2019, September). A Survey on Optical Technologies for IoT, Smart Industry, and Smart Infrastructures. *Journal of Sensor and Actuator Networks, 8*(3), 47. doi:10.3390/jsan8030047

Alfalou, A., & Brosseau, C. (2009). Optical image compression and encryption methods. Advances in Optics and Photonics, 1, 589. doi:10.1364/AOP.1.000589

Alfalou, A., & Brosseau, C. (2010). Exploiting root-mean-square time-frequency structure for multiple-image optical compression and encryption. Optics Letters, 35, 1914–1916.

Alfalou, A., Brosseau, C., & Alam, M. S. (2013). Smart pattern recognition. Proceedings of SPIE 8748, optical pattern recognition XXIV, 874809. doi:10.1117/12.2018249

Andersen, M. R., Jensen, T., Lisouski, P., Mortensen, A. K., Hansen, M. K., Gregersen, T., & Ahrendt, P. J. A. U. (2012). Kinect Depth Sensor Evaluation for Computer Vision Applications. Aarhus University.

Anderson, S. A., & Tallapally, L. K. (1996). Hydrologic response of a steep tropical slope to heavy rainfall. In 7th International Syposium on Landslides (pp. 1489-1498). Trondheim, Norway: Brookfield.

Andrews, H. C. (1990). *Computer Techniques for Image Processing.* Academic Press.

Apostolopoulos, Fallah, Folmer, & Bekris. (2014). Integrated Online Localization and Navigation for People with Visual Impairments Using Smart Phones. *ACM Transactions on Interactive Intelligent Systems, 3*(4), 1–28.

Assistive Technology Industry Association (ATIA). (2020). *What Is at?* Assistive Technology Industry Association (ATIA). https://www.atia.org/at-resources/what-is-at/

Bai, Lian, Liu, Wang, & Liu. (2017). Smart Guiding Glasses for Visually Impaired People in Indoor Environment. *IEEE Transactions on Consumer Electronics, 63*(3), 258–66.

Bajscy, R., Lee, S. W., & Leonardis, A. (1990). Color image segmentation with detection of highlights and local illumination induced by inter-reflections. In *Proceedings of the International Conference on Pattern Recognition* (pp. 785-790). Academic Press.

Banerjee, T. P., & Das, S. (2012). Multi-sensor data fusion using support vector machine for motor fault detection. *Information Sciences, 217*, 96–107. doi:10.1016/j.ins.2012.06.016

Baouche, F. Z., Hobar, F., & Hervé, Y. (2013). Level Modeling in Optoelectronic Systems; Transmission Line Application. *International Journal of Advances in Engineering and Technology, 6*(1), 392–404.

Bardoux, F., Bushko, J.-F., Aspe, A., & Cohen-Tannuji, K. (2006). *Levy Statistics and Laser Cooling. How rare events stop atoms*. FIZMATLIT.

Baron, I. R. S. (2010). State of the Art of Landslide Monitoring in Europe: Preliminary results of the Safeland Questionnaire. In Landslide Monitoring Technologies & Early Warning Systems (pp. 15-21). Vienna: Geological Survey of Austria.

Barros, Moura, Freire, Taleb, Valentim, & Morais. (2020). Machine Learning Applied to Retinal Image Processing for Glaucoma Detection: Review and Perspective. *BioMedical Engineering OnLine, 19*, 1–21.

Barthelemy, P., Bertolotti, J., & Weirsma, D. S. (2008). Levy flight for light. *Nature, 453*(7194), 495–498. doi:10.1038/nature06948 PMID:18497819

Basaca-Preciado, L. C.-Q.-L.-B., Sergiyenko, O. Y., Rodríguez-Quinonez, J. C., García, X., Tyrsa, V. V., Rivas-Lopez, M., Hernandez-Balbuena, D., Mercorelli, P., Podrygalo, M., Gurko, A., Tabakova, I., & Starostenko, O. (2014). Optical 3D laser measurement system for navigation of autonomous mobile robot. *Optics and Lasers in Engineering, 54*, 159–169. doi:10.1016/j.optlaseng.2013.08.005

Bass, M. (1995). Handbook of Optics, Vol. II- Devices, Measurements and Properties. New York: McGraw-Hill, Inc.

Bates, R.H.T. & McDonnel, M.J. (1989). *Image Restoration and Reconstruction*. Clarendon Press.

Bedard, G. (1967). Analysis of light fluctuations from photon counting statistics. *Journal of the Optical Society of America, 57*(10), 1201–1203. doi:10.1364/JOSA.57.001201

Beiser, L. (1995). Fundamental architecture of optical scanning systems. *Applied Optics, 31*(34), 7307–7317. doi:10.1364/AO.34.007307 PMID:21060601

Bekshaev A., Chernykh A., Khoroshun A., & Mikhaylovskaya L. (2017). Singular skeleton evolution and topological reactions in edge-diffracted circular optical-vortex beams. *Optics Communications, 397*, 72–83. doi:10.1016/j.optcom.2017.03.062

Belbachir, A. N. (2010). *Smart cameras*. Springer. doi:10.1007/978-1-4419-0953-4

Benet, G., Blanes, F., Simo, J., & Perez, P. (2002, September 30). Using infrared sensors for distance measurement in mobile robots. *Robotics and Autonomous Systems, 40*(4), 255–266. doi:10.1016/S0921-8890(02)00271-3

Bhatlawande, S. S., Mukhopadhyay, J., & Mahadevappa, M. (2012). Ultrasonic Spectacles and Waist-Belt for Visually Impaired and Blind Person. In *2012 National Conference on Communications (Ncc)*, (pp. 1–4). IEEE. 10.1109/NCC.2012.6176765

Bondarev, V. N., Trester, G., & Chernega, V. S. (1999). *Digital signal processing: Methods and tools.* Academic Press.

Born, M., & Wolf, E. (1999). *Principles of Optics.* Cambridge University Press. doi:10.1017/CBO9781139644181

Bouhamed, S. A., Kallel, I. K., & Masmoudi, D. S. (2013). Stair Case Detection and Recognition Using Ultrasonic Signal. In *2013 36th International Conference on Telecommunications and Signal Processing (Tsp),* (pp. 672–76). IEEE. 10.1109/TSP.2013.6614021

Bouman, C. A., & Shapiro, M. (1994, March). A Multiscale Random Field Model for Bayesian Image Segmentation. *IEEE Transactions on Image Processing, 3*(2), 162–177. doi:10.1109/83.277898 PMID:18291917

Bourne, Flaxman, Braithwaite, Cicinelli, Das, Jonas, Keeffe, Kempen, Leasher, & Limburg. (2017). Magnitude, Temporal Trends, and Projections of the Global Prevalence of Blindness and Distance and Near Vision Impairment: A Systematic Review and Meta-Analysis. *The Lancet Global Health, 5*(9), e888–e897.

Bousbia-Salah, Bettayeb, & Larbi. (2011). A Navigation Aid for Blind People. *Journal of Intelligent & Robotic Systems, 64*(3-4), 387–400.

Bouwmans, T., El Baf, F., & Vachon, B. (2008). Background Modeling using Mixture of Gaussians for Foreground Detection. A Survey. Recent Patents on Computer Science, 1(3), 219-237.

Boyat, A. K., & Joshi, B. K. (2015, April). A review paper: Noise model in digital image processing. *Signal and Image Processing: an International Journal, 6*(2), 63–75. doi:10.5121ipij.2015.6206

Brainard, D. H., & Wandell, B. A. (1986). Analysis of the retinex theory of color vision. Optical Society of America, 3(10), 1651-1660.

Breitenstein, M. D., Kuettel, D., Weise, T., Van Gool, L., & Pfister, H. (2008). Real-Time Face Pose Estimation from Single Range Images. In *2008 IEEE Conference on Computer Vision and Pattern Recognition,* (pp. 1–8). IEEE. 10.1109/CVPR.2008.4587807

Breve, F., & Fischer, C. N. (2020). *Visually Impaired Aid Using Convolutional Neural Networks, Transfer Learning, and Particle Competition and Cooperation.* arXiv Preprint arXiv:2005.04473

Brooks, D. R., & Mims, F. M. III. (2001). Development of an inexpensive handheld LED-based Sun photometer for the GLOBE program. *Journal of Geophysical Research, D, Atmospheres, 106*(D5), 4733–4740. doi:10.1029/2000JD900545

Caldini, A., Fanfani, M., & Colombo, C. (2015). Smartphone-Based Obstacle Detection for the Visually Impaired. In *International Conference on Image Analysis and Processing,* (pp. 480–88). Springer. 10.1007/978-3-319-23231-7_43

Campbell, G. A., & Foster, R. M. (1978). *Fourier Integrals for Practical Application.* Von Nostrand.

Carbonara, S., & Guaragnella, C. (2014). Efficient Stairs Detection Algorithm Assisted Navigation for Vision Impaired People. In *2014 IEEE International Symposium on Innovations in Intelligent Systems and Applications (Inista) Proceedings,* (pp. 313–18). IEEE. 10.1109/INISTA.2014.6873637

Carfagni, M. R., Furferi, R., Governi, L., Santarelli, C., Servi, M., Uccheddu, F., & Volpe, Y. (2019). Metrological and critical characterization of the Intel D415 stereo depth camera. *Sensors (Basel), 19*(3), 489. doi:10.339019030489 PMID:30691011

Castanedo, F. (2013). A review of data fusion techniques. *TheScientificWorldJournal, 2013,* 1–19. doi:10.1155/2013/704504 PMID:24288502

Castleman, K. R. (1979). *Digital Image Processing.* N.J. Prentice Hill.

Chandrasekhar, S. (1943). Stochastic Problems in Physics and Astronomy. *Reviews of Modern Physics, 15*(1), 1–89. doi:10.1103/RevModPhys.15.1

Chen, H., Wang, K., & Yang, K. (2018). Improving Realsense by Fusing Color Stereo Vision and Infrared Stereo Vision for the Visually Impaired. *Proceedings of the 2018 International Conference on Information Science and System*, 142–46. 10.1145/3209914.3209944

Chen, Y.-R., Chao, K., & Kim, M. S. (2002). Machine vision technology for agricultural applications. *Computers and Electronics in Agriculture, 36*(2-3), 173–191. doi:10.1016/S0168-1699(02)00100-X

Cokes, D., & Lewis, P. (1969). *Stochastic analysis of chains of events*. Academic Press.

Cook, A. M., & Polgar, J. M. (2014). *Assistive Technologies-E-Book: Principles and Practice*. Elsevier Health Sciences.

Corke, P. (2013). *Robotics, Vision and Control*. Springer.

Correal, R., Pajares, G., & Ruz, J. (2014). Automatic expert system for 3D terrain reconstruction based on stereo vision and histogram matching. *Expert System with Application, 41*, 2043-2051. Obtenido de https://www.sciencedirect.com/science/article/pii/S0957417413007227

Council, National Research. (1986). *Electronic Travel Aids: New Directions for Research*. National Research Council.

Cox, D., & Lewis, P. (1969). *Stochastic analysis of sequences of events*. Moscow. *WORLD (Oakland, Calif.)*, 310.

Cunninghom, I. A., & Shaw, R. (1999). Signal-to-noise optimizatiom of medical imaging systems. *Journal of the Optical Society of America. A, Optics, Image Science, and Vision, 19*(3), 621–632. doi:10.1364/JOSAA.16.000621

Dagenais, M. a. (1995). *Integrated optoelectronics*. Academic Press.

Dakopoulos, Boddhu, & Bourbakis. (2007). A 2D Vibration Array as an Assistive Device for Visually Impaired. In *2007 IEEE 7th International Symposium on Bioinformatics and Bioengineering*, (pp. 930–37). IEEE.

Damelin, S. B. (2012). *The mathematics of signal processing*. Cambridge University Press.

Dasarathy, B. V. (1997). Sensor fusion potential exploitation-innovative architectures and illustrative applications. *Proceedings of the IEEE, 1*(85), 24–38. doi:10.1109/5.554206

Davies, E. R. (2004). *Machine vision: theory, algorithms, practicalities*. Elsevier.

Davis, A. B., Marshak, A., & Pfeilsticker, K. P. (1999). Anomalous/Lévy Photon Diffusion Theory: Toward a New Parameterization of Shortwave Transport in Cloudy Columns. *Ninth ARM Science Team Meeting Proceedings*, 1-13.

Davis, A. B., Suszcynsky, D. M., & Marshak, A. (2000). Shortwave Transport in the Cloudy Atmosphere by Anomalous/Lévy Diffusion: New Diagnostics Using FORTÉ Lightning Data. *Tenth ARM Science Team Meeting Proceedings*, 1-18.

Deng, Z. H., Sun, H., Zhou, S., Zhao, J., Lei, L., & Zou, H. (2018). Multi-scale object detection in remote sensing imagery with convolutional neural networks. *ISPRS Journal of Photogrammetry and Remote Sensing, 145*, 3–22. doi:10.1016/j.isprsjprs.2018.04.003

Denisov, V. I., & Timofeev, V. S. (2011). Sustainable Distributions and Estimation of Parameters of Regression Dependencies. Izvestia, Tomsk Polytechnic University, 318(2), 10-15.

Dizdarević, J. F.-B., Carpio, F., Jukan, A., & Masip-Bruin, X. (2019). A survey of communication protocols for internet of things and related challenges of fog and cloud computing integration. *ACM Computing Surveys, 51*(6), 1–29. doi:10.1145/3292674

Dolezel, P. &. (2020). Neural Network for Smart Adjustment of Industrial Camera-Study of Deployed Application. In *International Conference on Advanced Engineering Theory and Applications* (pp. 101-113). Springer. 10.1007/978-3-030-14907-9_11

Don, H. (2013). *Johnson.* Statistical Signal Processing.

Eben Upton, G. H. (2014). *Raspberry Pi User Guide.* John Wiley & Sons.

Elfes, A. (1992). Multi-source spatial data fusion using Bayesian reasoning. *Data fusion in robotics and machine intelligence*, 137-163.

Ellison, S. L. R., & Williams, A. (Eds.). (2012). Quantifying Uncertainty in Analytical Measurement. EURACHEM/CITAC Guide CG 4.

Elmannai & Elleithy. (2017). Sensor-Based Assistive Devices for Visually-Impaired People: Current Status, Challenges, and Future Directions. *Sensors, 17*(3), 565.

Elmogy, M., & Zhang, J. (2009). Robust Real-time Landmark Recognition for Humanoid Robot Navigation. In *Proceedings of the 2008 IEEE International Conference on Robotics and Biomimetics* (pp. 572-577. 10.1109/ROBIO.2009.4913065

Engel, K. J., Steadman, R., & Herrmann, C. (2012). Pulse Temporal Splitting in Photon Counting X-Ray Detectors. *IEEE Transactions on Nuclear Science, 59*(4), 1480–1490. doi:10.1109/TNS.2012.2203610

Esfahani, A. G., Mantas, G., Matischek, R., Saghezchi, F. B., Rodriguez, J., Bicaku, A., Maksuti, S., Tauber, M. G., Schmittner, C., & Bastos, J. (2017, August 9). A lightweight authentication mechanism for M2M communications in industrial IoT environment. *IEEE Internet of Things Journal, 6*(1), 288–296. doi:10.1109/JIOT.2017.2737630

Faceli, K., de Carvalho, A. C. P. L. F., & Rezende, S. O. (2004). Combining intelligent techniques for sensor fusion. *Applied Intelligence, 3*(20), 199–213. doi:10.1023/B:APIN.0000021413.05467.20

Fang, W., Huang, X., Zhang, F., & Li, D. (2015). Intensity Correction of Terrestrial Laser Scanning Data by Estimating Laser Transmission Function. *IEEE Transactions on Geoscience and Remote Sensing, 53*(2), 942-951. doi:10.1109/TGRS.2014.2330852

Faugeras, O. D. (1979). Digital Color Image Processing within the Framework of Human Visual Model. *IEEE Transactions on Acoustics, Speech, and Signal Processing, ASSP-27*(Aug), 380–393. doi:10.1109/TASSP.1979.1163262

Fedoseev, V. I. (2011). *Priem prostranstvenno-vremennyh signalov v optiko-jelektronnyh sistemah* (puassonovskaja model'). Sp-b.: Universitetskaja kniga.

Fedoseev, V. I. (2011). Reception of spatio-temporal signals in optoelectronic systems (Poisson model). Moscow, Russia: University Book.

Fedoseev, V. I., & Kolosov, M. P. (2007). Optoelectronic devices for orientation and navigation of spacecraft. Academic Press.

Field, D. J. (1987). Relations between the statistics of natural images and the response properties of cortical cells. *Journal of the Optical Society of America. A, Optics and Image Science, 4*(12), 2379–2394. doi:10.1364/JOSAA.4.002379 PMID:3430225

Filipe, Fernandes, Fernandes, Sousa, Paredes, & Barroso. (2012). Blind Navigation Support System Based on Microsoft Kinect. *Procedia Computer Science, 14*, 94–101.

Firman, M. (2016). RGBD Datasets: Past, Present and Future. *Proceedings of the Ieee Conference on Computer Vision and Pattern Recognition Workshops*, 19–31.

Fischer, R. (2003). *Elektrische Maschinen* (12 ed.). Carl Hanser Verlag GmbH & Co. KG. Obtenido de https://www.amazon.de/Elektrische-Maschinen-Rolf-Fischer/dp/3446226931

Flores Fuentes, W., Rivas López, M., Srgiyenko, O., González Navarro, F. F., Rivera Castillo, J., & Hernández Balbuena, D. (2014). Machine Vision Supported by Artificial Intelligence Applied to Rotary Mirror Scanners. In *Proceedings of the 2014 IEEE 23th International Symposium on Industrial Electronics* (pp. 1949-1954). Istambul, Turkey: ISIE.

Flores-Fuentes, W. R.-L.-B.-Q.-P.-C. (2016). Online shm optical scanning data exchange. In *2016 IEEE 25th International Symposium on Industrial Electronics (ISIE)* (pp. 940-945). IEEE.

Flores-Fuentes, W., Rivas-Lopez, M., & Sergiyenko, O. (2013). Digital Signal Processing on Optoelectronic for SHM. *Proceedings of the World Congress on Engineering and Computer Science.*

Flores-Fuentes, W., Rivas-Lopez, M., Hernandez-Balbuena, D., Sergiyenko, O., Rodríguez-Quiñonez, J. C., Rivera-Castillo, J., . . . Basaca-Preciado, L. C. (2020). Applying Optoelectronic Devices Fusion in Machine Vision: Spatial Coordinate Measurement. In Natural Language Processing: Concepts, Methodologies, Tools, and Applications (pp. 184-213). IGI Global.

Flores-Fuentes, W., Rivas-Lopez, M., Hernandez-Balbuena, D., Sergiyenko, O., Rodriguez-Quinonez, J., Rivera-Castillo, J., . . . Basaca-Preciado, L. (2016). Applying Optoelectronic Devices Fusion in Machine Vision: Spatial Coordinate Measurement. In O. Sergiyenko, & J. Rodriguez-Quinonez (Eds.), Developing and Applying Optoelectronics in Machine Vision (p. 37). IGI Global. doi:10.4018/978-1-5225-0632-4.ch001

Flores-Fuentes, W., Rivas-Lopez, M., Sergiyenko, O., Gonzalez-Navarro, F., Rivera-Castillo, J., Hernandez-Balbuena, D., & Rodriguez-Quinonez, J. (2014). Combined application of Power Spectrum Centroid and Support Vector Machines for measurement improvement in Optical Scanning Systems. *Signal Processing, 98*, 37-51. Obtenido de https://www.sciencedirect.com/science/article/pii/S0165168413004337

Flores-Fuentes, W., Rodriguez-Quinonez, J., Hernandez-Balbuena, D., Rivas-Lopez, M., Sergiyenko, O., Rivera-Castillo, J., . . . Mayorga-Ortiz, P. (2015). Photodiode and charge-coupled device fusioned sensors. In *2015 IEEE 24th International Symposium on Industrial Electronics (ISIE)* (pp. 966-971). Buzios: IEEE. doi:10.1109/ISIE.2015.7281602

Flores-Fuentes, W. L.-N.-C.-B.-Q., Rivas-Lopez, M., Sergiyenko, O., Gonzalez-Navarro, F. F., Rivera-Castillo, J., Hernandez-Balbuena, D., & Rodríguez-Quiñonez, J. C. (2014). Combined application of power spectrum centroid and support vector machines for measurement improvement in optical scanning systems. *Signal Processing, 98*, 37–51. doi:10.1016/j.sigpro.2013.11.008

Flores-Fuentes, W. L.-Q.-B.-C., Rivas-Lopez, M., Sergiyenko, O., Rodriguez-Quinonez, J. C., Hernandez-Balbuena, D., & Rivera-Castillo, J. (2014). Energy Center Detection in Light Scanning Sensors for Structural Health Monitoring Accuracy Enhancement. *Sensors Journal, IEEE, 7*(14), 2355–2361. doi:10.1109/JSEN.2014.2310224

Flores-Fuentes, W., Sergiyenko, O., Gonzalez-Navarro, F., Rivas-Lopez, M., Rodriguez-Quinonez, J., Hernandez-Balbuena, D., Tyrsa, V., & Lindner, L. (2016, August). Multivariate outlier mining and regression feedback for 3D measurement improvement in opto-mechanical system. *Optical and Quantum Electronics, 48*(8), 21. doi:10.100711082-016-0680-1

Forsyth, D., & Pons, J. (2004). *Computer vision. The modern approach.* Moscow, Russia: "Williams" Publishing House.

Franca, J. G. (2005). A 3D scanning system based on laser triangulation and variable field of view. *Image Processing, 2005. ICIP 2005. IEEE International Conference on.* IEEE.

Fu, H. L. (2011). Home-Made 3-D image measuring instrument data process and analysis. In *2011 International Conference on Multimedia Technology* (págs. 6386-6389). Hangzhou, China: IEEE.

Fujiyoshi, H. T., Hirakawa, T., & Yamashita, T. (2019). Deep learning-based image recognition for autonomous driving. *IATSS Research, 43*(4), 244–252. doi:10.1016/j.iatssr.2019.11.008

Fukuda, M. (1999). *Optical semiconductor devices*. John Wiley and Sons.

Gal'jardi, R., & Karp, Sh. (1978). Opticheskaja svjaz. Moscow: Svjaz.

Ganz, A., Schafer, J., Gandhi, S., Puleo, E., Wilson, C., & Robertson, M. (2012). PERCEPT Indoor Navigation System for the Blind and Visually Impaired: Architecture and Experimentation. *International Journal of Telemedicine and Applications, 2012*, 2012. doi:10.1155/2012/894869 PMID:23316225

Garcia-Cruz, X., Sergiyenko, O., Tyrsa, V., Rivas-Lopez, M., Hernandez-Balbuena, D., Rodriguez-Quinonez, J., ... Mercorelli, P. (2014). Optimization of 3D laser scanning speed by use of combined variable step. *Optics and Lasers in Engineering, 54*, 141-151. Obtenido de https://www.sciencedirect.com/science/article/pii/S0143816613002546

Garcia-Talegon, J., Calabres, S., Fernandez-Lozano, J., Inigo, A., Herrero-Fernandez, H., Arias-Perez, B., & Gonzalez-Aguilera, D. (2015, February 25-27). Assessing Pathologies On Villamayor Stone (Salamanca, Spain) By Terrestrial Laser Scanner Intensity Data. *Remote Sensing and Spatial Information Sciences, XL-5/W4*, 445-451. Retrieved from https://www.int-arch-photogramm-remote-sens-spatial-inf-sci.net/XL-5-W4/445/2015/isprsarchives-XL-5-W4-445-2015.html

Gascón-Moreno, J. O.-G.-S.-T.-M.-F. (2011, June). Multi-parametric gaussian kernel function optimization for ε-SVMr using a genetic algorithm. In *International Work-Conference on Artificial Neural Networks* (pp. 113-120). Berlin: Springer.

Gasvik, K. (2002). *Optical Metrology* (3rd ed.). John Wiley & Sons, Ltd. doi:10.1002/0470855606

Gikhman, I. I., & Skorokhod, A. V. (1977). *Introduction to the theory of random processes*. Science.

Glauber, R. (1966). *Optical coherence and photon statistics*. World.

Gnedenko, B. V. (1950). Areas of attraction of the normal law. *Reports of the USSR Academy of Sciences, 71*(3), 425–428.

Gnedenko, B. V., & Kolmogorov, A. N. (1949). *Limit distributions for sums of independent random variables* (Vol. M). GITTL.

Gonzalez-Aguilera, D. L.-G. (2010). *Camera and laser robust integration in engineering and architecture applications*. INTECH Open Access Publisher. doi:10.5772/9959

Gonzalez, R. (2008). *Richard Woods*. Digital Image Processing.

Goodman, J. (1988). *Statistical Optics*. World.

Gracheva, M. E., Gurvich, A. S., & Kallistratova, M. A. (1970). Measurement of the dispersion of" strong "fluctuations in the intensity of laser radiation in the atmosphere. *Izv. Universities. Radiophysics, 13*, 56.

Greenwood, P. E., & Nikulin, M. S. (1996). *A guide to chi-squared testing*. Wiley.

Grubbs, F. (1969). Procedures for Detecting Outlying Observations in Samples. *Technometrics, 11*(1), 1–21.

Guerrero, J. J., Pérez-Yus, A., Gutiérrez-Gómez, D., Rituerto, A., & López-Nicolás, G. (2015). Human Navigation Assistance with a Rgb-d Sensor. *ACTAS V Congreso Internacional de Turismo Para Todos: VI Congreso Internacional de Diseno, Redes de Investigacion Y Tecnologia Para Todos Drt4all*, 285–312.

Gülcan, E. (2020). A novel approach for sensor based sorting performance determination. *Minerals Engineering, 146*, 106130. doi:10.1016/j.mineng.2019.106130

Gundecha, T. J. (2020). *Automation of Mechanical Press Machine Using Revolution Pi and PLC*. Academic Press.

Gunn, S. R. (1998). Support vector machines for classification and regression. *ISIS Technical Report*, 5-16.

Gustafsson, L., & Lanshammar, H. (1977). *Enoch - An integrated system for measurement and analysis of human gait* (Ph.D. thesis). UPTEC 7723 R.

Habasaki, J., Okada, I., & Hivatari, Y. (1997). Fraction excitation and Levy flight dynamics in alkali silicate glasses. *Physical Review*, *55*(10), 6309–6315. doi:10.1103/PhysRevB.55.6309

Habib, A., Islam, M. M., Kabir, M. N., Mredul, M. B., & Hasan, M. (2019). Staircase Detection to Guide Visually Impaired People: A Hybrid Approach. *Revue d'Intelligence Artificielle*, *33*(5), 327–334. doi:10.18280/ria.330501

Hall, D. L., & Llinas, J. (1997). An introduction to multisensor data fusion. *Proceedings of the IEEE*, *1*(85), 6–23. doi:10.1109/5.554205

Halmos, P. R. (2019). *Finite-Dimensional Vector Spaces* (2nd ed.). Dover Publications Inc.

Hamming, R. W. (1962). *Numerical Methods for Scientists and Engineers*. MC Gaw-Hill Book Company.

Harms, H., Rehder, E., Schwarze, T., & Lauer, M. (2015). Detection of Ascending Stairs Using Stereo Vision. In *2015 Ieee/Rsj International Conference on Intelligent Robots and Systems (Iros)*, (pp. 2496–2502). IEEE. 10.1109/IROS.2015.7353716

Harvey, J. E., Choi, N., Krywonos, A., Peterson, G., & Bruner, M. (2010). *Image degradation due to scattering effects in two-mirror telescopes*. https://opticalengineering.spiedigitallibrary.org/article.aspx?articleid=1096366

Hassen, R., Wang, Z., & Salama, M. (2010). No-reference image sharpness assessment based on local phase coherence measurement. *Proceedings of the IEEE International Conference on Acoustics, Speech, and Signal Processing (ICASSP '10)*. 10.1109/ICASSP.2010.5496297

He, Goodkind, & Kowal. (2016). *An Aging World: 2015*. U.S. Census Bureau, International Population Reports, P95/16-1. U.S. Government Publishing Office.

He, A. C. (2018). A twofold siamese network for real-time object tracking. In *Proceedings of the IEEE Conference on Computer Vision and Pattern Recognition* (pp. 4834-4843). Salt Lake City: CVF. 10.1109/CVPR.2018.00508

Henry, Krainin, Herbst, Ren, & Fox. (2012). RGB-d Mapping: Using Kinect-Style Depth Cameras for Dense 3D Modeling of Indoor Environments. *The International Journal of Robotics Research*, *31*(5), 647–63.

Hernandez, D. C., & Jo, K.-H. (2010). Outdoor Stairway Segmentation Using Vertical Vanishing Point and Directional Filter. In *International Forum on Strategic Technology 2010*, (pp. 82–86). IEEE. 10.1109/IFOST.2010.5667914

Hernández, D. C., & Jo, K.-H. (2011). Stairway Tracking Based on Automatic Target Selection Using Directional Filters. In *2011 17th Korea-Japan Joint Workshop on Frontiers of Computer Vision (Fcv)*, (pp. 1–6). IEEE. 10.1109/FCV.2011.5739752

Hernandez, D. C., & Jo, K.-H. (2011). Stairway Segmentation Using Gabor Filter and Vanishing Point. In *2011 IEEE International Conference on Mechatronics and Automation*, (pp. 1027–32). IEEE. 10.1109/ICMA.2011.5985801

Hernández, D. C., Kim, T., & Jo, K.-H. (2011). Stairway Detection Based on Single Camera by Motion Stereo. *International Conference on Industrial, Engineering and Other Applications of Applied Intelligent Systems*, 338–47. 10.1007/978-3-642-21822-4_34

Hersh & Johnson. (2010). *Assistive Technology for Visually Impaired and Blind People*. Springer Science & Business Media.

Hill & Bradfield. (1986). Electronic Travel Aids for Blind Persons. *Journal of Special Education Technology, 8*(3), 31–42.

Hinks, T., Carr, H., Gharibi, H., & Laefer, D. (2015). Visualisation of urban airborne laser scanning data with occlusion images. *ISPRS Journal of Photogrammetry and Remote Sensing, 104*, 77-87. doi:10.1016/j.isprsjprs.2015.01.014

Hiremath, S., van der Heijden, G., van Evert, F., Stein, A., & ter Braak, C. (2014). Laser range finder model for autonomous navigation of a robot in a maize field using a particle filter. *Computers and Electronics in Agriculture, 100*, 41-50. Obtenido de https://www.sciencedirect.com/science/article/pii/S0168169913002470

Hodge, V. J., & Austin, J. (2004). A Survey of Outlier Detection Methodologies. *Artificial Intelligence Review, 22*(2), 85–126. doi:10.1023/B:AIRE.0000045502.10941.a9

Hoon Sohn, C. F. (2002). *A Review of Structural Health Monitoring Literature 1996-2001.* www.lanl.gov.projects/damage

Hou, B. (2011). Charge-coupled devices combined with centroid algorithm for laser beam deviation measurements compared to a position-sensitive device. *Optical Engineering (Redondo Beach, Calif.), 50*(3), 033603–033603. doi:10.1117/1.3554379

Hoydal, T. O., & Zelano, J. A. (1991). An Alternative Mobility Aid for the Blind: The 'ultrasonic Cane'. In *Proceedings of the 1991 Ieee Seventeenth Annual Northeast Bioengineering Conference,* (pp. 158–59). IEEE. 10.1109/NEBC.1991.154627

Huang, N. E., Shen, Z., Long, S. R., Wu, M. C., Shih, H. H., Zheng, Q., Yen, N.-C., Tung, C. C., & Liu, H. H. (1998). The empirical mode decomposition and the Hilbert spectrum for nonlinear and non-stationary time series analysis. In Proceedings of R. Soc. London (pp. 903-995). doi:10.1098/rspa.1998.0193

Huang, X. Y., Yu, S., Xu, H., Aheto, J. H., Bonah, E., Ma, M., Wu, M., & Zhang, X. (2019). Rapid and nondestructive detection of freshness quality of postharvest spinaches based on machine vision and electronic nose. *Journal of Food Safety, 39*(6), e12708. doi:10.1111/jfs.12708

Hubbard, M. R. (2012). *Statistical quality control for the food industry.* Springer Science & Business Media.

Hubeli, A. a. (2001). *Multiresolution feature extraction for unstructured meshes. In Proceedings Visualization, 2001. VIS'01.* IEEE.

Hug, C., & Wehr, A. (1997, September 17-19). Detecting And Identifying Topographic Objects In Imaging Laser Altimeter Data. *IAPRS, 32*, 19-26. Retrieved from https://www.ifp.uni-stuttgart.de/publications/wg34/wg34_hug.pdf

Hu, M., Chen, Y., Zhai, G., Gao, Z., & Fan, L. (2019). An Overview of Assistive Devices for Blind and Visually Impaired People. *International Journal of Robotics and Automation, 34*(5). Advance online publication. doi:10.2316/J.2019.206-0302

Hussain, T., Muhammad, K., Ser, J. D., Baik, S. W., & de Albuquerque, V. H. C. (2019, August 27). Intelligent Embedded Vision for Summarization of Multiview Videos in IIoT. *IEEE Transactions on Industrial Informatics, 16*(4), 2592–2602. doi:10.1109/TII.2019.2937905

Ibragimov, I. A., & Linnik, Yu. V. (1965). *Independent and stationary related quantities.* Science.

Ignatov, A., Timofte, R., Kulik, A., Yang, S., Wang, K., Baum, F., Wu, M., Xu, L., & Van Gool, L. (2019). *AI Benchmark: All About Deep Learning on Smartphones in 2019.* arXiv Preprint arXiv:1910.06663

Imai. (1986). Statistical Properties of Optical Fiber Speckles. *Bulletin of the Faculty of Engineering, Hokkaido University, 130*, 89-10.

Islam, Sadi, Zamli, & Ahmed. (2019). Developing Walking Assistants for Visually Impaired People: A Review. *IEEE Sensors Journal, 19*(8), 2814–28.

Ivanov, M. O.-F.-Q. (2020). Influence of data clouds fusion from 3D real-time vision system on robotic group dead reckoning in unknown terrain. *IEEE/CAA Journal of Automatica Sinica, 7*(2), 368-385.

Janesick, J. R. (2001). *Scientific charge-coupled devices.* SPIE Press.

Jarret, M. O., Andrews, B. J., & Paul, J. P. (1974). Quantitative analysis of locomotion using television. *Proceedings of lSPO World Congress.*

Johnson, D. H. (n.d.). *Statistical Signal Processing.* http://www.ece.rice.edu/~dhj/courses/elec531 / notes.pdf

Johnson, G., Song, X., Montag, E. D., & Fairchild, M. D. (2010, December). Derivation of a Color Space for Image Color Difference Measurement. *Color Research and Application, 35*(6), 387–400. doi:10.1002/col.20561

Jouppi, Young, Patil, & Patterson. (2018). Motivation for and Evaluation of the First Tensor Processing Unit. *IEEE Micro, 38*(3), 10–19.

Kammoun, Macé, Oriola, & Jouffrais. (2011). Toward a Better Guidance in Wearable Electronic Orientation Aids. In *IFIP Conference on Human-Computer Interaction*, (pp. 624–27). Springer. 10.1007/978-3-642-23768-3_98

Kanade, T. (2012). Three-dimensional machine vision (Vol. 21). Boston: Springer Science & Business Media.

Kanwal, N., Bostanci, E., Currie, K., & Clark, A. F. (2015). A Navigation System for the Visually Impaired: A Fusion of Vision and Depth Sensor. *Applied Bionics and Biomechanics, 2015*, 2015. doi:10.1155/2015/479857 PMID:27057135

Katz, Kammoun, Parseihian, Gutierrez, Brilhault, Auvray, Truillet, Denis, Thorpe, & Jouffrais. (2012). NAVIG: Augmented Reality Guidance System for the Visually Impaired. *Virtual Reality, 16*(4), 253–69.

Keelan, B. (2002). *Handbook of Image Quality: Characterization and Prediction, Optical engineering.* Taylor & Francis. doi:10.1201/9780203910825

Kennedy, W. P. (2012). *The Basics of triangulation sensors.* Obtenido de http://archives.sensorsmag.com/articles/0598/tri0598/main.shtml

Kerle, N., Stump, A., & Malet, J.-P. (2010). Object oriented and cognitive methods for multidata event-based landslide detection and monitoring. *Landslide Monitoring Technologies & Early Warning Systems.*

Khoroshun, A. N. (2010). Optimal linear phase mask for the singular beam synthesis from a Gaussian beam and the scheme of its experimental realization. Journal of Modern Optics, 57(16), 1542–1549.

Khoroshun, G. (2019). Model of service providing on the use of optical laboratory in conditions of individual customer needs. Visnik of the Volodymyr Dahl East Ukrainian National University, 8(256), 118-122.

Khoroshun, G., Riazantsev, A., Ryazantsev, O., Popiolek-Masajada, A., & Strelkova, T. (2020). Estimation of the image quality and noise forinformation system. *Scientific and practical conference "Now Technologies In Nauts Ta Osviti",* 107.

Khoroshun, A. N., Chernykh, A. V., Tsimbaluk, A. N., Kirichenko, J. A., Yezhov, P. V., Kuzmenko, A. V., & Kim, J. T. (2016). Properties of an Axial Optical Vortex Generated with the use of a Gaussian Beam and Two Ramps. *Journal of Nanoscience and Nanotechnology, 16*(2), 1–3. doi:10.1166/jnn.2016.12029 PMID:27433739

Khoroshun, A., Ryazantsev, A., Ryazantsev, O., Sato, S., Kozawa, Y., Masajada, J., Popiołek-Masajada, A., Szatkowski, M., Chernykh, A., & Bekshaev, A. (2020). Formation of an optical field with regular singular-skeleton structure by the double-phase-ramp converter. *Journal of Optics, 22*(2), 025603. doi:10.1088/2040-8986/ab61c9

Khoroshun, G., Luniakin, R., Riazantsev, A., Ryazantsev, O., Skurydina, T., & Tatarchenko, H. (2020). The Development of an Application for Microparticle Counting Using a Neural Network. *Proceedings of the 4th International Conference on Computational Linguistics and Intelligent Systems (COLINS 2020)*, 1186-1195.

Kijima, T. N.-o. (2017, Oct. 18). Study on object recognition by active stereo camera for clean-up robot. In *2017 17th International Conference on Control, Automation and Systems (ICCAS)* (pp. 27-31). Jeju, South Korea: IEEE.

Klein, L. A. (2003). Sensor and data fusion: a tool for information assessment and decision making. Bellingham, WA: Spie Press.

Klyatskin, V. I. (1975). Statistical description of dynamical systems with fluctuating parameters. *Science*, 239.

Klyshko, D. N. (1996). The nonclassical light. *Успехи физических наук*, *166*(6), 613–638. doi:10.3367/UFNr.0166.199606b.0613

Köberlein, Beifus, Schaffert, & Finger. (2013). The Economic Burden of Visual Impairment and Blindness: A Systematic Review. *BMJ Open, 3*(11).

Kolyadin, V. L. (2002). Distributions with infinite dispersion and the limitations of classical statistics. *Wuxiandian Gongcheng*, (2), 4–11.

Korzhenevsky, A. L., & Kamzina, L. S. (1998). Anomalous diffusion of light in ferroelectrics with a diffuse phase transition. *Solid State Physics*, *40*(8), 1537–1541. doi:10.1134/1.1130566

Kowsari, K., & Alassaf, M. H. (2016). Weighted Unsupervised Learning for 3D Object Detection. *Information and Organization*. Advance online publication. doi:10.14569/IJACSA.2016.070180

Krantz, D. (1975). Color Measurement and Color Theory: Opponents Colors Theory. *Journal of Mathematical Psychology*, *12*(3), 304–327. doi:10.1016/0022-2496(75)90027-9

Kumar, N., Harbola, U., & Lindenberg, K. (2010). *Memory-induced anomalous dynamics: emergence of diffusion, subdiffusion, and superdiffusion from a single random walk model.* https://journals.aps.org/pre/abstract/10.1103/PhysRevE.82.021101

Kumar, P., McElhinney, C., Lewis, P., & McCarthy, T. (2013). An automated algorithm for extracting road edges from terrestrial mobile LiDAR data. *ISPRS Journal of Photogrammetry and Remote Sensing, 85*, 44-55. Obtenido de https://www.sciencedirect.com/science/article/pii/S0924271613001834

Kunhoth, J., AbdelGhani, K., Al-Maadeed, S., & Al-Ali, A. (2020). Indoor Positioning and Wayfinding Systems: A Survey. Human-Centric Computing and Information Sciences, 10, 1–41.

Kupchenko, L. F. (2009). *Acousto-optical effects with strong interaction. Theory and experiment (Continuous fraction method for solving acousto-optic problems).* EDENA.

Labayrade, R., Royere, C., Gruyer, D., & Aubert, D. (2005). Cooperative fusion for multi-obstacles detection with use of stereovision and laser scanner. *Autonomous Robots, 2*(19), 117–140. doi:10.100710514-005-0611-7

Lalos, A. S. (2019). Secure and safe ΠoT systems via machine and deep learning approaches. In W. F. Gilb (Ed.), Security and Quality in Cyber-Physical Systems Engineering (pp. 443-470). Springer. doi:10.1007/978-3-030-25312-7_16

Lanshammar, H. (1985). Measurement and analysis of displacement. *Gait Analysis in Theory and Practice Proceedings of the 1985 Uppsala Gait Analysis Meeting*, 29 – 45.

Lappa, A. V., Bakhvalov, E. V., & Anikina, A. S. (2004). The method of characteristic functions in estimating the mathematical expectation of random variables with infinite dispersion. *Izvestia Chelyabinsk Scientific Center*, 2, 1–6.

LeCun, Bengio, & Hinton. (2015). Deep Learning. *Nature, 521*(7553), 436–44.

Lee & Medioni. (2016). RGB-d Camera Based Wearable Navigation System for the Visually Impaired. *Computer Vision and Image Understanding, 149*, 3–20.

Lee, Y. H., & Medioni, G. (2014). Wearable Rgbd Indoor Navigation System for the Blind. In *European Conference on Computer Vision*, (pp. 493–508). Springer.

Lee, S.-Y. J.-Y.-H.-y. (2019). Educational Indoor Autonomous Mobile Robot System Using a LiDAR and a RGB-D Camera. *Journal of IKEEE, 23*(1), 44–52.

Levin, B. R. (1976). Theoretical foundations of statistical radio engineering [Lexical characteristics of Russian language]. Sov. radio, book.

Levin, B.R. (1968). *Theoretical foundations of statistical radio engineering.* Sov. radio.

Levy, P. (1972). *Stochastic processes and Brownian motion.* Science.

Li, Munoz, Rong, Chen, Xiao, Tian, Arditi, & Yousuf. (2018). Vision-Based Mobile Indoor Assistive Navigation Aid for Blind People. *IEEE Transactions on Mobile Computing, 18*(3), 702–14.

Liang, Q. Z., Zhu, W., Sun, W., Yu, Z., Wang, Y., & Zhang, D. (2019). In-line inspection solution for codes on complex backgrounds for the plastic container industry. *Measurement, 148*, 106965. doi:10.1016/j.measurement.2019.106965

Li, F. a. (2006). *CCD image sensors in deep-ultraviolet: degradation behavior and damage mechanisms.* Springer Science and Business Media.

Li, F. M. (2003). Degradation behavior and damage mechanisms of CCD image sensor with deep-UV laser radiation. *Electron Devices. IEEE Transactions on, 12*(51), 2229–2236.

Lifshits, M. A. (2007). *Stable distributions, random variables and processes.* SPb.

Li, G. Y., Hou, Y., & Wu, A. (2017). Fourth Industrial Revolution: Technological drivers, impacts and coping methods. *Chinese Geographical Science, 27*(4), 626–637. doi:10.100711769-017-0890-x

Li, L., Ma, G., & Du, X. (2013). Edge detection in potential-field data by enhanced mathematical morphology filter. *Pure and Applied Geophysics, 4*(179), 645–653. doi:10.100700024-012-0545-x

Lim, S. H. (2006). Characterization of Noise in Digital Photographs for Image Processing. *Proceedings of SPIE Digital Photography II, 6069*, 1–11. doi:10.1117/12.655915

Lin, H. Y., & Lin, J. H. (2006). A visual positioning system for vehicle or mobile robot navigation. IEICE Transactions on Information and Systems, 89(7), 2109-2116. doi:10.1093/ietisy/e89-d.7.2109

Lin, Y., Liu, Z., Huang, J., Wang, C., Du, G., Bai, J., & Lian, S. (2019). Deep Global-Relative Networks for End-to-End 6-Dof Visual Localization and Odometry. In *Pacific Rim International Conference on Artificial Intelligence*, (pp. 454–67). Springer. 10.1007/978-3-030-29911-8_35

Lindholm, L. E. (1974). An optoelectronic instrument for remote on-line movement monitoring. In Biomechanics IV. University Park Press.

Lindner, L. (2016). Laser Scanners. In O. Sergiyenko, J. Rodriguez-Quinonez, O. Sergiyenko, & J. Rodriguez-Quinonez (Eds.), Developing and Applying Optoelectronics in Machine Vision (p. 38). IGI Global. doi:10.4018/978-1-5225-0632-4.ch004

Lindner, L. O.-L.-Q.-B.-F.-R. (2017). Machine vision system errors for unmanned aerial vehicle navigation. In *2017 IEEE 26th international symposium on industrial electronics (ISIE)* (pp. 1615-1620). Edinburgh, UK: IEEE.

Lindner, L., Sergiyenko, O., Rivas-Lopez, M., Ivanov, M., Rodriguez-Quinonez, J., Hernandez-Balbuena, D., . . . Mercorelli, P. (2017). Machine vision system errors for unmanned aerial vehicle navigation. In *Industrial Electronics (ISIE), 2017 IEEE 26th International Symposium on* (pp. 1615-1620). Edinburgh: IEEE. doi:10.1109/ISIE.2017.8001488

Lindner, L., Sergiyenko, O., Rivas-Lopez, M., Rodriguez-Quinonez, J., Hernandez-Balbuena, D., Flores-Fuentes, W., . . . Kartashov, V. (2016). Issues of exact laser ray positioning using DC motors for vision-based target detection. In *2016 IEEE 25th International Symposium on Industrial Electronics (ISIE)* (pp. 929-934). Santa Clara: IEEE.

Lindner, L., Sergiyenko, O., Rivas-Lopez, M., Valdez-Salas, B., Rodriguez-Quinonez, J., Hernandez-Balbuena, D., . . . Kartashov, V. (2016). Machine vision system for UAV navigation. In *Electrical Systems for Aircraft, Railway, Ship Propulsion and Road Vehicles & International Transportation Electrification Conference (ESARS-ITEC), International Conference on.* Toulouse: IEEE.

Lindner, L., Sergiyenko, O., Rivas-Lopez, M., Valdez-Salas, B., Rodriguez-Quinonez, J., Hernandez-Balbuena, D., . . . Mercorelli, P. (2016). UAV remote laser scanner improvement by continuous scanning using DC motors. In Industrial Electronics Society, IECON 2016. Florence: IEEE.

Lindner, L., Sergiyenko, O., Rodriguez-Quinonez, J., Tyrsa, V., Mercorelli, P., Fuentes-Flores, W., . . . Nieto-Hipolito, J. (2015). Continuous 3D scanning mode using servomotors instead of stepping motors in dynamic laser triangulation. In *Industrial Electronics (ISIE), 2015 IEEE 24th International Symposium on* (pp. 944-949). Buzios: IEEE. doi:10.1109/ISIE.2015.7281598

Lindner, L., Sergiyenko, O., Tyrsa, V., & Mercorelli, P. (2014, June 01-04). An approach for dynamic triangulation using servomotors. In *Industrial Electronics (ISIE), 2014 IEEE 23rd International Symposium on* (pp. 1926-1931). Istanbul: IEEE. doi:10.1109/ISIE.2014.6864910

Lindner, L. S.-L.-B.-F.-Q.-P. (2017). Exact laser beam positioning for measurement of vegetation vitality. *Industrial Robot. International Journal (Toronto, Ont.).*

Lindner, L. S.-Q.-L.-B.-F. (2016). Mobile robot vision system using continuous laser scanning for industrial application. *Industrial Robot. International Journal (Toronto, Ont.).*

Lindner, L., Sergiyenko, O., Rivas-Lopez, M., Valdez-Salas, B., Rodriguez-Quinonez, J., Hernandez-Balbuena, D., & Mercorelli, P. (2016). *UAV remote laser scanner improvement by continuous scanning using DC motors. In Industrial Electronics Society, IECON 2016.* IEEE. doi:10.1109/IECON.2016.7793316

Lindner, L., Sergiyenko, O., & Rodriguez-Quinonez, J. (2017). *Theoretical Method to Increase the Speed of Continuous Mapping in a three-dimensional Laser Scanning System using Servomotors Control.* UABC, Editorial Universitario.

Lindner, L., Sergiyenko, O., Rodriguez-Quinonez, J., Rivas-Lopez, M., Hernandez-Balbuena, D., Flores-Fuentes, W., Natanael Murrieta-Rico, F., & Tyrsa, V. (2016). Mobile robot vision system using continuous laser scanning for industrial application. *The Industrial Robot, 43*(4), 360–369. doi:10.1108/IR-01-2016-0048

Lins, R. G., de Araujo, P. R. M., & Corazzim, M. (2020). In-process machine vision monitoring of tool wear for Cyber-Physical Production Systems. *Robotics and Computer-integrated Manufacturing, 61*, 101859. doi:10.1016/j.rcim.2019.101859

Lin, W., & Kuo, C. C. J. (2011). Perceptual visual quality metrics: A survey. *Journal of Visual Communication and Image Representation, 22*(4), 297–312. doi:10.1016/j.jvcir.2011.01.005

Li, S., Zhang, F., Ma, L., & Ngan, K. N. (2011). Image quality assessment by separately evaluating detail losses and additive impairments. *Proceedings of the IEEE Transactions on Multimedia*, 13(5). 10.1109/TMM.2011.2152382

Litex Electricals Pvt. Ltd. (2015). Obtenido de http://www.litexelectricals.com/halogen_lamp_characteristics.asp

Litomisky, K. (2012). Consumer Rgb-d Cameras and Their Applications. Rapport Technique, University of California 20.

Liu, A., Lin, W., & Narwaria, M. (2012). Image quality assessment based on gradient similarity. *Proceedings of the IEEE Transactions on Image Processing*, 21(4).

Liu, W., McGucken, E., Vitchiechom, K., Clements, M., De Juan, E., & Humayun, M. (1997). Dual Unit Visual Intra-ocular Prosthesis. *Proceedings of the 19th Annual International Conference of the Ieee Engineering in Medicine and Biology Society. 'Magnificent Milestones and Emerging Opportunities in Medical Engineering'(Cat. No. 97CH36136)*, 5, 2303–6. 10.1109/IEMBS.1997.758824

Li, X., Zhao, H., Liu, Y., Jiang, H., & Bian, Y. (2014). Laser scanning based three dimensional measurement of vegetation canopy structure. *Optics and Lasers in Engineering*, 54, 152–158. doi:10.1016/j.optlaseng.2013.08.010

Lodh, Subramaniam, & Paswan. (2016). Ultrasound Based Assistive Mobility Devices for the Visually-Impaired. In *2016 Ieee 7th Power India International Conference (Piicon)*, (pp. 1–6). IEEE.

Long, Y. Y., Gong, Y., Xiao, Z., & Liu, Q. (2017). Accurate object localization in remote sensing images based on convolutional neural networks. *IEEE Transactions on Geoscience and Remote Sensing*, 55(5), 2486–2498. doi:10.1109/TGRS.2016.2645610

Lorenser, D. a. (2003). Towards wafer-scale integration of high repetition rate passively mode-locked surface-emitting semiconductor lasers. *Applied Physics. B, Lasers and Optics*, 8(79), 927–932.

Luo, P. a. (2012). A moving average filter based method of performance improvement for ultraviolet communication system. In *2012 8th International Symposium on Communication Systems, Networks & Digital Signal Processing (CSNDSP)* (pp. 1- 4). Poznan, Poland: IEEE.

Luo, R. C.-C., & Chih-Chen Yih. (2002). Multisensor fusion and integration: Approaches, applications, and future research directions. *Sensors Journal, IEEE*, 2(2), 107–119. doi:10.1109/JSEN.2002.1000251

Lytyuga A. P. (2010). Algorithms for Detecting Optical Signals from Low-Orbit Space Objects in the Daytime. *Information Processing Systems*, 4(22), 41-46.

Lytyuga, A. (2009). Mathematical model of signals in television systems with low-orbit space objects observation in daytime. *Collected Works of Kharkiv University of Air Force*, 4(22), 41–46.

Lytyuga, A. P. (2009). A mathematical model of signals in television systems when observing low-orbit space objects in the daytime. *Scientific Works of Kharkiv National Air Force University*, 4(22), 41–46.

Mallick, Das, & Majumdar. (2014). Characterizations of Noise in Kinect Depth Images: A Review. *IEEE Sensors Journal*, 14(6), 1731–40.

Manjari, K., Verma, M., & Singal, G. (2020). A Survey on Assistive Technology for Visually Impaired. In Internet of Things. Elsevier. doi:10.1016/j.iot.2020.100188

Mankoff, K. D., Russo, T. A., Norris, B. K., Hossainzadeh, S., Beem, L., Walter, J. I., & Tulaczyk, S. M. (2011). Kinects as sensors in earth science: glaciological, geomorphological, and hydrological applications. *AGU Fall Meeting Abstracts*, C41D–0442.

Marston, R. M. (1999). *Optoelectronics circuits manual*. Butterworth-Heinemann.

Mathew, R. a. (2016). Trajectory tracking and control of differential drive robot for predefined regular geometrical path. *Procedia Technology*, 1273-1280.

Matveev, A. N. (1988). *Optics*. Mir Publishers.

McClure, W. F. (2003). 204 years of near infrared technology: 1800-2003. *Journal of Near Infrared Spectroscopy*, *6*(11), 487–518. doi:10.1255/jnirs.399

Merchan, Guerra, Poveda, Guzmán, & Sanchez-Galan. (2020). Bioacoustic Classification of Antillean Manatee Vocalization Spectrograms Using Deep Convolutional Neural Networks. *Applied Sciences, 10*(9), 3286.

Metzler, R., & Klafter, J. (n.d.). *The random walk's guide to anomalous diffusion: a fractional dynamics approach*. https://www.tau.ac.il/~klafter1/258.pdf

Mian, Bennamoun, & Owens. (2006). Three-Dimensional Model-Based Object Recognition and Segmentation in Cluttered Scenes. *IEEE Transactions on Pattern Analysis and Machine Intelligence, 28*(10), 1584–1601.

Miček, J., & Kapitulík, J. (2003). Median filter. *Journal of Information, Control and Management Systems*, 51-56.

Michoud, M. J.-H. (2012). New classification of landslide-inducing anthropogenic activities. *Geophysical Research Abstracts*.

Mills, Jalil, & Stanga. (2017). Electronic Retinal Implants and Artificial Vision: Journey and Present. *Eye, 31*(10), 1383–98.

Milyutin, E. R., & Gumbinas, A. Y. (2002). *Statistical Theory of the Atmospheric Channel of Optical Information Systems*. Radio and Communications.

Mims, F. M. (2000). Solar radiometer with light-emitting diodes as spectrally-selective detectors. *Applied Optics, 39*(34), 6517–6518. doi:10.1364/AO.39.006517

Ming-Chih, L., Su-Chin, C., Yu-Shen, L., & Cheng-Pei, T. (n.d.). *Machine Vision in the realization of Landslide monitoring applications*. Academic Press.

Miranda-Vega, J. E.-F.-L.-Q.-B., Flores-Fuentes, W., Sergiyenko, O., Rivas-López, M., Lindner, L., Rodríguez-Quiñonez, J. C., & Hernández-Balbuena, D. (2018). Optical cyber-physical system embedded on an FPGA for 3D measurement in structural health monitoring tasks. *Microprocessors and Microsystems, 56*, 121–133. doi:10.1016/j.micpro.2017.11.005

Miranda-Vega, J. E., Rivas-López, M., Flores-Fuentes, W., Sergiyenko, O., Lindner, L., & Rodríguez-Quiñonez, J. C. (2019). Implementación digital de filtros FIR para la minimización del ruido óptico y optoelectrónico de un sistema de barrido óptico. *Revista Iberoamericana de Automática e Informática., 16*(3), 344–357. doi:10.4995/riai.2019.10210

Miranda-Vega, J. E., Rivas-Lopez, M., Flores-Fuentes, W., Sergiyenko, O., Rodríguez-Quiñonez, J. C., & Lindner, L. (2019). Methods to reduce the optical noise in a real-world environment of an optical scanning system for structural health monitoring. In *Optoelectronics in machine vision-based theories and applications* (pp. 301–336). IGI Global. doi:10.4018/978-1-5225-5751-7.ch011

Mitchelson, D. (1975). Recording of movement without photography. In Techniques for the Analysis of Human Movement. London Lepus Books.

Mokyr, J. (1998). The second industrial revolution, 1870-1914. *Storia dell'economia Mondiale*, 1-18.

Moreno, D. a. (2012 Oct 13). Simple, accurate, and robust projector-camera calibration. In *2012 Second International Conference on 3D Imaging, Modeling, Processing, Visualization & Transmission* (pp. 464-471). Zurich, Switzerland: IEEE. 10.1109/3DIMPVT.2012.77

Mori, H., Kobayashi, K., Ohtuki, N., & Kotani, S. (1997). Color impression factor: an image understanding method for outdoor mobile robots. In *Proceedings of the IEEE/RSJ International Conference of Intelligent Robots and Systems* (pp. 380-387). 10.1109/IROS.1997.649088

Moru, D. K., & Borro, D. (2020). A machine vision algorithm for quality control inspection of gears. *International Journal of Advanced Manufacturing Technology, 106*(1-2), 105–123. doi:10.100700170-019-04426-2

Mosyagin, G. M., Nemtinov, V. B., & Lebedev, E. N. (1990). *Theory of Optoelectronic Systems*. Engineering.

Mousavi Hondori, H., & Khademi, M. (2014). A Review on Technical and Clinical Impact of Microsoft Kinect on Physical Therapy and Rehabilitation. *Journal of Medical Engineering*. doi:10.1155/2014/846514 PMID:27006935

Murali & Coughlan. (2013). Smartphone-Based Crosswalk Detection and Localization for Visually Impaired Pedestrians. In *2013 IEEE International Conference on Multimedia and Expo Workshops (Icmew)*, (pp. 1–7). IEEE. 10.1109/ICMEW.2013.6618432

Murrieta Rico, F., Petranovskii, V., Raymond-Herrera, O., Sergiyenko, O., Lindner, L., Hernandez-Balbuena, D., . . . Tyrsa, V. (2016). High resolution measurement of physical variables change for INS. In *2016 IEEE 25th International Symposium on Industrial Electronics (ISIE)* (pp. 912-917). Santa Clara: IEEE.

Murrieta-Rico, F., Hernandez-Balbuena, D., Rodriguez-Quinonez, J., Petranovskii, V., Raymond-Herrera, O., Nieto-Hipolito, J., . . . Melnyk, V. (2015). Instability measurement in time-frequency references used on autonomous navigation systems. In *2015 IEEE 24th International Symposium on Industrial Electronics (ISIE)* (pp. 956-961). Buzios: IEEE. doi:10.1109/ISIE.2015.7281600

Murrieta-Rico, F., Petranovskii, V., Raymond-Herrera, O., Sergiyenko, O., Lindner, L., Valdez-Salas, B., . . . Tyrsa, V. (2016). Resolution improvement of accelerometers measurement for drones in agricultural applications. In Industrial Electronics Society, IECON 2016. Florence: IEEE.

Murrieta-Rico, F., Sergiyenko, O., Petranovskii, V., Hernandez-Balbuena, D., Lindner, L., Tyrsa, V., Rivas-Lopez, M., Nieto-Hipolito, J. I., & Karthashov, V. (2016, May). Pulse width influence in fast frequency measurements using rational approximations. *Measurement, 86*, 67–78. doi:10.1016/j.measurement.2016.02.032

Nagata, Tokuno, Otsuka, Ochi, Ikeda, Watanabe, & Habib. (2020). Development of Design and Training Application for Deep Convolutional Neural Networks and Support Vector Machines. In *Machine Vision and Navigation*, (pp. 769–86). Springer.

Nehmetallah, G., & Banerjee, P. P. (2012). Applications of digital and analog holography in three-dimensional imaging. Advances in Optics and Photonics, 4, 472–553.

Nielsen, M. A. (2015). *Neural Networks and Deep Learning*. Determination Press.

Nikitin, V. M., Fomin, V. N., Borisenkov, A. I., Nikolaev, A. I., & Borisenkov, I. L. (2008). Adaptive noise protection of optical-electronic information systems. Academic Press.

Nikitin, V. M., Fomin, V. N., Nikolaev, A. I., & Borisenkov, I. L. (2008). *Adaptive Interference Protection of Optoelectronic Information Systems*. BelSU Publishing House.

Ningombam, D., Rai, P., & Chinghtham, T. S. (2018, December). Autonomous Mobile Robot Navigation and Localisation using Cloud, based on Landmark. *International Journal of Management. Technology And Engineering, 8*(12), 1265–1278.

Nishad, P. M., & Chezian, R. M. (2013, January). Various colour spaces and colour space conversion algorithms. *Journal of Global Research in Computer Science, 4*(1), 44–48.

NOAA. (2015). Obtenido de http://oceanservice.noaa.gov: https://oceanservice.noaa.gov/facts/lidar.html

Noguez, C., & Ulloa, S. E. (1997). First-principles calculations of optical properties: Application to silicon clusters. *Physical Review. B*, *56*(15), 9719–9725. Advance online publication. doi:10.1103/PhysRevB.56.9719

Nowak, R. D., & Kolaczyk, E. D. (2000). A Statistical Multiscale Framework for Poisson Inverse Problems. *IEEE Transactions on Information Theory*, *46*(5), 1811–1825. doi:10.1109/18.857793

Ohnishi, N., & Imiya, A. (2013). Appearance-based navigation and homing for autonomous mobile robot. *Image and Vision Computing*, *31*, 511-532. Obtenido de https://www.sciencedirect.com/science/article/pii/S0262885612002120

Onta, Y., Kanade, T., & Sakai, T. (1980). Color Information for Region Segmentation. *Computer Graphics and Image Processing*, *13*(3), 222–241. doi:10.1016/0146-664X(80)90047-7

Pallejà, Tresanchez, Teixidó, & Palacin. (2010). Bioinspired Electronic White Cane Implementation Based on a Lidar, a Tri-Axial Accelerometer and a Tactile Belt. *Sensors (Basel)*, *10*(12), 11322–11339.

Pardasani, Indi, Banerjee, Kamal, & Garg. (2019). Smart Assistive Navigation Devices for Visually Impaired People. In *2019 IEEE 4th International Conference on Computer and Communication Systems (Icccs)*, (pp. 725–29). IEEE.

Parfenov, V. I., & Kirillov, V. S. (2012). Detection of optical signals when receiving a stream of photoelectrons with an unknown density form. *Computer Optics*, *36*(4), 618–622.

Patel, K. K., Kar, A., Jha, S. N., & Khan, M. A. (2012). Machine vision system: A tool for quality inspection of food and agricultural products. *Journal of Food Science and Technology*, *49*(2), 123–141. doi:10.100713197-011-0321-4 PMID:23572836

Patro, P. P. & Panda, C. S. (2016). A review on: noise model in digital image processing. *International Journal of Engineering Sciences & Research Technology*, 891-897.

Pedersen, K. S., Duits, R., & Nielsen, M. (2005). On α Kernels, Levy Processes, and Natural Image Statistics. In *Scale Space and PDE Methods in Computer Vision* (pp. 468–479). SpringerVerlag. doi:10.1007/11408031_40

Peli, Luo, Bowers, & Rensing. (2009). Development and Evaluation of Vision Multiplexing Devices for Vision Impairments. *International Journal on Artificial Intelligence Tools*, *18*(3), 365–78.

Pérez, L. Í., Rodríguez, Í., Rodríguez, N., Usamentiaga, R., & García, D. (2016). Robot guidance using machine vision techniques in industrial environments: A comparative review. *Sensors (Basel)*, *16*(3), 335. doi:10.339016030335 PMID:26959030

Peters, J. F. (2017). *Foundations of Computer Vision: Computational Geometry, Visual Image Structures and Object Shape Detection* (Vol. 124). Springer. doi:10.1007/978-3-319-52483-2

Piprek, J. (2003). *Semiconductor optoelectronic devices: introduction to physics and simulation*. Academic Press.

Pisarenko, V. F., (1973), The Retrieval of Harmonics from a Covariance Function. *Geophys. J.R. astr. Soc.*, *33*.

Plaza, J. P., Plaza, A., & Barra, C. (2009). Multi-Channel Morphological Profiles for Classification of Hyperspectral Images Using Support Vector Machines. *Sensors (Basel)*, *9*(1), 197. doi:10.339090100196 PMID:22389595

Poggi, M., Nanni, L., & Mattoccia, S. (2015). Crosswalk Recognition Through Point-Cloud Processing and Deep-Learning Suited to a Wearable Mobility Aid for the Visually Impaired. In *International Conference on Image Analysis and Processing*, (pp. 282–89). Springer. 10.1007/978-3-319-23222-5_35

Poggi, M., & Mattoccia, S. (2016). A Wearable Mobility Aid for the Visually Impaired Based on Embedded 3d Vision and Deep Learning. In *2016 IEEE Symposium on Computers and Communication (Iscc)*, (pp. 208–13). IEEE. 10.1109/ISCC.2016.7543741

Poliarus, O., & Poliakov, Y. (2020a) The Methods of Radar Detection of Landmarks by Mobile Autonomous Robots. In O. Sergiyenko, W. Flores-Fuentes & P. Mercorelli (Eds.), Machine Vision and Navigation (pp. 171-196). Springer. doi:10.1007/978-3-030-22587-2_6

Poliarus, O., & Poliakov, Ye. (2020b) The Problem of Using Landmarks for the Navigation of Mobile Autonomous Robots on Unknown Terrain. In O. Sergiyenko, M. Rivas-Lopez, W. Flores-Fuentes, Ju. Rodríguez-Quiñonez, & L. Lindner (Eds.), Control and Signal Processing Applications for Mobile and Aerial Robotic Systems. IGI Global. doi:10.4018/978-1-5225-9924-1.ch008

Poliarus, O., Poliakov, Ye., & Lindner, L. (2018). Determination of landmarks by mobile robot's vision system based on detecting abrupt changes of echo signals parameters. In *The 44th Annual Conference of the IEEE Industrial Electronics Society* (pp. 3165-3170). 10.1109/IECON.2018.8591362

Poliarus, O., Poliakov, Ye., Sergiyenko, O., Tyrsa, V., & Hernandez, W. (2020). Azimuth estimation of landmarks by mobile autonomous robots using one scanning antenna. *Proceedings of IEEE 28th International Symposium on Industrial Electronics (ISIE 2019)*.

Pradeep, V., Medioni, G., & Weiland, J. (2010). Robot Vision for the Visually Impaired. In *2010 Ieee Computer Society Conference on Computer Vision and Pattern Recognition-Workshops*, (pp. 15–22). IEEE.

Prasad, D. K. (2016). *Challenges in video based object detection in maritime scenario using computer visio.* arXiv preprint arXiv:1608.01079

Prokhorov, Yu. V., & Rozanov, Yu. A. (1973). *Probability Theory.* Science.

Prokopenko, I. G. (2011). Statistical signal processing. Academic Press.

Pronina, O. I. (2018). Formalization of the order organization in the conditions of individual needs of the client. *Proceedings of All-Ukrainian scientific-practical conference of higher education seekers and young scientists "Computer Engineering and Cybersecurity: Achievements and Innovations"*, 93 – 95.

Pronina, O. I. (2018). Model of service presentation in the context of individual customer needs. Computer Integrated Technologies: Education, Science, Production, 33, 128– 133.

Ramík, D. M., Sabourin, C., Moreno, R., & Madani, K. (2014). A machine learning based intelligent vision system for autonomous object detection and recognition. *Applied Intelligence*, *40*(2), 358–375. doi:10.100710489-013-0461-5

Ramírez-Hernández, L. R., Rodríguez-Quiñonez, J. C., Castro-Toscano, M. J., Hernández-Balbuena, D., Flores-Fuentes, W., Rascón-Carmona, R., Lindner, L., & Sergiyenko, O. (2020). Improve three-dimensional point localization accuracy in stereo vision systems using a novel camera calibration method. *International Journal of Advanced Robotic Systems*, *17*(1), 1–15. doi:10.1177/1729881419896717

Ray, P. P. (2016). Internet of robotic things: Concept, technologies, and challenges. *IEEE Access: Practical Innovations, Open Solutions*, *4*, 9489–9500. doi:10.1109/ACCESS.2017.2647747

Real-Moreno, O. C.-T.-O.-B.-F.-L. (2018). Implementing k-nearest neighbor algorithm on scanning aperture for accuracy improvement. *IECON 2018-44th Annual Conference of the IEEE Industrial Electronics Society*, 3182-3186.

Rea, M. S. (2000). *The IESNA lighting handbook: reference & application.* Illuminating Engineering Society of North America.

Rentschler, K., & Moser, M. (1996). Geotechniques of claystone hillslopes-properties of weathered claystone and formation of sliding surfaces. In *7th International Symposium on Landslides* (pp. 1571-1578). Tronheim, Norway: Brookfield.

Reyes-García, M. O.-Q.-B.-F.-A.-R. (2019). Defining the Final Angular Position of DC Motor shaft using a Trapezoidal Trajectory Profile. In *2019 IEEE 28th International Symposium on Industrial Electronics (ISIE)* (pp. 1694-1699). Vancouver, Canada: IEEE.

Reyes-Garcia, M., Lindner, L., Rivas-Lopez, M., Ivanov, M., Rodriguez-Quinonez, J., Murrieta-Rico, F., . . . Melnyk, V. (2018). Reduction of Angular Position Error of a Machine Vision System Using the Digital Controller LM629. In *IECON 2018-44th Annual Conference of the IEEE Industrial Electronics Society* (pp. 3200-3205). Washington, DC: IEEE. doi:10.1109/IECON.2018.8592803

Riazantsev, A.O., Khoroshun, G.M., & Ryazantsev, O.I. (2019). Statistical image analysis for information system. *Visnik of the Volodymyr Dahl East Ukrainian National University, 5*(253), 84-86. Doi:10.33216/1998-7927-2019-253-5-84-86

Rifkin, J. (2011). *The third industrial revolution: how lateral power is transforming energy, the economy, and the world.* Macmillan.

Rivas López, M., Flores Fuentes, W., Rivera Castillo, J., Sergiyenko, O., & Hernández Balbuena, D. (2013). A Method and Electronic Device to Detect the Optoelectronic Scanning Signal Energy Centre. In S. L. Pyshkin, & J. M. Ballato (Eds.), Optoelectronics (pp. 389-417). Crathia: Intech.

Rivas-Lopez, M. C.-F.-Q.-B.-B. (2014). Scanning for light detection and Energy Centre Localization Methods assesment in vision systems for SHM. In *2014 IEEE 23rd International Symposium on* (pp. 1955-1960). Industrial Electronics (ISIE).

Rivas-Lopez, M., Sergiyenko, O., Flores-Fuentes, W., & Rodríguez-Quiñonez, J. C. (Eds.). (2018). *Optoelectronics in Machine Vision-Based Theories and Applications.* IGI Global.

Rivera-Castillo, J. F.-F.-L.-N.-Q.-P., Flores-Fuentes, W., Rivas-López, M., Sergiyenko, O., Gonzalez-Navarro, F. F., Rodríguez-Quiñonez, J. C., Hernández-Balbuena, D., Lindner, L., & Básaca-Preciado, L. C. (2017). Experimental image and range scanner datasets fusion in shm for displacement detection. *Structural Control and Health Monitoring, 24*(10), e1967. doi:10.1002tc.1967

Rodriguez-Quinonez, J., Sergiyenko, O., Basaca-Preciado, L., Tyrsa, V., Gurko, A., Podrygalo, M., ... Hernandez-Balbuena, D. (2014). Optical monitoring of scoliosis by 3D medical laser scanner. *Optics and Lasers in Engineering, 54*, 175-186. Obtenido de https://www.sciencedirect.com/science/article/pii/S014381661300242X

Rodriguez-Quiñonez, J. B.-L.-F.-P., Sergiyenko, O., Hernandez-Balbuena, D., Rivas-Lopez, M., Flores-Fuentes, W., & Basaca-Preciado, L. (2014). Improve 3D laser scanner measurements accuracy using a FFBP neural network with Widrow-Hoff weight/bias learning function. *Opto-Electronics Review, 4*(22), 224–235. doi:10.247811772-014-0203-1

Rodriguez-Quiñonez, J. C., Sergiyenko, O. Y., Preciado, L. C. B., Tyrsa, V. V., Gurko, A. G., Podrygalo, M. A., Lopez, M. R., & Balbuena, D. H. (2014). Optical monitoring of scoliosis by 3D medical laser scanner. *Optics and Lasers in Engineering, 54*, 175–186. doi:10.1016/j.optlaseng.2013.07.026

Rodriguez-Quinonez, J., Sergiyenko, O., Gonzalez-Navarro, F., Basaca-Preciado, L., & Tyrsa, V. (2013, February). Surface recognition improvement in 3D medical laser scanner using Levenberg–Marquardt method. *Signal Processing, 93*(2), 378–386. doi:10.1016/j.sigpro.2012.07.001

Roemer, G., & Bechthold, P. (2014). Electro-optic and Acousto-optic Laser Beam Scanners. *8th International Conference on Laser Assisted Net Shape Engineering, 56*, 29-39. doi:10.1016/j.phpro.2014.08.092

Rogalski, A. (2002). Infrared detectors: An overview. *Infrared Physics & Technology, 3*(43), 187–210. doi:10.1016/S1350-4495(02)00140-8

Romanovsky M. Yu. (2007). Analytical representations of non-Gaussian random walks. *Actual problems of statistical physics (Malakhovsky digest), 63*, 56-81.

Romanovsky, M. Yu. (2009). Levy Functional Walks. *Proceedings of the Institute of General Physics named after A.M. Prokhorova, 65*, 20-28. 10.3103/S1541308X09030078

Romero, J., Hernández-Andrés, J., Nieves, J. L., & García, J. A. (2003, February). Color Coordinates of Objects with Daylight Changes. *Color Research and Application, 28*(1), 25–35. doi:10.1002/col.10111

Rose, A. (1973). *Vision: Human and Electronic*. Plenum Press.

Rosen, J., Katz, B., & Brooker, G. (2009). Fresnel incoherent correlation hologram—a review. Chinese Optics Letters, 7, 1134–1141.

Rozovsky, L. V. (n.d.). Probabilities of large deviations of sums of independent random variables with a common distribution function from the domain of attraction of the normal law. *Probability Theory and Its Application, 4*, 686-705.

Ruderman, D. L., & Bialek, W. (1994). Statistics of natural images: Scaling in the woods. *Physical Review Letters, 73*(6), 814–817. doi:10.1103/PhysRevLett.73.814 PMID:10057546

Rudoi, Yu. G. (2007). Random walks and anomalous diffusion of Levi-Khinchin in the physical chemistry of polymers. *Science - the foundation for solving technological problems of the development of Russia, 2*, 74-101.

Ruelle, D. (2001). *Chance and Chaos*. Izhevsk: Research Center. *Regular and Chaotic Dynamics*.

Ryazantsev, O., Khoroshun, G., Riazantsev, A., Ivanov, V., & Baturin, A. (2019). Statistical Optical Image Analysis for Information System. In *Proceedings of 2019 7th International Conference on Future Internet of Things and Cloud Workshops (FiCloudW), Istanbul, Turkey*, (pp. 130-134). IEEE. 10.1109/FiCloudW.2019.00036

Saad, M. A., Bovik, A. C., & Charrier, C. (2012). Blind image quality assessment: A natural scene statistics approach in the DCT domain. *IEEE Transactions on Image Processing, 21*(8), 3339–3352. doi:10.1109/TIP.2012.2191563 PMID:22453635

Sabathil, M. (2005). *Opto-electronic and quantum transport properties of semiconductor nanostructures. In Selected Topics of Semiconductor Physics and Technology (* Vol. 67). Verein zur Förderung des Walter Schottky Instituts der Technischen Universität München.

Sáez, Muñoz, Canto, García, & Montes. (2019). Assisting Visually Impaired People in the Public Transport System Through Rf-Communication and Embedded Systems. *Sensors, 19*(6), 1282.

Sanchez-Galan, J. E., Jo, K.-H., & Cáceres-Hernández, D. (2020). Stairway Detection Based on Single Camera by Motion Stereo for the Blind and Visually Impaired. In *Machine Vision and Navigation* (pp. 657–673). Springer. doi:10.1007/978-3-030-22587-2_20

Šanda, F., & Mukamel, S. (2006). Anomalous continuous-time random-walk spectral diffusion in coherent third-order optical response. *Physical Review E: Statistical, Nonlinear, and Soft Matter Physics, 73*(1), 1–14. doi:10.1103/PhysRevE.73.011103 PMID:16486118

Sarkar, D. a. (2018). Practical Machine Learning with Python. In *A Problem-Solvers Guide To Building Real-World Intelligent Systems*. Berkeley: Apress.

Savvaidis, P. D. (2003). Existing Landslide Monitoring Systems aand Techniques. School of Rural and Surveying Engineering, The Aristotle University of Thessaloniki, 242-258.

Schauwecker, K. (2018). Real-time stereo vision on FPGAs with SceneScan. In T. P. Längle (Ed.), Forum Bildverarbeitung (p. 339). Academic Press.

Scolnik, M. I. (Ed.). (1990). *Radar handbook*. McGraw-Hill.

Seitz. (2011). *Single-Photon Imaging.* Springer – Verlag Berlin Heidelberg.

Seitz, P. (2011). *Fundamentals of Noise in Optoelectronics.* Single-Photon Imaging. *Springer Series in Optical Sciences, 160,* 1–25. doi:10.1007/978-3-642-18443-7_1

Sergiyenko, O. (2010). Optoelectronic System for Mobile Robot Navigation, Optoelectronics. *Instrumentation and Data Processing, 46,* 414-428. Obtenido de https://link.springer.com/article/10.3103/S8756699011050037

Sergiyenko, O., Tyrsa, V., Basaca-Preciado, L., Rodriguez-Quinonez, J., Hernandez, W., Nieto-Hipolito, J., . . . Starostenko, O. (2011). Electromechanical 3D Optoelectronic Scanners: Resolution Constraints and Possible Ways of Improvement. In Optoelectronic Devices and Properties. InTech. doi:10.5772/14263

Sergiyenko, O. (2010). Optoelectronic System for Mobile Robot Navigation. *Optoelectronics, Instrumentation and Data Processing, 46*(5), 414–428. doi:10.3103/S8756699011050037

Sergiyenko, O., Flores-Fuentes, W., & Mercorelli, P. (Eds.). (2019). *Machine Vision and Navigation.* Springer Nature.

Shalev-Shwartz, S. D. (2014). *Understanding machine learning: From theory to algorithms.* Cambridge University Press. doi:10.1017/CBO9781107298019

Shan, J., & Toth, C. (2008). *Topographic Laser Ranging And Scanning.* CRC Press. Obtenido de. https://www.amazon.com/Topographic-Laser-Ranging-Scanning-Principles/dp/1420051423/ref=tmm_hrd_title_0?_encoding=UTF8&qid=1441252814&sr=1-1

Shao, J. (1993). Linear model selection by cross-validation. *Journal of the American Statistical Association, 88*(422), 486–494. doi:10.1080/01621459.1993.10476299

Sharma, R. R., & Ingole, P. V. (2015, June). Review of Vision Based Robot Navigation System in Dynamic Environment. *International Journal for Research in Applied Science and Engineering Technology, 3*(VI), 393–397.

Shearman, J. D., & Manzhos, V. N. (1981). Theory and techniques of processing radar information on the background noise [Lexical characteristics of Russian language]. Radio and Communication.

Shih, C., & Ku, Y. (2016). Image-Based Mobile Robot Guidance System by Using Artificial Ceiling Landmarks. *Journal of Computer and Communications, 4*(11), 1–14. doi:10.4236/jcc.2016.411001

Shih, H.-C. P., Hwang, C., Barriot, J.-P., Mouyen, M., Corréia, P., Lequeux, D., & Sichoix, L. (2015). High-resolution gravity and geoid models in Tahiti obtained from new airborne and land gravity observations: Data fusion by spectral combination. *Earth, Planets, and Space, 1*(67), 1–16. doi:10.118640623-015-0297-9

Shimakawa, M., Akutagawa, R., Kiyota, K., & Nakano, M. (2017). A Study on Staircase Detection for Visually Impaired Person by Machine Learning Using RGB-d Images. In *Proceedings of the 5th IIAE International Conference on Industrial Application Engineering 2017.* The Institute of Industrial Application Engineers. 10.12792/icisip2017.080

Shimakawa, M., Matsushita, K., Taguchi, I., Okuma, C., & Kiyota, K. (2019). Smartphone Apps of Obstacle Detection for Visually Impaired and Its Evaluation. *Proceedings of the 7th Acis International Conference on Applied Computing and Information Technology,* 1–6. 10.1145/3325291.3325381

Shoval, S., Ulrich, I., & Borenstein, J. (2000). Computerized Obstacle Avoidance Systems for the Blind and Visually Impaired. Intelligent Systems and Technologies in Rehabilitation Engineering, 414–48.

Shoval, Ulrich, & Borenstein. (2003). NavBelt and the Guide-Cane [Obstacle-Avoidance Systems for the Blind and Visually Impaired]. *IEEE Robotics & Automation Magazine, 10*(1), 9–20.

Sibatov, R. T., & Uchaikin, V. V. (2007). Fractional differential kinetics of charge transfer in disordered semiconductors. *Physics and Technology of Semiconductors, 41*(3), 346–357.

Signal-to-noise ratio. (2020). In *Wikipedia*. https://en.wikipedia.org/wiki/Signal-to-noise_ratio

Singh, I., Centea, D., & Elbestawi, M. (2019). IoT, IIoT and Cyber-Physical Systems Integration in the SEPT Learning Factory. *Procedia Manufacturing, 31*, 116–122. doi:10.1016/j.promfg.2019.03.019

Sinha, P. K. (2012). *Image Acquisition and Preprocessing for Machine Vision Systems*. SPIE Press. doi:10.1117/3.858360

Sixta, Z., Linhart, J., & Nosek, J. (2013). Experimental Investigation of Electromechanical Properties of Amplified Piezoelectric Actuator. In *Electronics, Control, Measurement, Signals and their application to Mechatronics (ECMSM), 2013 IEEE 11th International Workshop of* (pp. 1-5). Toulouse: IEEE. doi:10.1109/ECMSM.2013.6648957

Skwarek, V. (2017 Dec 4). Blockchains as security-enabler for industrial IoT-applications. *Asia Pacific Journal of Innovation and Entrepreneurship*.

Smith, S. W. (1997). Digital Signal Processors. In The scientist and engineer's guide to digital signal processing (pp. 503-534). San Diego, CA: California Technical Pub.

Snyder, W. E. (2010). *Machine vision*. Cambridge University Press. doi:10.1017/CBO9781139168229

Sokolova, M. a. (2006). Beyond accuracy, F-score and ROC: a family of discriminant measures for performance evaluation. In *Australasian joint conference on artificial intelligence* (pp. 1015-1021). Berlin: Springer.

Solchenbach, T., & Plapper, P. (2013). Mechanical characteristics of laser braze-welded aluminium–copper connections. *Optics & Laser Technology, 54*, 249-256. doi:10.1016/j.optlastec.2013.06.003

Song, D., & Chang, J. (2012, February 29). Super wide field-of-regard conformal optical imaging system using liquid crystal spatial light modulator. *Optik (Stuttgart), 2455-2458*. Advance online publication. doi:10.1016/j.ijleo.2012.08.013

Sonka, M. a. (2014). *Image processing, analysis, and machine vision*. Cengage Learning.

Sontakke, M. D., & Kulkarni, M. S. (2015, January). Different types of noises in image and noise removing technique. *International Journal of Advanced Technology in Engineering and Science, 3*(1), 102–115.

Souto, L., Castro, A., Gonçalves, L., & Nascimento, T. P. (2017). Stairs and Doors Recognition as Natural Landmarks Based on Clouds of 3D Edge-Points from RGB-D Sensors for Mobile Robot Localization. Sensors, 17(8), 1824.

Stockman, A., & Sharpe, L. T. (2006). Physiologically based color matching functions. In *ISCC/CIE Expert Symposium '06: 75 Years of the CIE Standard Colorimetric Observer* (pp. 13-20) Vienna: CIE Central Bureau.

Stratonovich, R. L. (1968). *Conditional Markov Process and Their Application to the Theory of Optimal Control*. Elsevier.

Strelkov, A. I., Lytuga, A. P., & Strelkova, T. A. (2013). Prospects for the development of special-purpose optoelectronic systems using the methods of stochastic-deterministic processing of optical signals. *Int. scientific-practical conf. Actual problems and prospects of development of radio engineering and info-communication systems*, 50-53.

Strelkova, T. A. (1999). The Potentialities of Optical and Electronic Devices for Biological Objects Investigations. Telecommunications and Radio Engineering, 53, 190-194.

Strelkova, T. A. (2014a). Studies on the Optical Fluxes Attenuation Process in Optical-electronic Systems. *Semiconductor Physics, Quantum Electronics & Optoelectronics (SPQEO), 4*, 421–424.

Strelkova, T. A., & Sautkin, V. A. (2013). A stochastic approach to assessing the quality of optical glass. Scientific-practical. conf. Technologies for processing optical elements and applying vacuum coatings, 85-86.

Strelkova, T., Kartashov, V., Lytyuga, A., & Strelkov, A. (2016). Theoretical methods of images processing in optoelectronic systems. In *Developing and Applying Optoelectronics and Machine Vision* (pp. 181-206). https://www.igi-global.com/book/developing-applying-optoelectronics-machine-vision/147652

Strelkova, T.A. (2014c). Studies on the Optical Fluxes Attenuation Process in Optical-electronic Systems. *Semiconductor physics, quantum electronics & optoelectronics (SPQEO), 4*, 421-424.

Strelkov, A. I., Moskvitin, S. V., Lytyuga, A. P., & Strelkova, T. A. (2010). *Optical location. The theoretical basis of the reception and processing of optical signals.* Apostrophe.

Strelkov, A. I., Stadnik, A. M., Lytyuga, A. P., & Strelkova, T. A. (1998). Comparative Analysis of Probabilistic and Determinate Methods for Attenuating Light Flux. *Telecommunications and Radio Engineering, 52*(8), 54–57. doi:10.1615/TelecomRadEng.v52.i8.110

Strelkov, A., Zhilin, Ye., Lytyuga, A., & Lisovenko, S. (2007). Signal Detection in Technical Vision Systems. *Telecom. and Radio Engineering, 66*(4), 283–293. doi:10.1615/TelecomRadEng.v66.i4.10

Strelkova, T. (2014a). Influence of Video Stream Compression on Image Microstructure in Medical Systems. *Biomedical Engineering, 47*(6), 307–311. doi:10.100710527-014-9398-1

Strelkova, T. (2014b). Statistical properties of output signals in optical-television systems with limited dynamic range. *Eastern-European Journal of Enterprise Technologies, 9*(68), 38–44. doi:10.15587/1729-4061.2014.23361

Strelkova, T. A. (2014b). Statistical properties of the output signals of optical-television systems with a limited dynamic range. *East European Journal of Advanced Technologies, No., 2/9*(68), 38–44.

Strelkova, T. A. (2015). The use of stable distribution laws in evaluating the efficiency of signal processing in optoelectronic systems. *East European Journal of Advanced Technologies, No., 2/9*(74), 4–9.

Strelkova, T. A., Sozonov, Yu. I., & Yanovsky, Yu. A. (2012). Study of the statistics of spatio-temporal signals in optoelectronic systems. *All-Ukrainian Interdepartmental Scientific and Technical Journal Radiotehnika, 170*, 185–188.

Strikova, T.O., Lytyuga, O.P., Skorupski, K. & Bugubayeva, A. (2019). Stochastic deterministic methods for processing signals and images in optical electronic systems. *Proc. SPIE 11176, Photonics Applications in Astronomy, Communications, Industry, and High-Energy Physics Experiments.* doi:10.1117/12.2536605

Strilkova, T. O., & Lytyuga, O. P. (2018). *Stochastic-deterministic methods of signals and images processing in optpelectronic systems. Optoelectronic information technology PHOTONIKA-ODS-2018.* Vinnitsya.

Structural similarity. (2020). In *Wikipedia.* https://en.wikipedia.org/wiki/Structural_similarity

Sun, C. a. (2019 Nov 22). Design and Development of Modbus/MQTT Gateway for Industrial IoT Cloud Applications Using Raspberry Pi. In *2019 Chinese Automation Congress (CAC)* (pp. 2267-2271). Hangzhou, China: IEEE. 10.1109/CAC48633.2019.8997492

Suresh, S. (2017). A Survey on Types of Noise Model, Noise and Denoising Technique in Digital Image Processing. *International Journal of Innovative Research in Computer and Communication Engineering, 5*(2), 50-56.

Syms, R. R. (1992). *Optical guided waves and devices.* McGraw-Hill.

Tachaphetpiboon, S. a. (2006). Applying FFT features for fingerprint matching. In *1st International Symposium on Wireless Pervasive Computing* (pp. 1-5). Phuket, Thailand: IEEE.

Takao, A., Yasuda, M., & Sawada, K. (1995). Noise Suppression Effect in an Avalanche Multiplication Photodiode Operating in a Charge Accumulation Mode. *IEEE Transactions on Electron Devices, 42*(10), 1769–1774. doi:10.1109/16.464420

Tame, B., & Stutzke, N. (2010). Steerable Risley Prism antennas with low side lobes in the Ka band. In *Wireless Information Technology and Systems (ICWITS), 2010 IEEE International Conference on* (págs. 1-4). IEEE. doi:10.1109/ICWITS.2010.5611931

Tan & Jiang. (2020). Digital Signal Processing. 3rd Edition. Fundamentals and Applications. Academic Press.

Tang, H., Joshi, N., & Kapoor, A. (2011). Learning a blind measure of perceptual image quality. *Proceedings of the International Conference on Computer Vision and Pattern Recognition.* 10.1109/CVPR.2011.5995446

Tatarsky, V. I. (1967). *Wave propagation in a turbulent atmosphere.* Science.

Tavernier, F. a. (2011). *High-speed optical receivers with integrated photodiode in nanoscale CMOS.* Springer Science and Business Media. doi:10.1007/978-1-4419-9925-2

Testorf, M., & Lohmann, A. W. (2008). Holography in phase space. Applied Optics, 47, A70–A77.

The AICPA Assurance Services Executive Committee's SOC 2® Working Group Description. (2018). *Criteria for a Description of a Service Organization's System in a SOC 2.* https://www.aicpa.org/content/dam/aicpa/interestareas/frc/assuranceadvisoryservices/downloadabledocuments/dc-200.pdf

Thomas, J. B. (1970). Nonparametric signal detection methods. *TIER, 5,* 23-31.

Tikhonov, V. I. (1983). Optimal reception of signals [Lexical characteristics of Russian language]. Radio and Communication.

Tikhonov, V. I., & Mironov, M. A. (1977). Markov's processes [Lexical characteristics of Russian language]. Sov. radio.

Toth, T., & Zivcak, J. (2014). A Comparison of the Outputs of 3D Scanners. In *24th DAAAM International Symposium on Intelligent Manufacturing and Automation,* (pp. 393-401). Kosice: Elsevier. doi:10.1016/j.proeng.2014.03.004

Tribolet, P. a. (2008). Advanced HgCdTe technologies and dual-band developments. In *SPIE Defense and Security Symposium.* International Society for Optics and Photonics. 10.1117/12.779902

Tukey, J. W. (1977). *Exploratory Data Analysis.* Addison-Wesley.

Tuy, H. (2020). *Convex Analysis and Global Optimization.* Springer.

Uchaikin V. V. (2003). Self-similar anomalous diffusion and stable laws. *Uspekhi Fizicheskikh Nauk., 173*(8), 847-876.

Uchaikin, V. V., & Korobko, D. A. (1999). On the theory of multiple scattering in a fractal medium. *Letters in ZhTF, 25*(1), 34–40. doi:10.1134/1.1262508

Uchaikin, V. V., & Sibatov, R. T. (2004). One-Dimensional Fractal Walking with a Finite Speed of Free Motion. *Letters in ZhTF, 30*(8), 27–33.

Ulrich & Borenstein. (2001). The Guidecane-Applying Mobile Robot Technologies to Assist the Visually Impaired. *IEEE Transactions on Systems, Man, and Cybernetics-Part A: Systems and Humans, 31*(2), 131–36.

Upadhyay, D., & Sampalli, S. (2020). SCADA (Supervisory Control and Data Acquisition) systems: Vulnerability assessment and security recommendations. *Computers & Security, 89,* 1–18. doi:10.1016/j.cose.2019.101666

Vahelal, A. (2010). *Sensors on 3D Digitization*. HasmukhGoswami College of Engineering.

Vargas, S. S., Stivanello, M. E., Roloff, M. L., Stiegelmaier, É., & Stemmer, M. R. (2020). Development of an Online Automated Fabric Inspection System. *Journal of Control. Automation and Electrical Systems, 31*(1), 73–83. doi:10.100740313-019-00514-6

Varghese, D., Wanat, R., & Mantiuk, R. K. (2014) Colorimetric calibration of high dynamic range images with a ColorChecker chart. In *HDRi 2014 - Second International Conference and SME Workshop on HDR imaging* (pp. 1-6). Academic Press.

Vauhkonen, J. a. (2014). Tree species recognition based on airborne laser scanning and complementary data sources. In Forestry applications of airborne laser scanning (pp. 135-156). Springer. doi:10.1007/978-94-017-8663-8_7

Vejarano, Henriuez, & Montes. (2018). Sistema Para La Interacción Activa Con Autobuses de Rutas Urbanas de Panamá Para Personas Con Discapacidad Visual. *I+ D Tecnológico, 14*(2), 17–23.

Velazquez, Pissaloux, Guinot, & Maingreaud. (2006). Walking Using Touch: Design and Preliminary Prototype of a Non-Invasive Eta for the Visually Impaired. In *2005 IEEE Engineering in Medicine and Biology 27th Annual Conference*, (pp. 6821–4). IEEE.

Veneri, G., & Capasso, A. (2018). *Hands-On Industrial Internet of Things: Create a powerful Industrial IoT infrastructure using Industry 4.0*. Packt Publishing Ltd.

Vetnzel, E. S., & Ovcharov, L. A. (2000). *Theory of random processes and its engineering applications*. Higher. Shk.

Vidov, P. V., & Romanovsky, M. Yu. (2007). Analytical representations of non-Gaussian laws of random walks. Actual problems of statistical physics (Malakhovsky digest), 63, 3-19.

Volos, C., Kyprianidis, I., & Stouboulos, I. (2013, December). Experimental investigation on coverage performance of a chaotic autonomous mobile robot. *Robotics and Autonomous Systems, 61*(12), 1314–1322. doi:10.1016/j.robot.2013.08.004

Voulodimos, A., Doulamis, N., Doulamis, A., & Protopapadakis, E. (2018). Deep learning for computer vision: A brief review. *Computational Intelligence and Neuroscience, 2018*, 1–14. doi:10.1155/2018/7068349 PMID:29487619

Wahab, Talib, Kadir, Johari, Noraziah, Sidek, & Mutalib. (2011). *Smart Cane: Assistive Cane for Visually-Impaired People*. arXiv Preprint arXiv:1110.5156

Wang, J., Chandler, D. M., & Le Callet, P. (2010), Quantifying the relationship between visual salience and visual importance. In Human Vision and Electronic Imaging, (vol. 7527). doi:10.1117/12.845231

Wang, L., Liu, L., Sun, J., Zhou, Y., Luan, Z., & Liu, D. (2010). Large-aperture double-focus laser collimator for PAT performance testing of inter-satellite laser communication terminal. *Optik, 121*(17), 1614-1619. doi:10.1016/j.ijleo.2009.03.004

Wang, W. S.-S., Capitaneanu, S. L., Marinca, D., & Lohan, E.-S. (2019, July 8). Comparative analysis of channel models for industrial IoT wireless communication. *IEEE Access: Practical Innovations, Open Solutions, 7*, 91627–91640. doi:10.1109/ACCESS.2019.2927217

Waske, B. a. (2007). Fusion of support vector machines for classification of multisensor data. *Geoscience and Remote Sensing. IEEE Transactions on, 122*(45), 3858–3866.

Weckenmann, A. D.-R. (2009). Multisensor data fusion in dimensional metrology. *CIRP Annals-Manufacturing Technology, 2*(58), 701–721.

Weiland, Liu, & Humayun. (2005). Retinal Prosthesis. *Annu. Rev. Biomed. Eng., 7*, 361–401.

Wendy Flores-Fuentes, J. E.-V.-L.-Q. (2018). Comparison between different types of sensors used in the real operational environment based on optical scanning system. *Sensors (Basel)*, *18*(6), 1684. doi:10.339018061684 PMID:29882912

Westfechtel, Ohno, Mertsching, Hamada, Nickchen, Kojima, & Tadokoro. (2018). Robust Stairway-Detection and Localization Method for Mobile Robots Using a Graph-Based Model and Competing Initializations. *The International Journal of Robotics Research, 37*(12), 1463–83.

Wetzel, B., Blow, K. J., Turitsyn, S. K., Millot, G., Larger, L., & Dudle, J. M. (2012). Random walks and random numbers from supercontinuum generation. *OSA Optics Express*, *20*(10), 11143–11152. doi:10.1364/OE.20.011143 PMID:22565737

World Health Organization. (2013). *Vision Impairment and Blindness*. https://www.who.int/en/news-room/fact-sheets/detail/blindness-and-visual-impairment

World Health Organization. (2018). *Assistive Technology*. https://www.who.int/news-room/fact-sheets/detail/assistive-technology

Xiang, S., Chen, S., Wu, X., Xiao, D., & Zheng, X. (2010). Study on fast linear scanning for a new laser scanner. *Optics & Laser Technology, 42*(1), 42-46. doi:10.1016/j.optlastec.2009.04.019

Xu, L.-Y. Z.-Q., Cao, Z.-Q., Zhao, P., & Zhou, C. (2017). A new monocular vision measurement method to estimate 3D positions of objects on floor. *International Journal of Automation and Computing, 14*(2), 159–168. doi:10.100711633-016-1047-6

Xu, Q. (2014). Impact detection and location for a plate structure using least squares support vector machines. *Structural Health Monitoring, 13*(1), 5–18. doi:10.1177/1475921713495083

Yallup, K. a. (2014). *Technologies for smart sensors and sensor fusion*. CRC Press.

Yang, K., Wang, K., Lin, S., & Bai, J. (2018). Long-Range Traversability Awareness and Low-Lying Obstacle Negotiation with Realsense for the Visually Impaired. *Proceedings of the 2018 International Conference on Information Science and System*, 137–41.

Yang, S., Lee, K., Xu, Z., Zhang, X., & Xu, X. (2001). An accurate method to calculate the negative dispersion generated by prism pairs. *Optics and Lasers in Engineering, 36*(4), 381-387. doi:10.1016/S0143-8166(01)00055-0

Yang, Wang, Hu, & Bai. (2016). Expanding the Detection of Traversable Area with Realsense for the Visually Impaired. *Sensors, 16*(11), 1954.

Yang, Wang, Zhao, Cheng, Bai, Yang, & Liu. (2017). IR Stereo Realsense: Decreasing Minimum Range of Navigational Assistance for Visually Impaired Individuals. *Journal of Ambient Intelligence and Smart Environments, 9*(6), 743–55.

Yang, F., Lu, Y. M., Sbaiz, L., & Vetterli, M. (2012). Oversampled Image Acquistion Using Binary Poisson Statistics. *IEEE Transactions on Image Processing, 21*(4), 1421–1436. doi:10.1109/TIP.2011.2179306 PMID:22180507

Yang, Y. (2008, November 21). Analytic Solution of Free Space Optical Beam Steering Using Risley Prisms. *Journal of Lightwave Technology, 26*(21), 3576–3583. doi:10.1109/JLT.2008.917323

Zafar, M. N., & Mohanta, J. C. (2018). Methodology for path planning and optimization of mobile robots: A review. *Procedia Computer Science, 133*, 141–152. doi:10.1016/j.procs.2018.07.018

Zeuch, N. (2000). *Understanding and applying machine vision, revised and expanded*. CRC Press. doi:10.1201/b16927

Zhang, F.-m. a.-h.-h. (2008). Multiple sensor fusion in large scale measurement. *Optics and Precision Engineering*, (7), 18.

Zhang, Z. (2012). Microsoft Kinect Sensor and Its Effect. *IEEE Multimedia, 19*(2), 4–10.

Zhang, C. (2019). Automatic drill pipe emission control system based on machine vision. *Journal of Petroleum Exploration and Production Technology, 9*(4), 2737–2745.

Zhang, Q. A. (2012). 3-D shape measurement based on complementary Gray-code light. *Optics and Lasers in Engineering, 4*(50), 574–579.

Zhengt, J. Y., Barth, M., & Tsuji, S. (1991). Autonomous Landmark Selection for Route Recognition by A Mobile Robot. In *Proceedings of the 1991 IEEE International Conference on Robotics and Automation* (pp. 2004-2009). 10.1109/ROBOT.1991.131922

Zhongdong, Y., Peng, W., Xiaohui, L., & Changku, S. (2014, March). 3D laser scanner system using high dynamic range imaging. *Optics and Lasers in Engineering, 54*, 31–41. doi:10.1016/j.optlaseng.2013.09.003

Zhong, J. M., Li, M., Liao, X., & Qin, J. (2020). A Real-Time Infrared Stereo Matching Algorithm for RGB-D Cameras' Indoor 3D Perception. *SPRS International Journal of Geo-Information, 9*(8), 472. doi:10.3390/ijgi9080472

Zhong, X., Zhou, Y., & Liu, H. (2017). Design and recognition of artificial landmarks for reliable indoor self-localization of mobile robots. *International Journal of Advanced Robotic Systems, 14*(January-February), 1–13. doi:10.1177/1729881417693489

Zhou, X. W. (2019). Automated visual inspection of glass bottle bottom with saliency detection and template matching. *IEEE Transactions on Instrumentation and Measurement, 68*(11), 4253–4267.

Zimek, A., & Schubert, E. (2017). *Outlier Detection. Encyclopedia of Database Systems*. Springer. doi:10.1007/978-1-4899-7993-3_80719-1

Zolotarev, V. M. (1983). *One-dimensional stable distributions*. Science.

Zolotarev, V. M. (1984). *Stable laws and their application*. Knowledge.

Zolotarev, V. M., Uchaykin, V. V., & Saenko, V. V. (1999). Superdiffusion and Stable Laws. *Soviet Physics, JETP, 115*(4), 1411–1425.

Zou, Z. a. (2019). *Object detection in 20 years: A survey*. arXiv preprint arXiv:1905.05055

About the Contributors

Oleg Sergiyenko received the B.S., and M.S., degrees in Kharkiv National University of Automobiles and Highways, Kharkiv, Ukraine, in 1991, 1993, respectively. He received the Ph.D. degree in Kharkiv National Polytechnic University on specialty "Tools and methods of non-destructive control" in 1997. He received the DSc. (habit.) degree in Kharkiv National University of Radioelectronics in 2018. He has been an author of 1 book and editor of 8 books (in Springer, IGI Global, etc.), written 31 book chapters, over 140 papers indexed in Scopus and holds 3 patents (of Ukraine and Mexico). Since 1994 till the present time he was represented by his research works in several International Congresses of IEEE, ICROS, SICE, IMEKO in USA, England, Japan, Italy, Austria, Ukraine, Canada, Portugal, Brazil and Mexico. In many of them, he was a Session Chair. He gets the "Best presentation award" on IEEE conferences IECON2014 in Dallas, USA, on IECON2016 in Florence, Italy, ISIE2019 in Vancouver, Canada. He also did receive the Emarald Literati Award of 2017 Outstanding paper for his article published in Industrial Robot: an International Journal. Dr. Sergiyenko in December 2004 was invited by Engineering Institute of Baja California Autonomous University for researcher position. He is currently Head of Applied Physics Department of Engineering Institute of Baja California Autonomous University, Mexico, director of several Master's and Doctorate thesis. He is a full-member (Academician) of Academy of Applied Radioelectronics of Belorussia, Ukraine and Russia. His current research interests include automated metrology, machine vision systems, fast electrical measurements, control systems, robot navigation and 3D laser scanners.

Julio C. Rodríguez-Quiñonez received the B.S. degree in CETYS, Mexico in 2007. He received the Ph.D. degree from Baja California Autonomous University, México, in 2013. He is currently Head of the extension and liaison department in the Engineering Faculty of the Autonomous University of Baja California, and member of the National Research System Level 1. Since 2016 he is Senior Member of IEEE. He is involved in the development of optical scanning prototype in the Applied Physics Department and research leader in the development of a new stereo vision system prototype. He has been thesis director of 2 Doctor's Degree students and 3 Master's degree students. He holds two patents referred to dynamic triangulation method, has been editor of 2 books, Guest Editor of IEEE Sensors Journal, International Journal of Advanced Robotic Systems and Journal of Sensors, written over 60 papers, 8 book chapters and has been a reviewer for IEEE Sensors Journal, Optics and Lasers in Engineering, IEEE Transaction on Mechatronics and Neural Computing and Applications of Springer; he participated as a reviewer and Session Chair of IEEE ISIE conferences in 2014 (Turkey), 2015 (Brazil), 2016 (USA), 2017 (UK), 2019 (Canada), IECON 2018 (USA), IECON 2019 (Portugal), ISIE 2020 (Netherlands). His current research interests include automated metrology, stereo vision systems, control systems, robot navigation and 3D laser scanners.

Wendy Flores-Fuentes received the bachelor's degree in electronic engineering from the Autonomous University of Baja California in 2001, the master's degree in engineering from Technological Institute of Mexicali in 2006, and the Ph.D. degree in science, applied physics, with emphasis on Optoelectronic Scanning Systems for SHM, from Autonomous University of Baja California in June 2014. By now she is the author of 30 journal articles in Elsevier, IEEE Emerald and Springer, 13 book chapters and 5 books in Springer, Intech, IGI global Lambert Academic and Springer, 35 proceedings articles in IEEE ISIE 2014-2019, IECON 2014, 2018, 2019, the World Congress on Engineering and Computer Science (IAENG 2013), IEEE Section Mexico IEEE ROCC2011, and the VII International Conference on Industrial Engineering ARGOS 2014. Recently, she has organized and participated as Chair of Special Session on "Machine Vision, Control and Navigation" at IEEE ISIE 2015, 2016, 2017, 2019, 2020 and IECON 2018, 2019. She holds 1 patent of Mexico and 1 patent of Ukraine. She has been a reviewer of several articles in Taylor and Francis, IEEE, Elsevier, and EEMJ. Currently, she is a full-time professor at Universidad Autónoma de Baja California, at the Faculty of Engineering, and is the coordinator of the physics area in the basic science department. She has been incorporated into CONACYT National Research System in 2015. She did receive the award of "Best session presentation" in WSECS2013 in San-Francisco, USA. She did receive as coauthor the award of "Outstanding Paper in the 2017 Emerald Literati Network Awards for Excellence".

* * *

Luis Carlos Básaca Preciado was born October 25, 1985. He received a B.S. degree in Cybernetics and Electronics Engineering from CETYS University, Mexicali, Baja California, Mexico in 2007. In 2013 he obtained his Ph.D. in Optoelectronics from the Institute of Engineering of the Autonomous University of Baja California, México. As of 2014 he became a full-time professor, teaching automation, robotics and instrumentation at the Graduate School of Engineering and in the Electronics and Mechatronics majors at the School of Engineering of CETYS University, Mexicali Campus. From 2015 to 2018 he coordinated the Program of Cybernetics Electronics Engineering Major. He currently focuses on research in 3D vision systems, intelligent transportation by autonomous vehicles, autonomous UAV's, underwater ROV's, VR/AR and also participates in applied engineering projects linked to the industry. He has also been a mentor to student robotics teams since 2012, to promote science and engineering among young people. He has written three book chapters, nine journal articles, fourteen conference documents, including international conferences such as IEEE ISIE, IECON, EEEIC, ROC&C, PAHCE, and IPC; in the United States, Italy, Brazil and Mexico, where he holds a patent for a 3D vision system.

Danilo Cáceres-Hernández received his BS in Electrical and Electronic Engineering from the Technological University of Panamá (UTP), in 2004. From January 1, 2002, to March 14, 2005, he was working at the Computer Laboratory at the Department of Electrical Engineering. In 2011, he completed his MSc of Science in Electrical Engineering at the University of Ulsan. He received his PhD in Electrical Engineering from the University of Ulsan, South Korea, in February 2017. Since 2017, he is currently a full-time professor in the Electrical Engineering Department at Universidad Tecnológica de Panamá. His research interests are focused on intelligent systems, robotics, autonomous navigation, and computer vision.

Felix Fernando Gonzalez-Navarro holds a Ph.D. in Artificial Intelligence (2011) from the Software Department at the Universitat Politecnica de Catalunya (UPC). Currently, he is full-professor at Universidad Autonoma de Baja California leading the Artificial Intelligence Lab, where machine learning techniques are applied in research fields such as Bioinformatics, Image Processing, Human Movements Recognition, Data Mining and Custom-Made Software. He is reviewer of several international journals and conferences. He has been recognized by Mexican Government as National Researcher of Mexico Excellence Level 1 and is member of the Mexican Society of Artificial Intelligence and The Mexican Academy of Computation (AMEXCOMP).

Daniel Hernández-Balbuena was born on July 25, 1971. He received the BS degree from Puebla Autonomous University, Puebla, México, in 1996, and the MS degree from Ensenada Center for Scientific Research and Higher Education, Baja California, México, in 1999. He received his PhD degree in Baja California Autonomous University in 2010. His research interests are in the areas of time and frequency metrology, design and characterization of microwave devices and systems RF measurements, research applications of unmanned aerial vehicles, and image digital processing.

Mykhailo Ivanov received the B.S. and M.S. degrees at Kharkiv National Aerospace University "KhAI", Ukraine, in 2010 and 2012, respectively. In 2012 he joined the Kharkiv National Aerospace University where he held positions of Assistant and Vice Dean in questions of students recruitment. Since 2016 he is working to get a Ph.D. degree in Universidad Autónoma de Baja California, Mexico, and from 2019 he holds the position of Professor in the Faculty of Administrative Sciences. Till the present time, he has written 8 publications in international congresses of IEEE in the USA, England, Italy, etc., 3 papers and 3 book chapters. He has been a Reviewer for International Journal of Advanced Robotic Systems, IEEE ISIE 2017 (UK), 2019 (Canada), IECON 2018 (USA), and IECON 2019 (Portugal). His current research interests include robot navigation and communication, robotic group behavior models, vision systems, and 3D laser scanners.

Alexander S. Kalmykov is a Postgraduate at the Department of Microelectronics, Electronic Devices and Appliances (NURE). Assistant of Microelectronics, Electronic Devices and Appliances (NURE).

Vladimir Kartashov graduated from Kharkov Institute of Radio Electronics in 1980. Since 2006, he is Head of the Department of Media Engineering and Information Electronic Systems of Kharkiv National University of Radio Electronics. He received Degree of Dr. Sc. (Technology) in 2004 in the field of acoustic and radio-acoustic sounding systems of the atmosphere. He is the author of more than 250 scientific papers and 30 patents. His research interests include signal and image processing, navigation methods for mobile ground and flying robots.

Ganna Khoroshun, is currently an associated Professor of V. Dahl East-Ukrainian National University, Oct 2018-Sept 2020 doctorate courses on the specialty 05.13.06 - Information technologies. Previous positions at the same University Head of Department, 2012-2014 Associated Professor, 2008-2012 Lecturer 2007 Assistant of Lecturer 2005-2007 Junior research Scientist 2003-2005 Engineer 2002-2003Education: Ass. Prof. degree in Physics Department of East-Ukrainian National University, Ph.D. degree in Optics and Laser Physics from Institute of Physics National Academy of Sciences of Ukraine, in Quantum Electronics Department, Kiev, Ukraine. The supervisor was Prof. Vasnetsov M.V., the consultant was Prof. Gorshkov V.N. Thesis "Phase singularities analysis in a diffraction light field". He has taught postgraduate courses at the Physics and Engineering Education and Research Center of NAS of Ukraine, Kyiv, Ukraine, 2002-2006. His research skills are experimental research of diffraction problems. Experimental interferometric investigations. Experimental microparticles guiding. Theoretical research of interference and diffraction problems and the intensity pattern analysis by LG mode decomposition of the perturbed field. Statistical analysis of the image. Machine vision system. He achievements and awards are 2018-2019 Two years Joint Program on Scientific and Technological Cooperation between the Ministry of Education and Science of Ukraine and the Ministry of Science and Higher Education of the Republic of Poland. 2015 September – 2015 November internship as a researcher in the Institute of Multidisciplinary Research and Advanced Materials, Tohoku University, Japan. 2011 November - 2012 February Internship at Theoretical Phys. Dept. of Bristol University with Prof. Mark Dennis, UK. 2010 Grant of the President of Ukraine for young scientists of National Academy of Science of Ukraine for young scientists «The improvement of known methods of the optical vortices synthesis and new high-efficient ones development». 2002 Prize for student scientific work "The theory of the synthesis of the optical vortices by the technique of a phase wedge" from National Academy of Science of Ukraine. He is a member of the Ukrainian Physical Society.

Lars Lindner was born on July 20th 1981 in Dresden, Germany. He received his M.S. degree in mechatronics engineering from the TU Dresden University in January 2009. He was working as graduate assistant during his studies at the Fraunhofer Institute for Integrated Circuits EAS in Dresden and also prepared his master thesis there. After finishing his career, he moved to Mexico and started teaching engineering classes at different universities in Mexicali. Since August 2013 he began his PhD studies at the Engineering Institute of Autonomous University of Baja California in Mexicali with the topic "Theoretical Method to Increase the Speed of Continuous Mapping of a Three-dimensional Laser Scanner using Servomotor Control", in which he worked in the development of an optoelectronic prototype for the measurement of 3D coordinates, using laser dynamic triangulation. Its academic products include 13 original research articles, 3 articles in national congresses, 21 articles in international congresses and 10 book chapters. In September 2017, he was appointed as a Level 1 National Researcher by the National System of Researchers CONACYT for the period 2018-2020. Right now he is working as a technician assistant at the Engineering Institute of Autonomous University of Baja California for the department of applied physics.

Alexander P. Lytyuga graduated in 1988 from Kharkov State University named after M.Gorky, Radiophysics Department, specialty Radiophysics. He attended graduate school (NURE) in 1995. He received a Ph.D. in 2008 at the Kharkov National University of Radio Electronics. His specialty "Radio Electronic and Television Systems". in 2010 he was awarded the scientific title of Senior Scientific Worker. His Research interests are statistical methods for signals and images processing in optoelectronic systems; stochastic processes; photons statistics.

Fernando Merchán received a BSc. in Electrical and Electronic Engineering from Universidad Tecnológica de Panamá (UTP) in 2003. Obtained his MSc from the University of Bordeaux I (2005). He received a Diplome d'Ingenieur-Grade Master, telecommunications option, from the École Nationale Supérieure d'Electronique, d'Informatique et de Radiocommunications de Bordeaux (ENSEIRB), France. In 2009, he completed a PhD in Signal and Image Processing Université Bordeaux I, in France

(2009). He developed his doctoral research at Laboratoire d'intégration du Matériau au Système (IMS) in Bordeaux, France. His doctoral research addressed parametric modeling of signals and images based on Wold's decomposition and techniques of signal and image decomposition based on local regularity. Since 2011, he is a full-time professor in the Electrical Engineering Department at Universidad Tecnológica de Panamá. His research interests are focused on parametric modeling of random processes, bioacoustic signal processing and applications of image processing.

Jesús Elias Miranda-Vega was born in 1984 and received a BE degree in Electrical and Electronic Engineering from Los Mochis Institute Technology in Sinaloa, Mexico, in 2007, and a Master's degree in Electronic Engineering from the Mexicali Institute of Technology, Mexicali, Mexico, in 2014. He received his PhD degree with honors in Engineering Institute at the Autonomous University of Baja California (UABC) in December 2019. His current research interest includes machine vision, data signal processing, the theory and opto-electronics devices, and their applications. Until now he is the author of 5 impact journal articles and 3 book chapters, and 3 proceedings articles in IEEE.

Fabian N. Murrieta-Rico received B.Eng. and M.Eng. degrees from Mexicali Institute of Technology (ITM) in 2004 and 2013 respectively. In 2017, he received his PhD in Materials Physics at Ensenada Center for Scientific Research and Higher Education (CICESE). He has worked as an automation engineer, systems designer, as a university professor, and as postdoctoral researcher at Facultad de Ingeniería, Arquitectura y Diseño from Universidad Autónoma de Baja California (UABC). Currently he works as professor of bioengineering topics at UABC and as postdoctoral researcher at the Centro de Nanociencias y Nanotecnología from Universidad Nacional Autónoma de México (CNyN-UNAM). His research has been published in different journals and presented at international conferences since 2009. He has served as reviewer for different journals, some of them include IEEE Transactions on Industrial Electronics, IEEE Transactions on Instrumentation, Measurement and Sensor Review. His research interests are focused on the field of time and frequency metrology, the design of wireless sensor networks, automated systems and highly sensitive chemical detectors.

Yevhen Polyakov was born in 1985 in Pavlograd, Dnipropetrovsk region, Ukraine. In 2003 he graduated from Pavlograd secondary school number 9 and entered the Kharkiv National Automobile and Highway University. In 2008 he graduated from Kharkiv National Automobile and Highway University and received a full higher education in the specialty "Automated Control of Technological Processes". From 2008 to 2012 he was graduate student of the Department of Metrology and Life Safety of Kharkiv National Automobile and Highway University. From 2012 to present he works at the Kharkiv National Automobile and Highway University as Associate Professor of the Department of Metrology and Life Safety. In 2014 he defended his Ph.D thesis on the theme "Improvement of methods of sensors dynamic errors decrease" (specialty 05.01.02 "Standardization, certification and metrological support"). Married, has a daughter. Lives in Kharkiv, Ukraine. Research interests: inverse problems of measurement, estimation of parameters of quick changing processes, identification of dynamic characteristics of technical objects, correction of dynamic measurement errors.

Oleksandr Poliarus was born on February 18, 1950 in Gadyach town of Poltava region (Ukraine). From 1967 to 1973 he studied at the Moscow Higher Technical School named after M. E. Bauman, after which he worked as an engineer at a research institute for one year. From 1974 to 1999 he served in the

Armed Forces of the USSR and then Ukraine. In 1980 graduated from the Military Radio Engineering Academy of Air Defense in Kharkiv and in 1985 defended his PhD thesis. He was in teaching positions of Academy, and since 1996 - the head of the department of antenna-feeder devices. In 1994 he defended the doctoral thesis and in 1999 he retired from the Armed Forces of Ukraine. Since September 2007 he is the head of the Department of Metrology and Life Safety of the Kharkiv National Automobile and Highway University. Married, has two sons. Lives in Kharkiv, Ukraine. Research interests: signal processing, remote measurements, system identification, inverse problems of measurement.

Martin Poveda received a BSc. in Electronic and Telecommunications Engineering from Universidad Tecnológica de Panama (UTP) in 2016. He Is working for Cable Onda since 2015, where he currently serves as Data Center infrastructure engineer. He is currently in the Electrical Engineering Master Program of the UTP. His research interest includes applications of computer vision, image processing applications and electronics.

Andrii Riazantsev is a Postgraduate at the Department of Computer Science and Engineering of V. Dahl East-Ukrainian National University Severodonetsk, Ukraine. Assistant of Department of Computer Science and Engineering of V. Dahl East-Ukrainian National University, Severodonetsk, Ukraine.

Moisés Rivas-López was born on June 1, 1960. He received his BS and MS degrees in Autonomous University of Baja California, México, in 1985 and 1991, respectively. He received his PhD degree in the same university, on specialty "Optical Scanning for Structural Health Monitoring," in 2010. He was the editor of 3 books and has written 5 book chapters and 48 journal and proceedings conference papers. Since 1992 until the present time, he has presented different works in several International Congresses of IEEE, ICROS, SICE, and AMMAC in the USA, England, Japan, Turkey, and Mexico. In 1997–2005, he was dean of the Engineering Institute of Autonomous University Baja California; also in 2006–2010, he was rector of the Polytechnic University of Baja California. Since 2013 to date , he is a member of the National Researcher System.

Javier Rivera Castillo received the M.S. degrees in Baja California Autonomous University, México in 2006. He has written 2 book chapter and 9 Journal and Proceedings Conference papers. M.S. Rivera was in charge of Calibration Laboratory at Engineering Institute of Baja California Autonomous University and then he was Head of Mechatronic Program at Baja California Polytechnic University Since 2006 to 2011. He is now full researcher in Applied Physics Department of Engineering Institute of Baja California Autonomous University, Mexico.

Oleksandr Ryazantsev is Vice-Rector for Scientific and Educational Affairs and International Cooperation, Volodymyr Dahl East Ukrainian National University, Doctor of Engineering Sciences, Professor.

Javier E. Sanchez-Galan received his BSc degree in Computer Systems Engineering at Universidad Tecnológica de Panamá (UTP), in 2006. He completed a MSc in Computer Science (focusing on bioinformatics) and a PhD in Experimental Medicine (focusing the use of Near-Infrared spectroscopy of biofluids for the determination of maternal-fetal health status), both from McGill University (Montreal, Canada). Since 2013, he has been a researcher and coordinator of the Research Group in Biotechnology, Bioinformatics, and Systems Biology (GIBBS) at UTP. In 2018, he was appointed as a Level I national researcher by the Sistema Nacional de Investigación of Senacyt (Panama). Since 2019, he is a full-time professor-researcher in the Department of Computer Systems Engineering at UTP. His research interests include the development of innovative approaches applying computational data analysis, machine learning and deep learning to massive data sets coming from biological, medical, environmental, and agricultural fields.

Tatyana Strelkova graduated in 1992 from Kharkov State University named after M.Gorky, Radio-physics Department, specialty Biophysics. She attended graduate school (KNURE) in 1995. in 1999 she became an assistant of the Department of Radioelectronic Devices (NURE). She earened her Ph.D. in 2001 at the Kharkov National University of Radio Electronics. Her specialty is 05.11.17 "Medical Devices and Systems". In 2002 she became the Associate Professor of the Department of Radioelectronic Devices (NURE). In 2004 she was awarded the scientific title of associate professor in the Department of Radioelectronic Devices.2004 – Associate Professor of the Department of Physical Basics of Electronic Technology (NURE).2012 – Doctoral (NURE). Doctor Sci. (Engineering) – 2017 in National Technical University of Ukraine "Igor Sikorsky Kyiv Polytechnic Institute" Her specialty 05.11.07 "Optical Instruments and Systems". Since 2018 she is a Professor of the Department of Microelectronics, Electronic Devices and Appliances (NURE). Her research interests are theory and practice of optical signals reception and processing; random fluxes theory; photons statistics. High school organization and management. Methods of technical education including dynamic innovation of modern high school education. She is an author of over 100 publications, including 4 monographs, 1 patent (Ukraine).

Alexander Strelkov was born in 1939 is a Prof., Dr.Sci. (Engineering), Honored Inventor of Ukraine. He is an author of over 450 printed works including over 150 patents (USSR, Ukraine). Director of over 100 Master's and more than 40 Ph.D. thesis. His research interest area theory of optoelectronic systems, theory of optical signals detection and processing, statistical methods of signals reception.

Oleg Sytnik was born in Dneprodzerginsk, Ukraine, on May 17, 1958. He received the M.S. degree in radio engineering from the Kharkov University of Radio Electronics, Kharkov in 1980. He worked in the Design Bureau of Machine-Building Plant of Dnipropetrovsk (1980-1982), and then at the Radio Engineering Department of the Kharkov University of Radio Electronics (1982-1986). He received the Ph.D. degree in radars and navigation from the Kharkov University of Radio Electronics, Kharkov, in 1986. Since 1986 he has worked as a senior researcher at the A. Ya. Usikov Institute for Radio Physics and Electronics under the National Academy of Sciences of Ukraine in the Department of Radiophysical Introscopy and in the A.I. Kalmykov Centre for Remote Sensing of the Earth under the Space Agency of Ukraine. During this time, he received the Dr. Sci. degree in Radio Physics and Mathematics (2005) and Professor (2014). He has published over 174 science papers in radars theory, digital signal processing and applications. His fields of research interests are radars, digital signal and image processing, pattern recognition and stochastic non-stationary processes.

Index

A

Absolute Position Error 37
artificial intelligence 1-2, 17, 28-31, 124, 131, 163, 167, 180, 184, 220-221, 226, 255, 258-259, 265, 278-279, 286

B

Blindness 187-188, 217, 220, 222
brightness 94, 98, 140, 174, 177-178, 289-292, 299-300, 305-307, 310

C

Classical and Nonparametric Detection 224
computer vision 4-5, 18, 101, 159, 187, 192, 195-197, 213-222, 259, 268, 284, 286, 310
confusion matrix 276-277, 282-283

D

Data post-processing 37
databases 14, 163, 170, 183
DC motor 37, 50-53, 58, 60-61, 64-65, 125, 286
deconvolution 294, 309-310
depth sensors 187-188, 197, 213-214
digital controller 37, 60, 65, 69
digital filter 103, 126, 129-130, 294
distortion 73, 146, 232, 289, 291-294, 308

E

Euclidean distance 258, 275

F

fluctuations 74, 79-80, 135, 140, 145-150, 153, 155, 157-158, 164, 173, 178, 229

H

Hilbert-Huang modes 224, 249-252, 254
human visual system 288-289, 300-301, 310

I

image analysis 26, 166, 171-172, 185-187, 199, 212, 218, 221
IMAGE PREPROCESSING 288, 301
image processing 2, 4, 8, 14, 25, 27-28, 33, 36, 72, 75-76, 86, 100-102, 111, 161, 172-173, 183, 214, 217, 226-228, 231-232, 254-255, 257, 264, 267, 282, 288-290, 293-294, 300, 302, 304, 307-310
image quality 71, 74, 158, 163, 166, 171-172, 176-177, 289, 309-310
image quality assessment 71, 309-310
image recognition 258-259, 268, 278, 283-284
Industrial Internet of Things 258-259, 265, 286
interference 19, 74-75, 77, 80, 82, 93, 105, 128, 135, 138, 142, 159, 167, 194, 233, 241, 253, 288, 291-292, 309

K

Kinect 187-189, 197-200, 204, 208, 210, 213-214, 216-218, 220-221, 223, 268

L

Laser Beam Positioning 34, 37, 53, 60

M

machine learning 28, 120, 124, 214, 216-217, 222, 258, 260, 267, 274-276, 278-279, 285-286
machine vision 1-5, 18-19, 21, 23-24, 28, 31, 34-36, 46, 49, 66-67, 69, 71-72, 100-101, 103-104, 131-132, 134-136, 160, 163-164, 167, 183-184,

Ensure Quality Research is Introduced to the Academic Community

Become an IGI Global Reviewer for Authored Book Projects

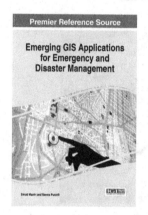

Premier Reference Source

Emerging GIS Applications for Emergency and Disaster Management

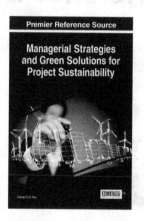

Premier Reference Source

Managerial Strategies and Green Solutions for Project Sustainability

Premier Reference Source

Comparative Approaches to Using R and Python for Statistical Data Analysis

Premier Reference Source

Solutions for High-Touch Communications in a High-Tech World

The overall success of an authored book project is dependent on quality and timely reviews.

In this competitive age of scholarly publishing, constructive and timely feedback significantly expedites the turnaround time of manuscripts from submission to acceptance, allowing the publication and discovery of forward-thinking research at a much more expeditious rate. Several IGI Global authored book projects are currently seeking highly-qualified experts in the field to fill vacancies on their respective editorial review boards:

Applications and Inquiries may be sent to:
development@igi-global.com

Applicants must have a doctorate (or an equivalent degree) as well as publishing and reviewing experience. Reviewers are asked to complete the open-ended evaluation questions with as much detail as possible in a timely, collegial, and constructive manner. All reviewers' tenures run for one-year terms on the editorial review boards and are expected to complete at least three reviews per term. Upon successful completion of this term, reviewers can be considered for an additional term.

If you have a colleague that may be interested in this opportunity, we encourage you to share this information with them.

Printed in the United States
By Bookmasters